SOCIOLOGY

Twenty-Sixth Edition

Editor

Kurt Finsterbusch

University of Maryland, College Park

Kurt Finsterbusch received his bachelor's degree in history from Princeton University in 1957, and his bachelor of divinity degree from Grace Theological Seminary in 1960. His Ph.D. in sociology, from Columbia University, was conferred in 1969. He is the author of several books, including *Understanding Social Impacts* (Sage Publications, 1980), *Social Research for Policy Decisions* (Wadsworth Publishing, 1980, with Annabelle Bender Motz), and *Organizational Change as a Development Strategy* (Lynne Rienner Publishers, 1987, with Jerald Hage). He is currently teaching at the University of Maryland, College Park, and, in addition to serving as editor for *Annual Editions: Sociology,* he is also coeditor for Dushkin/McGraw·Hill's *Taking Sides: Clashing Views on Controversial Social Issues.*

A Library of Information from the Public Press

Dushkin/McGraw·Hill

Sluice Dock, Guilford, Connecticut 06437

Visit us on the Internet—http://www.dushkin.com

The Annual Editions Series

ANNUAL EDITIONS is a series of over 65 volumes designed to provide the reader with convenient, low-cost access to a wide range of current, carefully selected articles from some of the most important magazines, newspapers, and journals published today. ANNUAL EDITIONS are updated on an annual basis through a continuous monitoring of over 300 periodical sources. All ANNUAL EDITIONS have a number of features that are designed to make them particularly useful, including topic guides, annotated tables of contents, unit overviews, and indexes. For the teacher using ANNUAL EDITIONS in the classroom, an Instructor's Resource Guide with test questions is available for each volume.

VOLUMES AVAILABLE

Abnormal Psychology
Adolescent Psychology
Africa
Aging
American Foreign Policy
American Government
American History, Pre-Civil War
American History, Post-Civil War
American Public Policy
Anthropology
Archaeology
Biopsychology
Business Ethics
Child Growth and Development
China
Comparative Politics
Computers in Education
Computers in Society
Criminal Justice
Criminology
Developing World
Deviant Behavior
Drugs, Society, and Behavior
Dying, Death, and Bereavement
Early Childhood Education
Economics
Educating Exceptional Children
Education
Educational Psychology
Environment
Geography
Global Issues
Health
Human Development
Human Resources
Human Sexuality
India and South Asia
International Business
Japan and the Pacific Rim
Latin America
Life Management
Macroeconomics
Management
Marketing
Marriage and Family
Mass Media
Microeconomics
Middle East and the Islamic World
Multicultural Education
Nutrition
Personal Growth and Behavior
Physical Anthropology
Psychology
Public Administration
Race and Ethnic Relations
Russia, the Eurasian Republics, and Central/Eastern Europe
Social Problems
Social Psychology
Sociology
State and Local Government
Urban Society
Western Civilization, Pre-Reformation
Western Civilization, Post-Reformation
Western Europe
World History, Pre-Modern
World History, Modern
World Politics

Cataloging in Publication Data
Main entry under title: Annual Editions: Sociology. 1997/98.
1. Sociology—Periodicals. 2. United States—Social Conditions—1960—Periodicals. I. Finsterbusch, Kurt, *comp.* II. Title: Sociology.
0-697-37361-4 301′.05 72–76876

Twenty-Sixth Edition

Cover image © 1996 PhotoDisc, Inc.

Printed in the United States of America

Printed on Recycled Paper

Editors/Advisory Board

Members of the Advisory Board are instrumental in the final selection of articles for each edition of ANNUAL EDITIONS. Their review of articles for content, level, currentness, and appropriateness provides critical direction to the editor and staff. We think that you will find their careful consideration well reflected in this volume.

To the Reader

In publishing ANNUAL EDITIONS we recognize the enormous role played by the magazines, newspapers, and journals of the *public press* in providing current, first-rate educational information in a broad spectrum of interest areas. Many of these articles are appropriate for students, researchers, and professionals seeking accurate, current material to help bridge the gap between principles and theories and the real world. These articles, however, become more useful for study when those of lasting value are carefully *collected, organized, indexed,* and *reproduced* in a *low-cost format,* which provides easy and permanent access when the material is needed. That is the role played by ANNUAL EDITIONS. Under the direction of each volume's *academic editor,* who is an expert in the subject area, and with the guidance of an *Advisory Board,* each year we seek to provide in each ANNUAL EDITION a current, well-balanced, carefully selected collection of the best of the public press for your study and enjoyment. We think that you will find this volume useful, and we hope that you will take a moment to let us know what you think.

Both the 1990s and the 1980s are full of crises, changes, and challenges. Crime is running rampant. The public is demanding more police, more jails, and tougher sentences, but less government spending. The economy suffers from foreign competition, trade deficits, budget deficits, and economic uncertainties. Government social policies seem to create almost as many problems as they solve. Laborers, women, blacks, and many other groups complain of injustices and victimization. The use of toxic chemicals has been blamed for increases in cancer, sterility, and other diseases. Marriage and the family have been transformed, in part by the women's movement and in part by the stress that current conditions create for women who try to combine family and careers. Schools, television, and corporations are commonly vilified. Add to all this the problems of population growth, ozone depletion, and the greenhouse effect, and it is easy to despair. Nevertheless, crises also provide opportunities.

The present generation may determine the course of history for the next 200 years. Great changes are taking place, and new solutions are being sought where old answers no longer work. The issues the current generation faces are complex and must be interpreted within a sophisticated framework. The sociological perspective provides such a framework. It expects people to act in terms of their position in the social structure; the political, economic, and social forces operating on them; and the norms that govern the situation. *Annual Editions: Sociology 97/98* should help you develop the sociological perspective that will enable you to determine how the issues of the day relate to the way society is structured. The essays provide not only information but also models of interpretation and analysis that will guide you as you form your own views.

Annual Editions: Sociology 97/98 emphasizes social change, institutional crises, and prospects for the future. It provides intellectual preparation for acting for the betterment of humanity in times of crucial change. The sociological perspective is needed more than ever as humankind tries to find a way to peace, prosperity, health, and well-being that can be maintained for generations in an improving environment. The obstacles that lie in the path of these important goals seem to increase yearly. The aims of this edition are to communicate to students the excitement and importance of the study of the social world and to provoke interest in, and enthusiasm for, the study of sociology.

Annual Editions depends upon reader response to develop and change. You are encouraged to return the postage-paid *article rating form* at the back of the book with your opinions about existing readings, recommendations of articles you think have sociological merit for subsequent editions, and advice on how the anthology can be made more useful as a teaching and learning tool.

Kurt Finsterbusch
Editor

Dedicated to Josephine for her 87 years of concern about social issues.

Contents

UNIT 1

Culture

Five selections consider what our culture can learn from primitive peoples, what forces are shaping today's cultures and lifestyles, and what impact crises have on culture.

UNIT 2

Socialization and Social Control

Six articles examine the effects of social influences on childhood, personality, and human behavior with regard to the socialization of the individual.

The concepts in bold italics are developed in the article. For further expansion please refer to the Topic Guide, the Glossary, and the Index.

UNIT 3

Groups and Roles in Transition

Nine articles discuss some of the social roles and group relationships that are in transition in today's society. Topics include primary and secondary groups and the reevaluation of social choices.

UNIT 4

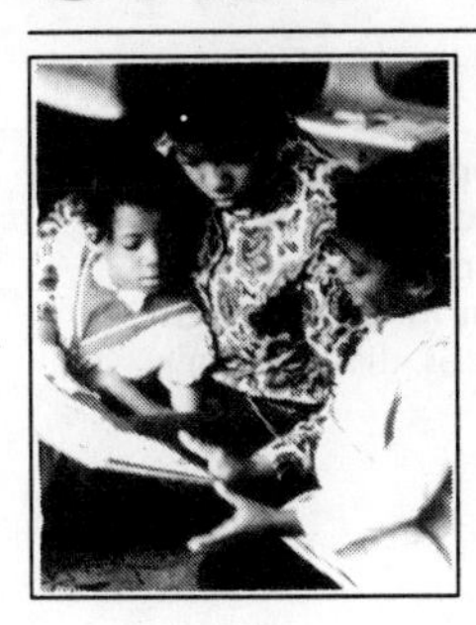

Stratification and Social Inequalities

Nine selections discuss the social stratification and inequalities that exist in today's society with regard to the rich, the poor, blacks, and women.

UNIT 5

Social Institutions: Issues, Crises, and Changes

Nine articles examine several social institutions that are currently in crisis. Selections focus on the political, economic, and social spheres, as well as the overall state of the nation.

UNIT 6

Social Change and the Future

Six selections discuss the impact that population, technology, environmental stress, and social values will have on society's future.

Selected World Wide Web Sites for Sociology

(Some Web sites are continually changing their structure and content, so the information listed here may not always be available.)

BUBL (BUlletin Board for Libraries) Information Server

http://www.bubl.bath.ac.uk/BUBL/Sociology.html This site includes links to U.S. National Data Archives on population studies, child abuse, disaster research center, men's issues, United Nations Gopher server, *CIA World Fact Book,* postmodern culture, and many other topics. It also links to the WWW Virtual Library in Sociology and the Virtual Library of Demographic and Population Studies.

Humanities HUB and Census Data

http://www.gu.edu.au/gwis/hub/hub.socio.html This site offers a wide range of references for the sociological researcher: the European Sociological Association, the American Communication Association, population study links, data from the U.S. Census Bureau, Eurostat, and data archives from many other countries. Other links offer access to discussion groups on conversation analysis, globalization, and a range of other topics.

American Sociological Association (ASA)

http://www.asanet.org This is the home page for the American Sociological Association that includes many of its journals: *American Sociological Review, Contemporary Sociology, Journal of Health and Social Behavior, Social Psychology Quarterly, Sociological Methodology, Sociological Theory, Sociology of Education, Teaching Sociology,* and many others. It also provides membership information, convention and conference announcements, and links to sites operated by special interest sections of the ASA.

Society for Applied Sociology

http://www.indiana.edu/~appsoc// This site was developed for professionals involved in applying sociological knowledge to a variety of settings. It includes access to *Journal of Applied Sociology Information* and *Social Insight,* and it lists relevant conferences.

Other Links of Interest

http://www.soc.surrey.ac.uk/Other_links.html This British site offers an enormous number of links to a world-wide collection of journals and magazines. It also offers links to libraries, data archives, and sociological associations, and it provides information on new sites that have appeared on the WWW.

Topic Guide

This topic guide suggests how the selections in this book relate to topics of traditional concern to students and professionals involved with the study of sociology. It is useful for locating articles that relate to each other for reading and research. The guide is arranged alphabetically according to topic. Articles may, of course, treat topics that do not appear in the topic guide. In turn, entries in the topic guide do not necessarily constitute a comprehensive listing of all the contents of each selection.

TOPIC AREA	TREATED IN
Population Growth	3. Overworked Americans or Overwhelmed Americans? 39. Carrying Capacity 43. Coming Anarchy
Poverty	2. Mountain People 23. Poverty's Children 43. Coming Anarchy
Race/Ethnic Relations	18. Crisis of Community 26. Whites' Myths about Blacks 27. Affirmative Action
Regulations	32. Death of Common Sense
Religion	38. Can Churches Save America?
Roles	6. Guns and Dolls 13. Where's Papa? 16. Ending the Battle between the Sexes 17. Men
Sex Roles	6. Guns and Dolls 16. Ending the Battle between the Sexes 17. Men 28. Longest Climb
Sexism	29. Violence against Women
Sexual Practices	15. Now for the Truth about Americans and Sex
Social Change	2. Mountain People 3. Overworked Americans or Overwhelmed Americans? 4. De-Moralized Society 5. Decline of Bourgeois America 12. The Way We Weren't 13. Where's Papa? 14. Modernizing Marriage 17. Men 19. Strange Disappearance of Civic America 31. Hyper Democracy 33. Reinventing the Corporation 34. Revolution in the Workplace 36. Health Unlimited 43. Coming Anarchy 44. Operating in a Period of Punctuated Equilibrium
Social Class/ Stratification	21. Winner Take All 22. Working Harder, Getting Nowhere 23. Poverty's Children 24. Upside-Down Welfare 25. Dismantling the Welfare State 26. Whites' Myths about Blacks 27. Affirmative Action 28. Longest Climb 29. Violence against Women 31. Hyper Democracy
Social Control	4. De-Moralized Society 7. Children of the Universe 8. Crime in America 9. Moral Credibility and Crime 10. When Violence Hits Home 11. Legalization Madness 43. Coming Anarchy
Social Relationships	2. Mountain People 4. De-Moralized Society 10. When Violence Hits Home 12. The Way We Weren't 13. Where's Papa? 14. Modernizing Marriage 16. Ending the Battle between the Sexes 18. Crisis of Community 29. Violence against Women
Socialization	6. Guns and Dolls 7. Children of the Universe 13. Where's Papa?
Technology	41. Alone Together 42. Price of Fanaticism
Terrorism	42. Price of Fanaticism
Underclass	23. Poverty's Children 25. Dismantling the Welfare State
Unemployment	*See* Work/Unemployment
Upper Class	24. Upside-Down Welfare 28. Longest Climb
Values	1. Tribal Wisdom 2. Mountain People 4. De-Moralized Society 5. Decline of Bourgeois America 7. Children of the Universe 9. Moral Credibility and Crime 15. Now for the Truth about Americans and Sex 18. Crisis of Community 19. Strange Disappearance of Civic America 20. Are Today's Suburbs Really Family-Friendly? 35. Seeking Abortion's Middle Ground
Violence	4. De-Moralized Society 8. Crime in America 10. When Violence Hits Home 42. Price of Fanaticism
Volunteerism	19. Strange Disappearance of Civic America
Wealth	22. Working Harder, Getting Nowhere
Welfare	24. Upside-Down Welfare 25. Dismantling the Welfare State 38. Can Churches Save America?
Women	6. Guns and Dolls 7. Children of the Universe 10. When Violence Hits Home 16. Ending the Battle between the Sexes 28. Longest Climb 29. Violence against Women
Work\ Unemployment	3. Overworked Americans or Overwhelmed Americans? 22. Working Harder, Getting Nowhere 25. Dismantling the Welfare State 33. Reinventing the Corporation 34. Revolution in the Workplace 41. Alone Together

Culture

- **Tribal Cultures and Their Lessons for Us (Articles 1 and 2)**
- **Forces Shaping Cultures and Lifestyles (Articles 3–5)**

The ordinary, everyday objects of living and the daily routines of life provide a structure to social life that is regularly punctuated by festivals, celebrations, and other special events (both happy and sad). Both routine and special times are the "stuff" of culture, for culture is the sum total of all the elements of one's social inheritance. Culture includes language, tools, values, habits, science, religion, literature, and art.

It is easy to take one's own culture for granted, so it is useful to pause and reflect on the shared beliefs and practices that form the foundations for our social life. Students share beliefs and practices and thus have a student culture. Obviously, the faculty has one also. Students, faculty, and administrators share a university culture. At the national level, Americans share an American culture. These cultures change over time and especially between generations. As a result, there is much variety among cultures across time and across nations, tribes, and groups. It is fascinating to study these differences and to compare the dominant values and signature patterns of different groups.

The two articles in the first section deal with primitive cultures that are under considerable stress today. David Maybury-Lewis challenges our sense of cultural superiority by demonstrating the wisdom of tribal patterns compared to our modern lifestyles. Tribal societies value people, but modern societies value things. The reader probably will not abandon his or her lifestyle after reading this article, but he or she should have a lot more respect for tribal societies. The report by Colin Turnbull tells how the Ik tribe suffered the loss of their tribal lands and were forced to live in a harsh environment. When a society's technology is very primitive, its environment has a profound impact on its social structure and culture. We would expect, therefore, that this momentous change in the tribe's environment would require some interesting adaptations. The change that occurred, however, was shocking. Literally all aspects of life changed for the tribe's members, in a disturbingly sinister way. Moreover, the experience of this tribe leads Turnbull to question some of the individualistic tendencies of America.

In the next section, Jeff Davidson looks at some major trends in American society and explores their impact on American culture and lifestyles. Longer working hours, population growth, the knowledge explosion, the growth of the media, the steady growth of paper trails, and the overabundance of choices are overwhelming Americans. Their leisure time is reduced, their attention is more fractured, and they feel more overwhelmed. Davidson recommends that the only way to live more fully is to exclude more from one's life and live more simply.

Gertrude Himmelfarb warns about a more ominous trend, the increase in such negative social indicators as the crime, illegitimacy, and divorce rates. These increases cannot be explained by worsening economic conditions, but rather by two trends. First, much that once was considered as abnormal or illegitimate has become accepted as normal and legitimate. Second, American society has abandoned moral judgments. "Most of us are uncomfortable with the idea of making moral judgments even in our private lives, let alone with the intrusion of moral judgments into public affairs," Himmelfarb declares.

In the last unit article, Stanley Rothman also analyses the cultural changes taking place in America, but he focuses on the values that are very consequential in the workplace. He notes the decline of bourgeois values of hard work, self-restraint, and ego control, and the rise of values such as expressive individualism. The decline of restraint and of the superego is evidenced in the crime, divorce, and deviancy statistics and does not bode well for the institutions of America.

Looking Ahead: Challenge Questions

What do you think are the core values in American society?

What are the strengths and weaknesses of cultures that emphasize either cooperation or individualism?

What is the relationship between culture and identity?

What might a visitor from a primitive tribe describe as shocking and barbaric about American society?

UNIT 1

Tribal wisdom

Is it too late for us to reclaim the benefits of tribal living?

David Maybury-Lewis

Tribal people hold endless fascination for us moderns. We imagine them as exotics trapped in a lyrical past, or as charming anachronisms embarking on the inevitable course toward modernity. What few of us realize is that tribal peoples have not tried (and failed) to be like us, but have actually chosen to live differently. It is critical that we examine the roads they took that we did not; only then can we get a clear insight into the choices we ourselves make and the price we pay for them—alienation, loneliness, disintegrating families, ecological destruction, spiritual famishment. Only then can we consider the possibility of modifying some of those choices to enrich our lives.

In studying tribal societies, as I have for 30 years, we learn that there is no single "tribal" way of life—I use the word here as a kind of shorthand to refer to small-scale, preindustrial societies that live in comparative isolation and manage their affairs without a central authority such as the state. But however diverse, such societies do share certain characteristics that make them different from "modern" societies. By studying the dramatic contrasts between these two kinds of societies, we see vividly the consequences of modernization and industrialization. Modernization has changed our thinking about every facet of our lives, from family relationships to spirituality to our importance as individuals. Has ours been the road best traveled?

Strange relations

The heart of the difference between the modern world and the traditional one is that in traditional societies people are a valuable resource and the interrelations between them are carefully tended; in modern society things are the valuables and people are all too often treated as disposable.

In the modern world we shroud our interdependency in an ideology of independence. We focus on individuals, going it alone in the economic sphere, rather than persons, interconnected in the social sphere. As French anthropologist Marcel Mauss put it, "It is our Western societies that have recently turned man into an economic animal." What happened?

A truly revolutionary change—a social revolution centering on the rights of the individual—swept Western Europe during the Renaissance and eventually came to dominate and define the modern world. While traditional societies had denounced individualism as anti-social, in Western Europe a belief in the rights and dignity of the individual slowly came to be regarded as the most important aspect of society itself.

The glorification of the individual, this focus on the dignity and rights of the individual, this severing of the obligations to kin and community that support and constrain the individual in traditional societies—all this was the sociological equivalent of splitting the atom. It unleashed the human energy and creativity that enabled people to make extraordinary technical advances and to accumulate undreamed-of wealth.

But we have paid a price for our success. The ever-expanding modern economy is a driven economy, one that survives by creating new needs so that people will consume more. Ideally, under the mechanics of this system, people should have unlimited needs so that the economy can expand forever, and advertising exists to convince them of just that.

The driven economy is accompanied by a restless and driven society. In the United States, for example, the educational system teaches children to be competitive and tries to instill in them the hunger for personal achievement. As adults, the most driven people are rewarded by status. Other human capabilities—for kindness, generosity, patience, tolerance, cooperation, compassion—all the qualities one might wish for in one's family and friends, are literally undervalued: Any job that requires such talents usually has poor pay and low prestige.

The tendency of modern society to isolate the individual is nowhere more clearly evident than in the modern family. In the West we speak of young people growing up, leaving their parents, and "starting a family." To most of the world, including parts of Europe, this notion seems strange. Individuals do not start families, they are born into them and stay in them until death or even beyond. In those societies you cannot leave your family without becoming a social misfit, a person of no account.

When the modern system works, it provides a marvelous release for individual creativity and emotion; when it does not, it causes a lot of personal pain and social stress. It is, characteristically, an optimistic system, hoping for and betting on the best. In contrast, traditional societies have settled for more cautious systems, designed to make life tolerable and to avoid the worst. Americans, in their version of the modern family, are free to be themselves at the risk of ultimate loneliness. In traditional family systems the individual may be suffocated but is never unsupported. Is there a middle way?

Finding that middle way is not a problem that tribal societies have to face, at least not unless they find their way of life overwhelmed by the outside world. They normally get on with the business of bringing up children against a background of consensus about what should be done and how, which means that they can also be more relaxed about who does the bringing up. Children may spend as much time with other adults as they do with their parents, or, as in the Xavante tribe of central Brazil, they may wander around in a flock that is vaguely supervised by whichever adults happen to be nearby. As soon as Xavante babies are old enough to toddle, they attach themselves to one of the eddies of children that come and go in the village. There they are socialized by their peers. The older kids keep an eye on the younger ones and teach them their place in the pecking order. Of course there are squabbles and scraps, and one often sees a little child who has gotten the worst of it wobbling home and yelling furiously. The child's parents never do what parents in our society often do—go out and remonstrate with the children in an attempt to impose some kind of adult justice (often leaving the children with a burning sense of unfairness). Instead they simply comfort the child and let her return to the fold as soon as her bruised knee or battered ego permits. At the same time, there is never any bullying among the Xavante children who are left to police themselves.

In traditional societies, people are valuable; in modern society, things are the valuables.

The Xavante system represents an informal dilution of parents' everyday responsibilities. In many societies these responsibilities are formally transferred to other relatives. In the Pacific Islands, for example, it is quite common for children to be raised by their parents' kin. Among the Trobriand Islanders, this is seen as useful for the child, since it expands his or her network of active kin relationships without severing ties to the biological parents. If children are unhappy, they can return to their true parents. If they are contented, they remain with their adoptive parents until adulthood.

Tribal societies also differ from the modern in their approach to raising teenagers. The tribal transition to maturity is made cleanly and is marked with great ceremony. In Western societies families dither over their often resentful young, suggesting that they may be old enough but not yet mature enough, mature enough but not yet secure enough, equivocating and putting adolescents through an obstacle course that keeps being prolonged.

Tribal initiation rites have always held a special interest for outside observers, who have been fascinated by their exotic and especially by their sexual aspects. It is the pain and terror of such initiations that make the deepest impression, and these are most frequently inflicted on boys, who are in the process of being taken out of the women's world and brought into that of the men. Some Australian Aboriginal groups peel the penis like a banana and cut into the flesh beneath the foreskin. Some African groups cut the face and forehead of the initiate in such a way as to leave deep scars.

Circumcision is, of course, the commonest of all initiation procedures. Its effect on the boy is, however, intensified in some places by an elaborate concern with his fortitude during the operation. The Maasai of East Africa, whose *moran* or warriors are world famous as epitomes of courage and bravado, closely watch a boy who is being circumcised for the slightest sign of cowardice. Even an involuntary twitch could make him an object of condemnation and scorn.

The tribal initiation gives girls a strong sense of the powers of women.

Initiation rituals are intended to provoke anxiety. They act out the death and rebirth of the initiate. His old self dies, and while he is in limbo he learns the mysteries of his society—instruction that is enhanced by fear and deprivation and by the atmosphere of awe that his teachers seek to create. In some societies that atmosphere is enhanced by the fact that the teachers are anonymous, masked figures representing the spirits. The lesson is often inscribed unforgettably on his body as well as in his mind. Later (the full cycle of ceremonies may last weeks or even months) he is reborn as an adult, often literally crawling between the legs of his sponsor to be reborn of man into the world of men.

Girls' initiation ceremonies are as dramatically marked in some societies as those of boys. Audrey Richards' account of the *chisungu,* a month-long initiation ceremony among the Bemba of Zambia, describes the complex ritual that does not so much add to the girl's practical knowledge as inculcate certain attitudes—a respect for age, for senior women and men, for the mystical bonds between husband and wife, for what the Bemba believe to be the dangerous potentials of sex, fire, and blood. The initiate learns the secret names of things and the songs and dances known only to women. She is incorporated into the group of women who form her immediate community, since this is a society that traces descent in the female line and a husband moves to his wife's village when they marry. Western writers tend to assume that it is more important for boys to undergo separation from their mothers as they mature than it is for girls. But the Bemba stress that mothers must surrender their daughters in the *chisungu* to the community at large (and to the venerable mistress of ceremonies in particular) as part of a process through which they will eventually gain sons-in-law.

The ceremony Richards observed for the initiation of three girls included 18 separate events, some 40 different pottery models (shaped for the occasion and destroyed immediately afterward), nearly a hundred songs, and numerous wall paintings and dances, all used to instruct the girls in their new status. All of this represents a large investment of time and resources. The initiation gives girls a strong sense of the solidarity and powers of women in a society that also stresses male authority and female submissiveness.

Ever since the influential work of Margaret Mead, there has been a tendency in the West to assume that, if growing up is less stressful in tribal societies, it is because they are less puritanical about sex. The modern world has, however, undergone a sexual revolution since Mead was writing in the 1930s and 1940s, and it does not seem to have made growing up much easier. I think that, in our preoccupation with sex, we miss the point. Take the case of tribal initiations. They not only make it clear to the initiates (and to the world at large) that they are now mature enough to have sex and to have children; the clarity also serves to enable the individual to move with a fair degree of certainty through clearly demarcated stages of life.

A moral economy

Since earliest times, the exchange of gifts has been the central mechanism through which human beings relate to one another. The reason is that the essence of a gift is obligation. A person who gives a gift compels the recipient either to make a return gift or to reciprocate in some other way. Obligation affects the givers as well. It is not entirely up to them whether or when to bestow a gift. Even in the modern world, which prides itself on its pragmatism, people are expected to give gifts on certain occasions—at weddings, at childbirth, at Christmas, and so on. People are expected to invite others to receive food and drink in their houses and those so invited are expected to return the favor.

In traditional societies, it is gifts that bond people to one another and make society work. It follows that in such societies a rich person is not somebody who accumulates wealth in money and goods but rather somebody who has a large network of people beholden to him. Such networks are the instrument through which prominent people can demonstrate their prestige. They are also the safety net that sees an individual through the crises of life.

In modern societies these networks have shrunk, just as the family continues to shrink. There are fewer and fewer people to whom we feel obligated and, more ominously, fewer and fewer who feel obligated to us. When we think of a safety net, when our politicians speak of it, we refer to arrangements made by abstract entities—the state, the corporation, the insurance company, the pension fund—entities we would not dream of giving presents to; entities we hope will provide for us (and fear they will not).

Traditional societies operate a moral economy, that is, an economy permeated by personal and moral considerations. In such a system, exchanges of goods in the "market" are not divorced from the personal relationships between those who exchange. On the contrary, the exchanges define those relationships. People who engage in such transactions select exchange partners who display integrity and reliability so that they can go back to them again and again. Even when cash enters such an economy, it does not automatically transform it. People still look for just prices, not bargain prices, and the system depends on trust and interdependence. In traditional societies the motto is "seller beware," for a person who gouges or shortchanges will become a moral outcast, excluded from social interaction with other people.

An ecology of mind

The sense of disconnection so characteristic of modern life affects not only the relations between people but equally importantly the relations between people and their environment. As a result, we may be gradually making the planet uninhabitable. The globe is warming up and is increasingly polluted. We cannot take fresh air or clean water for granted anymore. Even our vast oceans are starting to choke on human garbage. The rain forests are burning. The ozone layer is being depleted at rates that constantly exceed our estimates.

How have we come to this? A hundred years ago science seemed to hold such promising possibilities. But the scientific advances of the 19th century were built

Gift exchanges form the safety net that sees an individual through life's crises.

on the notion that human beings would master nature and make it produce more easily and plentifully for them. Medieval Christianity also taught that human beings, although they might be sinners, were created in God's image to have dominion over this earth. Whether human dominion was guaranteed by the Bible or by science, the result was the same—the natural world was ours to exploit.

Tribal societies, by contrast, have always had a strong sense of the interconnectedness of things on this earth and beyond. For example, human beings have, for the greater part of the history of our species on this earth, lived by hunting and gathering. Yet peoples who lived by hunting and gathering did not—and do not to this day—consider themselves the lords of creation. On the contrary, they are more likely to believe in (and work hard to maintain) a kind of reciprocity between human beings and the species they are obliged to hunt for food.

The reciprocity between hunter and hunted is elaborately expressed in the ideas of the Makuna Indians of southeastern Colombia. The Makuna believe that human beings, animals, and all of nature are parts of the same One. Their ancestors were fish people who came ashore along the rivers and turned into people. Out of their bodies or by their actions these ancestors created everything in the world, the hills and forests, the animals and the people. They carved out river valleys by pushing their sacred musical instruments in front of them.

People, animals, and fish all share the same spiritual essence and so, the Makuna say, animals and fish live in their own communities, which are just like human communities. They have their chiefs, their shamans, their dance houses, birth houses, and "waking up houses" (places where they originally came into being as species). They have their songs and dances and their material possessions. Above all, animals and fish are just like humans because they wear ritual ornaments, consume spirit foods—coca, snuff, and the hallucinogenic brew called *yage*—and use the sacred *yurupari* instruments in their ceremonies. When shamans blow over coca, snuff, and other spirit foods during human ceremonies, they are offering them to the animal people. When human beings dance in this world, the shaman invites the animal people to dance in theirs. If humans do not dance and shamans do not offer spirit food to the animal people, the animals will die out and there will be no more game left in this world.

Thus when the fish are spawning, they are actually dancing in their birth houses. That is why it is particularly dangerous to eat fish that have been caught at the spawning places, for then one eats a person who is ceremonially painted and in full dance regalia. A human being who does this or enters a fish house by mistake will sicken and die, for his soul will be carried away to the houses of the fish people.

Tribal people maintain a reciprocity with the species they must hunt for food.

It is clear that Makuna beliefs have specific ecological consequences. The sacredness of salt licks and fish-spawning places, the careful reciprocity between humans and their fellow animals and fish, all mediated by respected shamans, guarantee that the Makuna manage their environment and do not plunder it. The Swedish anthropologist Kaj Arhem, an authority on the Makuna, describes their ecological practices and cosmological speculations as an "ecosophy," where the radical division between nature and culture, humans and animals—so characteristic of Western thought—dissolves.

Arhem suggests that we need an ecosophy of our own, imbued with moral commitment and emotional power, if we are to protect the resources on which we depend and ensure not only our own survival but also that of our fellow creatures on this earth.

We, on the other hand, tend to forget our environment except when we want to extract wealth from it or use it as the backdrop for a scenic expedition. Then we take what we want. There is no compact, none of the reciprocity so characteristic of tribal societies. For the most part we mine the earth and leave it, for we do not feel we belong to it. It belongs to us. This rootlessness and the waste that goes with it are particularly shocking to traditional societies.

The Indians of the western United States were outraged by the way in which the invaders of their territories squandered the resources that they themselves used so sparingly. The Indians on the plains lived off the buffalo, killing only as many as they needed and using every bit of the dead animals. They ate the meat, made tents and clothes from the hides, and used the bones to make arrow straighteners, bows, mallets, even splints for setting fractures. They made butter from the marrow fat and cords from the sinews. When the white buffalo hunters came, it was more than an invasion. It was a sacrilege. These men slaughtered the herds with their powerful rifles, often taking only the tongue to eat and leaving the rest of the animal to rot.

The deep sadness of the Indians over this slaughter was expressed in a speech attributed to Chief Seattle, after whom the city of Seattle is named,believed to have been delivered in 1854 to an assembly of tribes preparing to sign away their lands under duress to the white man. Some contend the speech was actually written by a Texas speechwriter in 1971. Whatever their origin, these moving words convey an environmental and spiritual ethic that most tribal people share. They speak as much to us about our own predicament as they did to Chief Seattle's fellow chiefs about their defeated civilization. "What is man without the beasts?" he asked. "If all the beasts were gone, man would die from a great loneliness of spirit. For whatever happens to the beasts, soon happens to man. All things are connected. . . . We know that the white man does not understand our ways. One portion of the land is the same to him as the next, for he is a stranger who comes in the night and takes from the land whatever he needs. The earth is not his brother, but his enemy, and when he has conquered it, he moves on. He leaves his fathers' graves behind, and he does not care. He kidnaps the earth from his children. He does not care. His fathers' graves and his children's birthright are forgotten. He treats his mother, the earth, and his brother, the sky, as things to be bought, plundered, sold like sheep or bright beads. His appetite will devour the earth and leave behind only a desert."

Touching the timeless

Modern society is intensely secular. Even those who regret this admit it. Social theorists tend to assume that modernization is itself a process of secularization that has not only undermined people's religious beliefs but has also deprived them of their spirituality. In the industrial nations of the West many of the people who believe in God do not expect to come into close contact with the divine, except after death—and some of them are not too sure about it even then.

Indeed, it seems that those who live in the secular and industrialized West are already searching for ways to fill the vacuum in their lives left by "organized" religion and the numbing delights of mass society. We live in a world that prides itself on its modernity yet is hungry for wholeness, hungry for meaning. At the same time it is a world that marginalizes the very impulses that might fill this void. The pilgrimage toward the divine, the openness to knowledge that transcends ordinary experience, the very idea of feeling at one with the universe are impulses we tolerate only at the fringes of our society.

It seems that we denigrate our capacity to dream and so condemn ourselves to live in a disenchanted world. Shorn of the knowledge that we are part of something greater than ourselves, we also lose the sense of responsibility that comes with it. It is this connectedness that tribal societies cherish. Yet for modern society, this is a bond we cannot bring ourselves to seek. But if we do not listen to other traditions, do not even listen to our inner selves, then what will the future hold for our stunted and overconfident civilization?

The tightrope of power

Meanwhile, this civilization of ours, at once so powerful and so insecure, rolls like a juggernaut over societies that have explored the very solutions that might help us save ourselves. We do so in the name of progress, insisting all too often that we offer science, truth, plenty, and social order to peoples who lack these things. Yet the contrast between tribal societies and the centralized states that prey on them is not one of order and disorder, violence and peace. It is instead a contrast between societies in which no one has a monopoly on the legitimate use of force and others in which those rights are vested in a state. The 20th century has been one of the bloodiest in history, not only because of the wars between countries employing weapons of mass destruction but also because modern technology has been used by ruthless rulers to cow their own subjects. Hitler and Stalin are only the most notorious examples of dictators who directed violence against their own people in the name of the state. There are literally scores of shooting wars going on at the moment, most of them between states and their own subjects.

The state guarantees order, or is supposed to. Force, the monopoly of the government, is applied massively but, once the system is in place, relatively invisibly. Its victims are hidden in concentration camps or banished to Siberias. In many places today, the victims simply disappear.

It seems that people will often acquiesce in despotism for fear of anarchy. Recent history seems to indicate that the most advanced countries are more afraid of anarchy than they are of oppression. The Russians, whose whole history is a struggle to create order on the open steppes of Eurasia, have a fear of disorder (which they call *besporyadok*, the condition of not being "lined up") that has frequently led them to accept tyranny. At the other extreme, the United States, whose whole history is a determination to avoid despotism, allows more internal chaos than most other industrial nations. It values individual freedom to the point of allowing private citizens to own arsenals of weapons and puts up with a rate of interpersonal violence that would be considered catastrophic in other countries.

It seems that human beings are everywhere searching for the right balance between the mob and the dictator, between chaos and tyranny, between the individual and society. Industrial societies give a monopoly of power to the state in exchange for a guarantee of peace. We take this social order for granted to the extent that we tend to assume that there is anarchy and perpetual warfare in tribal societies. What we do not realize is that such societies are acutely conscious of the fragility of the social order and of the constant effort needed to maintain it. Paradoxically, the people who live in societies that do not have formal political institutions are more political than those who do since it is up to each individual to make sure that the system works, indeed to ensure that the system continues to exist at all. Tribal people avoid the perils of anarchy only through constant and unremitting effort.

Elijah Harper, an Ojibwa-Cree who is a member of parliament in the Canadian province of Manitoba, contrasted the democratic procedures of the native Canadians he represented with those of the Canadian government that was trying to push through a revision of Canada's constitution. The new constitution was designed to respond to Quebec's demand to be considered a distinct society within Canada, with appropriate protection for its own language and culture. Harper used parliamentary procedure to block the constitutional change, on the grounds that native Canadians had been asking for similar consideration for years without getting a hearing. A new round of discussions concerning the revision of Canada's constitution is now taking place and this time the rights of Canada's "first nations," the aboriginal peoples, are also on the agenda.

The Canadian crisis makes clear what is only dimly perceived in other countries, namely that the destiny of the majority in any state is intimately linked to the fate of its minorities. The failure of the first attempt to change their constitution has forced Canadians to think about what kind of society they want theirs to be. These are the same questions that the Aborigines are trying to put on the Australian agenda and that the Indians are forcing Brazilians to think about as they protest against the rape of Amazonian regions.

It is not only in authoritarian states that questions arise about how people within a state are allowed to go about their business. The dramatic events in Eastern Europe, however, have led some people to think so. Once the heavy hand of Communist dictatorship was lifted, the nations of Eastern Europe started to unravel. Old ethnic loyalties surfaced and ethnic rivalries threaten to dismember one nation after another. The problem in Eastern Europe is not that it is made up of more peoples than states, but rather that the states have not been successful in working out political solutions that could enable those peoples to live together amicably. But neither do democratic regimes find it easy to create more imaginative solutions that allow diverse groups of people to live together.

The reason for this failure is that such solutions require us to have a different idea of the state, a kind of new federalism which, after the manner of the League of the Iroquois, permits each people in the nation to keep its council fire alight. This requires more than rules; it requires commitment. The Great Law of the Iroquois was remarkable because it was a constitution that had the force of a religion. People were willing, indeed eager, to subscribe to it because they saw it and revered it as the source of peace. Is it too much to hope that in a world riven with ethnic conflict we might search for political solutions more energetically than we have in the past? That we will not continue to expect strong states to iron out ethnicity, even if it means wiping out the "ethnics"? A new federalism is in our own interest, for it offers the hope of peace and the prospect of justice. Nations that trample on the rights of the weak are likely to end up trampling on everybody's rights. As we wring our hands over the fate of tribal peoples in the modern world, we would do well to remember John Donne's words: "Never send to know for whom the bell tolls; it tolls for thee."

Serious consideration of tribal ways of life should lead us to think carefully and critically about our own. What would it take for us to try to live in harmony with nature or to rehumanize our economic systems? How can we mediate between the individual and the family, between genders and generations? Should we strive for a less fragmented view of physical reality or of our place in the scheme of things? These questions revolve around wholeness and harmony, around tolerance and pluralism. The answers are still emerging, but they too are variations on a grand theme that can be summed up in E. M. Forster's famous phrase: "Only connect." The project for the new millennium will be to re-energize civil society, the space between the state and the individual where those habits of the heart that socialize the individual and humanize the state flourish.

The Mountain People

Colin M. Turnbull

Anthropologist Colin M. Turnbull, author of The Forest People *and* The Lonely Africans, *went to study the Ik of Uganda, who he believed were still primarily hunters, in order to compare them with other hunting-and-gathering societies he had studied in totally different environments. He was surprised to discover that they were no longer hunters but primarily farmers, well on their way to starvation and something worse in a drought-stricken land.*

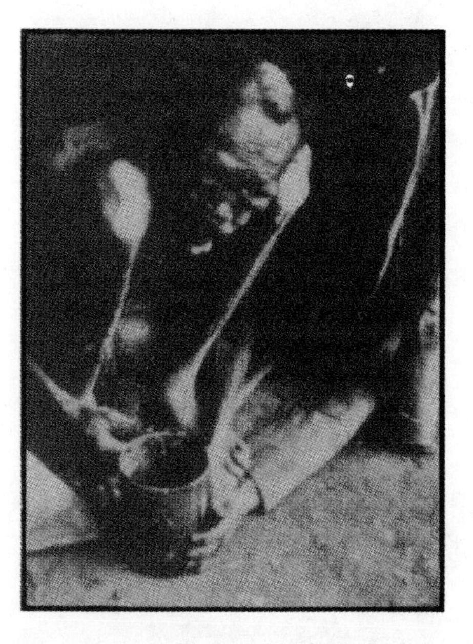

In what follows, there will be much to shock, and the reader will be tempted to say, "how primitive, how savage, how disgusting," and, above all, "how inhuman." The first judgments are typical of the kind of ethno- and egocentricism from which we can never quite escape. But "how inhuman" is of a different order and supposes that there are certain values inherent in humanity itself, from which the people described here seem to depart in a most drastic manner. In living the experience, however, and perhaps in reading it, one finds that it is oneself one is looking at and questioning; it is a voyage in quest of the basic human and a discovery of his potential for inhumanity, a potential that lies within us all.

Just before World War II the Ik tribe had been encouraged to settle in northern Uganda, in the mountainous northeast corner bordering on Kenya to the east and Sudan to the north. Until then they had roamed in nomadic bands, as hunters and gatherers, through a vast region in all three countries. The Kidepo Valley below Mount Morungole was their major hunting territory. After they were confined to a part of their former area, Kidepo was made a national park and they were forbidden to hunt or gather there.

The concept of family in a nomadic society is a broad one; what really counts most in everyday life is community of residence, and those who live close to each other are likely

to see each other as effectively related, whether there is any kinship bond or not. Full brothers, on the other hand, who live in different parts of the camp may have little concern for each other.

It is not possible, then, to think of the family as a simple, basic unit. A child is brought up to regard any adult living in the same camp as a parent, and age-mate as a brother or sister. The Ik had this essentially social attitude toward kinship, and it readily lent itself to the rapid and disastrous changes that took place following the restriction of their movement and hunting activities. The family simply ceased to exist.

It is a mistake to think of small-scale societies as "primitive" or "simple." Hunters and gatherers, most of all, appear simple and straightforward in terms of their social organization, yet that is far from true. If we can learn about the nature of society from a study of small-scale societies, we can also learn about human relationships. The smaller the society, the less emphasis there is on the formal system and the more there is on interpersonal and intergroup relations. Security is seen in terms of these relationships, and so is survival. The result, which appears so deceptively simple, is that hunters frequently display those characteristics that we find so admirable in man: kindness, generosity, consideration, affection, honesty, hospitality, compassion, charity. For them, in their tiny, close-knit society, these are necessities for survival. In our society anyone possessing even half these qualities would find it hard to survive, yet we think these virtues are inherent in man. I took it for granted that the Ik would possess these same qualities. But they were as unfriendly, uncharitable, inhospitable and generally mean as any people can be. For those positive qualities we value so highly are no longer functional for them; even more than in our own society they spell ruin and disaster. It seems that, far from being basic human qualities, they are luxuries we can afford in times of plenty or are mere mechanisms for survival and security. Given the situation in which the Ik found themselves, man has no time for such luxuries, and a much more basic man appears, using more basic survival tactics.

Turnbull had to wait in Kaabong, a remote administration outpost, for permission from the Uganda government to continue to Pirre, the Ik water hole and police post. While there he began to learn the Ik language and became used to their constant demands for food and tobacco. An official in Kaabong gave him, as a "gift," 20 Ik workers to build a house and a road up to it. When they arrived at Pirre, however, wages for the workers were negotiated by wily Atum, "the senior of all the Ik on Morungole."

The police seemed as glad to see me as I was to see them. They hungrily asked for news of Kaabong, as though it were the hub of the universe. They had a borehole and pump for water, to which they said I was welcome, since the water holes used by the Ik were not fit for drinking or even for washing. The police were not able to tell me much about the Ik, because every time they went to visit an Ik village, there was nobody there. Only in times of real hunger did they see much of the Ik, and then only enough to know that they were hungry.

The next morning I rose early, but even though it was barely daylight, by the time I had washed and dressed, the Ik were already outside. They were sitting silently, staring at the Land Rover. As impassive as they seemed, there was an air of expectancy, and I was reminded that these were, after all, hunters, and the likelihood was that I was their morning's prey. So I left the Land Rover curtains closed and as silently as possible prepared a frugal breakfast.

Atum was waiting for me. He said that he had told all the Ik that Iciebam [friend of the Ik] had arrived to live with them and that I had given the workers a "holiday" so they could greet me. They were waiting in the villages. They were very hungry, he added, and many were dying. That was probably one of the few true statements he ever made, and I never even considered believing it.

There were seven villages in all. Village Number One was built on a steep slope, and even the houses tilted at a crazy angle. Atum rapped on the outer stockade with his cane and shouted a greeting, but there was no response. This was Giriko's village, he said, and he was one of my workers.

"But I thought you told them to go back to their villages," I said.

"Yes, but you gave them a holiday, so they are probably in their fields," answered Atum, looking me straight in the eye.

At Village Number Two there was indisputably someone inside, for I could hear loud singing. The singing stopped, a pair of hands gripped the stockade and a craggy head rose into view, giving me an undeniably welcoming smile. This was Lokelea. When I asked him what he had been singing about, he answered, "Because I'm hungry."

Village Number Three, the smallest of all, was empty. Village Number Four had only 8 huts, as against the 12 or so in Lokelea's village and the 18 in Giriko's. The outer stockade was broken in one section, and we walked right in. We ducked through a low opening and entered a compound in which a woman was making pottery. She kept on at her work but gave us a cheery welcome and laughed her head off when I tried to speak in Icietot. She willingly showed me details of her work and did not seem unduly surprised at my interest. She said that everyone else had left for the fields except old Nangoli, who, on hearing her name mentioned, appeared at a hole in the stockade shutting off the next compound. Nangoli mumbled toothlessly at Losike, who told Atum to pour her some water.

As we climbed up to his own village, Number Five, Atum said that Losike never gave anything away. Later I remembered that gift of water to Nangoli. At the time I did not stop to think that in this country a gift of water could be a gift of life.

Atum's village had nearly 50 houses, each within its compound within the stout outer stockade. Atum did not invite me in.

A hundred yards away stood Village Number Six. Kauar, one of the workers, was sitting on a rocky slab just outside the village. He had a smile like Losike's, open and warm, and he said he had been waiting for me all morning. He offered us water and showed me his own small compound and that of his mother.

Coming up from Village Number Seven, at quite a respectable speed, was a blind man. This was Logwara, emaciated but alive and remarkably active. He had heard us and had come to greet me, he said, but he added the inevitable demand for tobacco in the

same breath. We sat down in the open sunlight. For a brief moment I felt at peace.

After a short time Atum said we should start back and called over his shoulder to his village. A muffled sound came from within, and he said, "That's my wife, she is very sick—and hungry." I offered to go and see her, but he shook his head. Back at the Land Rover I gave Atum some food and some aspirin, not knowing what else to give him to help his wife.

I was awakened well before dawn by the lowing of cattle. I made an extra pot of tea and let Atum distribute it, and then we divided the workers into two teams. Kauar was to head the team building the house, and Lokelatom, Losike's husband, was to take charge of the road workers.

While the Ik were working, their heads kept turning as though they were expecting something to happen. Every now and again one would stand up and peer into the distance and then take off into the bush for an hour or so. On one such occasion, after the person had been gone two hours, the others started drifting off. By then I knew them better; I looked for a wisp of smoke and followed it to where the road team was cooking a goat. Smoke was a giveaway, though, so they economized on cooking and ate most food nearly raw. It is a curious hangover from what must once have been a moral code that Ik will offer food if surprised in the act of eating, though they now go to enormous pains not to be so surprised.

I was always up before dawn, but by the time I got up to the villages they were always deserted. One morning I followed the little *oror* [gulley] up from *oror a pirre'i* [Ravine of Pirre] while it was still quite dark, and I met Lomeja on his way down. He took me on my first illicit hunt in Kidepo. He told me that if he got anything he would share it with me and with anyone else who managed to join us but that he certainly would not take anything back to his family. "Each one of them is out seeing what he can get for himself, and do you think they will bring any back for me?"

Lomeja was one of the very few Ik who seemed glad to volunteer information. Unlike many of the others, he did not get up and leave as I approached. Apart from him, I spent most of my time, those days, with Losike, the potter. She told me that Nangoli, the old lady in the adjoining compound, and her husband, Amuarkuar, were rather peculiar. They helped each other get food and water, and they brought it back to their compound to eat together.

I still do not know how much real hunger there was at that time, for most of the younger people seemed fairly well fed, and the few skinny old people seemed healthy and active. But my laboriously extracted genealogies showed that there were quite a number of old people still alive and allegedly in these villages, though they were never to be seen. Then Atum's wife died.

Atum told me nothing about it but kept up his demands for food and medicine. After a while the beady-eyed Lomongin told me that Atum was selling the medicine I was giving him for his wife. I was not unduly surprised and merely remarked that that was too bad for his wife. "Oh no," said Lomongin, "she has been dead for weeks."

It must have been then that I began to notice other things that I suppose I had chosen to ignore before. Only a very few of the Ik helped me with the language. Others would understand when it suited them and would pretend they did not understand when they did not want to listen. I began to be forced into a similar isolationist attitude myself, and although I cannot say I enjoyed it, it did make life much easier. I even began to enjoy, in a peculiar way, the company of the silent Ik. And the more I accepted it, the less often people got up and left as I approached. On one occasion I sat on the *di* [sitting place] by Atum's rain tree for three days with a group of Ik, and for three days not one word was exchanged.

The work teams were more lively, but only while working. Kauar always played and joked with the children when they came back from foraging. He used to volunteer to make the two-day walk into Kaabong and the even more tiring two-day climb back to get mail for me or to buy a few things for others. He always asked if he had made the trip more quickly than the last time.

Then one day Kauar went to Kaabong and did not come back. He was found on the last peak of the trail, cold and dead. Those who found him took the things he had been carrying and pushed his body into the bush. I still see his open, laughing face, see him giving precious tidbits to the children, comforting some child who was crying, and watching me read the letters he carried so lovingly for me. And I still think of him probably running up that viciously steep mountainside so he could break his time record and falling dead in his pathetic prime because he was starving.

Once I settled down into my new home, I was able to work more effectively. Having recovered at least some of my anthropological detachment, when I heard the telltale rustling of someone at my stockade, I merely threw a stone. If when out walking I stumbled during a difficult descent and the Ik shrieked with laughter, I no longer even noticed it.

Anyone falling down was good for a laugh, but I never saw anyone actually trip anyone else. The adults were content to let things happen and then enjoy them; it was probably conservation of energy. The children, however, sought their pleasures with vigor. The best game of all, at this time, was teasing poor little Adupa. She was not so little—in fact she should have been an adult, for she was nearly 13 years old—but Adupa was a little mad. Or you might say she was the only sane one, depending on your point of view. Adupa did not jump on other people's play houses, and she lavished enormous care on hers and would curl up inside it. That made it all the more jump-on-able. The other children beat her viciously.

Children are not allowed to sleep in the house after they are "put out," which is at about three years old, four at the latest. From then on they sleep in the open courtyard, taking what shelter they can against the stockade. They may ask for permission to sit in the doorway of their parents' house but may not lie down or sleep there. "The same thing applies to old people," said Atum, "if they can't build a house of their own and, of course, *if* their children let them stay in their compounds."

I saw a few old people, most of whom had taken over abandoned huts. For the first time I realized that there really was starvation and saw why I had never known it before: it was confined to the aged. Down in Giriko's village the old ritual priest, Lolim, confidentially told me that he was sheltering an old man who had been refused shelter by his son. But Lolim did not have enough food for himself, let alone his guest; could I . . . I liked old Lolim, so, not believing that Lolim had a visitor at all, I brought him a double ration that evening. There was a rustling in the back of the hut, and Lolim helped ancient Lomeraniang to the entrance. They shook with delight at the sight of the food.

When the two old men had finished eating, I left; I found a hungry-looking and disapproving little crowd clustered outside. They muttered to each other about wasting food. From then on I brought food daily, but in a very short time Lomeraniang was dead, and his son refused to come down from the village above to bury him. Lolim scratched a hole and covered the body with a pile of stones he carried himself, one by one.

Hunger was indeed more severe than I knew, and, after the old people, the children were the next to go. It was all quite impersonal—even to me, in most cases, since I had been immunized by the Ik themselves against sorrow on their behalf. But Adupa was an exception. Her madness was such that she did not know just how vicious humans could be. Even worse, she thought that parents were for loving, for giving as well as receiving. Her parents were not given to fantasies. When she came for shelter, they drove her out; and when she came because she was hungry, they laughed that Icien laugh, as if she had made them happy.

Adupa's reactions became slower and slower. When she managed to find food—fruit peels, skins, bits of bone, half-eaten berries—she held it in her hand and looked at it with wonder and delight. Her playmates caught on quickly; they put tidbits in her way and watched her simple drawn little face wrinkle in a smile. Then as she raised her hand to her mouth, they set on her with cries of excitement, fun and laughter, beating her savagely over the head. But that is not how she died. I took to feeding her, which is probably the cruelest thing I could have done, a gross selfishness of my part to try to salve my own rapidly disappearing conscience. I had to protect her, physically, as I fed her. But the others would beat her anyway, and Adupa cried, not because of the pain in her body but because of the pain she felt at the great, vast, empty wasteland where love should have been.

It was *that* that killed her. She demanded that her parents love her. Finally they took her in, and Adupa was happy and stopped crying. She stopped crying forever because her parents went away and closed the door tight behind them, so tight that weak little Adupa could never have moved it.

The Ik seem to tell us that the family is not such a fundamental unit as we usually suppose, that it is not essential to social life. In the crisis of survival facing the Ik, the family was one of the first institutions to go, and the Ik as a society have survived.

The other quality of life that we hold to be necessary for survival—love—the Ik dismiss as idiotic and highly dangerous. But we need to see more of the Ik before their absolute lovelessness becomes truly apparent.

In this curious society there is one common value to which all Ik hold tenaciously. It is *ngag,* "food." That is the one standard by which they measure right and wrong, goodness and badness. The very word for "good" is defined in terms of food. "Goodness" is "the possession of food," or the "*individual* possession of food." If you try to discover their concept of a "good man," you get the truly Icien answer: one who has a full stomach.

We should not be surprised, then, when the mother throws her child out at three years old. At that age a series of *rites de passage* begins. In this environment a child has no chance of survival on his own until he is about 13, so children form age bands. The junior band consists of children between three and seven, the senior of eight- to twelve-year-olds. Within the band each child seeks another close to him in age for defense against the older children. These friendships are temporary, however, and in-

evitably there comes a time when each turns on the one that up to then had been the closest to him; that is the *rite de passage,* the destruction of that fragile bond called friendship. When this has happened three or four times, the child is ready for the world.

The weakest are soon thinned out, and the strongest survive to achieve leadership of the band. Such a leader is eventually driven out, turned against by his fellow band members. Then the process starts all over again; he joins the senior age band as its most junior member.

The final *rite de passage* is into adulthood, at the age of 12 or 13. By then the candidate has learned the wisdom of acting on his own, for his own good, while acknowledging that on occasion it is profitable to associate temporarily with others.

One year in four the Ik can count on a complete drought. About this time it began to be apparent that there were going to be two consecutive years of drought and famine. Men as well as women took to gathering what wild fruits and berries they could find, digging up roots, cutting grass that was going to seed, threshing and eating the seed.

Old Nangoli went to the other side of Kidepo, where food and water were more plentiful. But she had to leave her husband, Amuarkuar, behind. One day he appeared at my *odok* and asked for water. I gave him some and was going to get him food when Atum came storming over and argued with me about wasting water. In the midst of the dispute Amuarkuar quietly left. He wandered over to a rocky outcrop and lay down there to rest. Nearby was a small bundle of grass that evidently he had cut and had been dragging painfully to the ruins of his village to make a rough shelter. The grass was his supreme effort to keep a home going until Nangoli returned. When I went over to him, he looked up and smiled and said that my water tasted good. He lay back and went to sleep with a smile on his face. That is how Amuarkuar died, happily.

There are measures that can be taken for survival involving the classical institutions of gift and sacrifice. These are weapons, sharp and aggressive. The object is to build up a series of obligations so that in times of crisis you have a number of debts you can recall; with luck one of them may be repaid. To this end, in the circumstances of Ik life, considerable sacrifice would be justified, so you have the odd phenomenon of these otherwise singularly self-interested people going out of their way to "help" each other. Their help may very well be resented in the extreme, but is done in such a way that it cannot be refused, for it has already been given. Someone may hoe another's field in his absence or rebuild his stockade or join in the building of a house.

The danger in this system was that the debtor might not be around when collection was called for and, by the same token, neither might the creditor. The future was too uncertain for this to be anything but one additional survival measure, though some developed it to a fine technique.

There seemed to be increasingly little among the Ik that could by any stretch of the imagination be called social life, let alone social organization. The family does not hold itself together; economic interest is centered on as many stomachs as there are people; and cooperation is merely a device for furthering an interest that is consciously selfish. We often do the same thing in our so-called "altruistic" practices, but we tell ourselves it is for the good of others. The Ik have dispensed with the myth of altruism. Though they have no centralized leadership or means of physical coercion, they do hold together with remarkable tenacity.

In our world, where the family has also lost much of its value as a social unit and where religious belief no longer binds us into communities, we maintain order only through coercive power that is ready to uphold a rigid law and through an equally rigid penal system. The Ik, however, have learned to do without coercion, either spiritual or physical. It seems that they have come to a recognition of what they accept as man's basic selfishness, of his natural determination to survive as an individual before all else. This they consider to be man's basic right, and they allow others to pursue that right without recrimination.

In large-scale societies such as our own, where members are individual beings rather than social beings, we rely on law for order. The absence of both a common law and a common belief would surely result in lack of any community of behavior; yet Ik society is not anarchical. One might well expect religion, then, to play a powerful role in Icien life, providing a source of unity.

The Ik, as may be expected, do not run true to form. When I arrived, there were still three ritual priests alive. From them and from the few other old people, I learned something of the Ik's belief and practice as they had been before their world was so terribly changed. There had been a powerful unity of belief in Didigwari—a sky god—and a body of ritual practice reinforcing secular behavior that was truly social.

Didigwari himself is too remote to be of much practical significance to the Ik. He created them and abandoned them and retreated into his domain somewhere in the sky. He never came down to earth, but the *abang* [ancestors] have all known life on earth; it is only against them that one can sin and only to them that one can turn for help, through the ritual priest.

While Morungole has no legends attached to it by the Ik, it nonetheless figures in their ideology and is in some ways regarded by them as sacred. I had noticed this by the almost reverential way in which they looked at it—none of the shrewd cunning and cold appraisal with which they regarded the rest of the world. When they talked about it, there was a different quality to their voices. They seemed incapable of talking about Morungole in any other way, which is probably why they talked about it so very seldom. Even that weasel Lomongin became gentle the only time he talked about it to me. He said, "If Atum and I were there, we would not argue. It is a good place." I asked if he meant that it was full of food. He said yes. "Then why do Ik never go there?" "They do go there." "But if hunting is good there, why not live there?" "We don't hunt there, we just go there." "Why?" "I told you, it is a good place." If I did not understand him, that was my fault; for once he was doing his best to communicate something to me. With others it was the same. All agreed that it was "a good place." One added, "That is the Place of God."

Lolim, the oldest and greatest of the ritual priests, was also the last. He was not much in demand any longer, but he was still held in awe, which means kept at a distance. Whenever he approached a *di*, people cleared a space for him, as far away from themselves as possible. The Ik rarely called on his services, for they had little to pay him with, and he had equally little to offer them. The main things they did try to get out of him were certain forms of medicine, both herbal and magical.

Lolim said that he had inherited his power from his father. His father had taught him well but could not give him the power to hear the *abang*—that had to come from the *abang* themselves. He had wanted his oldest son to inherit and had taught him everything he could. But his son, Longoli, was bad, and the *abang* refused to talk to him. They talked instead to his oldest daughter, bald Nangoli. But there soon came the time when all the Ik needed was food in their stomachs, and Lolim could not supply that. The time came when Lolim was too weak to go out and collect the medicines he needed. His children all refused to go except Nangoli, and then she was jailed for gathering in Kidepo Park.

Lolim became ill and had to be protected while eating the food I gave him. Then the children began openly ridiculing him and teasing him, dancing in front of him and kneeling down so that he would trip over them. His grandson used to creep up behind him and with a pair of hard sticks drum a lively tattoo on the old man's bald head.

I fed him whenever I could, but often he did not want more than a bite. Once I found him rolled up in his protective ball, crying. He had had nothing to eat for four days and no water for two. He had asked his children, who all told him not to come near them.

The next day I saw him leaving Atum's village, where his son Longoli lived. Longoli swore that he had been giving his father food and was looking after him. Lolim was not shuffling away; it was almost a run, the run of a drunken man, staggering from side to side. I called to him, but he made no reply, just a kind of long, continuous and horrible moan. He had been to Longoli to beg him to let him into his compound because he knew he was going to die in a few hours, Longoli calmly told me afterward. Obviously Longoli could not do a thing like that: a man of Lolim's importance would have called for an enormous funeral feast. So he refused. Lolim begged Longoli then to open up Nangoli's *asak* for him so that he could die in *her* compound. But Longoli drove him out, and he died alone.

Atum pulled some stones over the body where it had fallen into a kind of hollow. I saw that the body must have lain parallel with the *oror*. Atum answered without waiting for the question: "He was lying looking up at Mount Meraniang."

Insofar as ritual survived at all, it could hardly be said to be religious, for it did little or nothing to bind Icien society together. But the question still remained: Did this lack of social behavior and communal ritual or religious expression mean that there was no community of belief?

Belief may manifest itself, at either the individual or the communal level, in what we call morality, when we behave according to certain principles supported by our belief even when it seems against our personal interest. When we call ourselves moral, however, we tend to ignore that ultimately our morality benefits us even as individuals, insofar as we are social individuals and live in a society. In the absence of belief, law takes over and morality has little role. If there was such a thing as an Icien morality, I had not yet perceived it, though traces of a moral past remained. But it still remained a possibility, as did the existence of an unspoken, unmanifest belief that might yet reveal itself and provide a basis for the reintegration of society. I was somewhat encouraged in this hope by the unexpected flight of old Nangoli, widow of Amuarkuar.

When Nangoli returned and found her husband dead, she did an odd thing: she grieved. She tore down what was left of their home, uprooted the stockade, tore up whatever was growing in her little field. Then she fled with a few belongings.

Some weeks later I heard that she and her children had gone over to the Sudan and built a village there. This migration was so unusual that I decided to see whether this runaway village was different.

Lojieri led the way, and Atum came along. One long day's trek got us there. Lojieri pulled part of the brush fence aside, and we went in and wandered around. He and Atum looked inside all the huts, and Lojieri helped himself to tobacco from one and water from another. Surprises were coming thick and fast. That households should be left open and untended with such wealth inside . . . That there should have been such wealth, for as well as tobacco and jars of water there were baskets of food, and meat was drying on racks. There were half a dozen or so compounds, but they were separated from each other only by a short line of sticks and brush. It was a village, and these were homes, the first and last I was to see.

The dusk had already fallen, and Nangoli came in with her children and grandchildren. They had heard us and came in with warm welcomes. There was no hunger here, and in a very short time each kitchen hearth had a pot of food cooking. Then we sat around the central fire and talked until late, and it was another universe.

There was no talk of "how much better it is here than there"; talk revolved around what had happened on the hunt that day. Loron was lying on the ground in front of the fire as his mother made gentle fun of him. His wife, Kinimei, whom I had never seen even speak to him at Pirre, put a bowl of fresh-cooked berries and fruit in front of him. It was all like a nightmare rather than a fantasy, for it made the reality of Pirre seem all the more frightening.

The unpleasantness of returning was somewhat alleviated by Atum's suffering on the way up the stony trail. Several times he slipped, which made Lojieri and me laugh. It was a pleasure to move rapidly ahead and leave Atum gasping behind so that we could be sitting up on the *di* when he finally appeared and could laugh at his discomfort.

The days of drought wore on into weeks and months and, like everyone else, I became rather bored with sickness and death. I survived rather as did the young adults, by diligent attention to my own needs while ignoring those of others.

More and more it was only the young who could go far from the village as hunger became starvation. Famine relief had been initiated down at Kasile, and those fit enough to make the trip set off. When they came back, the contrast between them and the others was that between life and death. Villages were villages of the dead and dying, and there was little difference between the two. People crawled rather than walked. After a few feet some would lie down to rest, but they could not be sure of ever being able to sit up again, so they mostly stayed upright until they reached their destination. They were going nowhere, these semianimate bags of skin and bone; they just wanted to be with others, and they stopped whenever they met. Perhaps it was the most important demonstration of sociality I ever saw among the Ik. Once they met, they neither spoke nor did anything together.

Early one morning, before dawn, the village moved. In the midst of a hive of activity were the aged and crippled, soon to be abandoned, in danger of being trampled but seemingly unaware of it. Lolim's widow, Lo'ono, whom I had never seen before, also had been abandoned and had tried to make her way down the mountainside. But she was totally blind and had tripped and rolled to the bottom of the *oror a pirre'i;* there she lay on her back, her legs and arms thrashing feebly, while a little crowd laughed.

At this time a colleague was with me. He kept the others away while I ran to get medicine and food and water, for Lo'ono was obviously near dead from hunger and thirst as well as from the fall. We treated her and fed her and asked her to come back with us. But she asked us to point her in the direction of her son's new village. I said I did not think she would get much of a welcome there, and she replied that she knew it but wanted to be near him when she died. So we gave her more food, put her stick in her hand and pointed her the right way. She suddenly cried. She was crying, she said, because we had reminded her that there had been a time when people had helped each other, when people had been kind and good. Still crying, she set off.

The Ik up to this point had been tolerant of my activities, but all this was too much. They said that what we were doing was wrong. Food and medicine were for the living, not the dead. I thought of Lo'ono. And I thought of other old people who had joined in the merriment when they had been teased or had a precious morsel of food taken from their mouths. They knew that it was silly of them to expect to go on living, and, having watched others, they knew that the spectacle really was quite funny. So they joined in the laughter. Perhaps if we had left Lo'ono, she would have died laughing. But we prolonged her misery for no more than a few brief days. Even worse, we reminded her of when things had been different, of days when children had cared for parents and parents for children. She was already dead, and we made her unhappy as well. At the time I was sure we were right, doing the only "human" thing. In a way we *were*—we were making life more comfortable for ourselves. But now I wonder if the Ik way was not right, if I too should not have laughed as Lo'ono flapped about, then left her to die.

Ngorok was a man at 12. Lomer, his older brother, at 15 was showing signs of strain; when he was carrying a load, his face took on a curious expression of pain that was no physical pain. Giriko, at 25 was 40, Atum at 40 was 65, and the very oldest, perhaps a bare 50, were centenarians. And I, at 40, was younger than any of them, for I still enjoyed life, which they had learned was not "adult" when they were 3. But they retained their will to survive and so offered grudging respect to those who had survived for long.

Even in the teasing of the old there was a glimmer of hope. It denoted a certain intimacy that did not exist between adjacent generations. This is quite common in small-scale societies. The very old and the very young look at each other as representing the future and the past. To the child, the aged represent a world that existed before their own birth and the unknown world to come.

And now that all the old are dead, what is left? Every Ik who is old today was thrown out at three and has survived, and in consequence has thrown his own children out and knows that they will not help him in his old age any more than he helped his parents. The system has turned one full cycle and is now self-perpetuating; it has eradicated what we know as "humanity" and has turned the world into a chilly void where man does not seem to care even for himself, but survives. Yet into this hideous world Nangoli and her family quietly returned because they could not bear to be alone.

For the moment abandoning the very old and the very young, the Ik as a whole must be searched for one last lingering trace of humanity. They appear to have disposed of virtually all the qualities that we normally think of as differentiating us from other primates, yet they survive without seeming to be greatly different from ourselves in terms of behavior. Their behavior is more extreme, for we do not start throwing our children out until kindergarten. We have shifted responsibility from family to state, the Ik have shifted it to the individual.

It has been claimed that human beings are capable of love and, indeed, are dependent upon it for survival and sanity. The Ik offer us an opportunity for testing this cherished notion that love is essential to survival. If it is, the Ik should have it.

Love in human relationships implies mutuality, a willingness to sacrifice the self that springs from a consciousness of identity. This seems to bring us back to the Ik, for it implies that love is self-oriented, that even the supreme sacrifice of one's life is no more than selfishness, for the victim feels amply rewarded by the pleasure he feels in making the sacrifice. The Ik, however, do not value emotion above survival, and they are without love.

But I kept looking, for it was the one thing that could fill the void their survival tactics had created; and if love was not there in some form, it meant that for humanity love is not a necessity at all, but a luxury or an illusion. And if it was not among the Ik, it meant that mankind can lose it.

The only possibility for any discovery of love lay in the realm of interpersonal relationships. But they were, each one, simply alone, and seemingly content to be alone. It was this acceptance of individual isolation that made love almost impossible. Contact, when made, was usually for a specific practical purpose having to do with food and the filling of a stomach, a single stomach. Such contacts did not have anything like the permanence or duration required to develop a situation in which love was possible.

The isolation that made love impossible, however, was not completely proof against loneliness. I no longer noticed normal behavior, such as the way people ate, running as they gobbled, so as to have it all for themselves. But I did notice that when someone was making twine or straightening a spear shaft, the focus of attention for the spectators was not the person but the action. If they were caught watching by the one being watched and their eyes met, the reaction was a sharp retreat on both sides.

When the rains failed for the second year running, I knew that the Ik as a society were almost certainly finished and that the monster they had created in its place, that passionless, feelingless association of individuals, would spread like a fungus, contaminating all it touched. When I left, I too had been contaminated. I was not upset when I said good-bye to old Loiangorok. I told him I had left a sack of *posho* [ground corn meal] with the police for him, and I said I would send money for more when that ran out. He dragged himself slowly toward the *di* every day, and he always clutched a knife. When he got there, or as far as he could, he squatted down and whittled at some wood, thus proving that he was still alive and able to do things. The *posho* was enough to last him for months, but I felt no emotion when I estimated that he would last one month, even with the *posho* in the hands of the police. I underestimated his son, who within two days had persuaded the police that it would save a lot of bother if he looked after the *posho*. I heard later that Loiangorok died of starvation within two weeks.

So, I departed with a kind of forced gaiety, feeling that I should be glad to be gone but having forgotten how to be glad. I certainly was not thinking of returning within a year, but I did. The following spring I heard that rain had come at last and that the fields of the

Ik had never looked so prosperous, nor the country so green and fertile. A few months away had refreshed me, and I wondered if my conclusions had not been excessively pessimistic. So, early that summer, I set off to be present for the first harvests in three years.

I was not surprised too much when two days after my arrival and installation at the police post I found Logwara, the blind man, lying on the roadside bleeding, while a hundred yards up other Ik were squabbling over the body of a hyena. Logwara had tried to get there ahead of the others to grab the meat and had been trampled on.

First I looked at the villages. The lush outer covering concealed an inner decay. All the villages were like this to some extent, except for Lokelea's. There the tomatoes and pumpkins were carefully pruned and cleaned, so that the fruits were larger and healthier. In what had been my own compound the shade trees had been cut down for firewood, and the lovely hanging nests of the weaver birds were gone.

The fields were even more desolate. Every field without exception had yielded in abundance, and it was a new sensation to have vision cut off by thick crops. But every crop was rotting from sheer neglect.

The Ik said that they had no need to bother guarding the fields. There was so much food they could never eat it all, so why not let the birds and baboons take some? The Ik had full bellies; they were good. The *di* at Atum's village was much the same as usual, people sitting or lying about. People were still stealing from each other's fields, and nobody thought of saving for the future.

It was obvious that nothing had really changed due to the sudden glut of food except that interpersonal relationships had deteriorated still further and that Icien individualism had heightened beyond what I thought even Ik to be capable of.

The Ik had faced a conscious choice between being humans and being parasites and had chosen the latter. When they saw their fields come alive, they were confronted with a problem. If they reaped the harvest, they would have to store grain for eating and planting, and every Ik knew that trying to store anything was a waste of time. Further, if they made their fields look too promising, the government would stop famine relief. So the Ik let their fields rot and continued to draw famine relief.

The Ik were not starving any longer; the old and infirm had all died the previous year, and the younger survivors were doing quite well. But the famine relief was administered in a way that was little short of criminal. As before, only the young and well were able to get down from Pirre to collect the relief; they were given relief for those who could not come and told to take it back. But they never did—they ate it themselves.

The facts are there, though those that can be read here form but a fraction of what one person was able to gather in under two years. There can be no mistaking the direction in which those facts point, and that is the most important thing of all, for it may affect the rest of mankind as it has affected the Ik. The Ik have "progressed," one might say, since the change that has come to them came with the advent of civilization to Africa. They have made of a world that was alive a world that is dead—a cold, dispassionate world that is without ugliness because it is without beauty, without hate because it is without love, and without any realization of truth even, because it simply is. And the symptoms of change in our own society indicate that we are heading in the same direction.

Those values we cherish so highly may indeed be basic to human society but not to humanity, and that means that the Ik show that society itself is not indispensable for man's survival and that man is capable of associating for purposes of survival without being social. The Ik have replaced human society with a mere survival system that does not take human emotion into account. As yet the system if imperfect, for although survival is assured, it is at a minimal level and there is still competition between individuals. With our intellectual sophistication and advanced technology we should be able to perfect the system and eliminate competition, guaranteeing survival for a given number of years for all, reducing the demands made upon us by a social system, abolishing desire and consequently that ever-present and vital gap between desire and achievement, treating us, in a word, as individuals with one basic individual right—the right to survive.

Such interaction as there is within this system is one of mutual exploitation. That is how it already is with the Ik. In our own world the mainstays of a society based on a truly social sense of mutuality are breaking down, indicating that perhaps society as we know it has outworn its usefulness and that by clinging to an outworn system we are bringing about our own destruction. Family, economy, government and religion, the basic categories of social activity and behavior, no longer create any sense of social unity involving a shared and mutual responsibility among all members of our society. At best they enable the individual to survive as an individual. It is the world of the individual, as is the world of the Ik.

The sorry state of society in the civilized world today is in large measure due to the fact that social change has not kept up with technological change. This mad, senseless, unthinking commitment to technological change that we call progress may be sufficient to exterminate the human race in a very short time even without the assistance of nuclear warfare. But since we have already become individualized and desocialized, we say that extermination will not come in our time, which shows about as much sense of family devotion as one might expect from the Ik.

Even supposing that we can avert nuclear holocaust or the almost universal famine that may be expected if population keeps expanding and pollution remains unchecked, what will be the cost if not the same already paid by the Ik? They too were driven by the need to survive, and they succeeded at the cost of their humanity. We are already beginning to pay the same price, but we not only still have the choice (though we may not have the will or courage to make it), we also have the intellectual and technological ability to avert an Icien end. Any change as radical as will be necessary is not likely to bring material benefits to the present generation, but only then will there be a future.

The Ik teach us that our much vaunted human values are not inherent in humanity at all but are associated only with a particular form of survival called society and that all, even society itself, are luxuries that can be dispensed with. That does not make them any less wonderful, and if man has any greatness, it is surely in his ability to maintain these values, even shortening an already pitifully short life rather than sacrifice his humanity. But that too involves choice, and the Ik teach us that man can lose the will to make it. That is the point at which there is an end to truth, to goodness and to beauty, an end to the struggle for their achievement, which gives life to the individual and strength and meaning to society. The Ik have relinquished all luxury in the name of individual survival, and they live on as a people without life, without passion, beyond humanity. We pursue those trivial, idiotic technological encumbrances, and all the time we are losing our potential for social rather than individual survival, for hating as well as loving, losing perhaps our last chance to enjoy life with all the passion that is our nature.

Overworked Americans or Overwhelmed Americans?

YOU CANNOT HANDLE EVERYTHING

JEFF DAVIDSON

Author and Management Consultant

Delivered before the U.S. Treasury Executive Institute, Washington, D.C., November 12, 1992

HERE IS A multiple-choice quiz question. Which word best describes the typical working American today: A) Overworked, B) Underworked, C) Energetic, D) Lazy.

While much has been written of late as to whether A, B, C, or D, is correct, the most appropriate answer may well be: "None of the above." Powerful social forces have the potential to turn each of us into human whirlwinds charging about in "fast forward." *Work, time away from work,* and *everything in between* appear as if they are all part of a never-ending, ever-lengthening to-do list, to be handled during days that race by.

To say that Americans work too many hours, and that too much work is at the root of the time-pressure we feel and the leisure we lack, is to miss the convergence of larger, more fundamental issues. We could handle the longer hours (actually less than 79 minutes more per day) that we work compared to the Europeans. It's everything else competing for our attention that leaves us feeling overwhelmed. Once overwhelmed, the feeling of overworked quickly follows.

Nearly every aspect of American society has become more complex even since the mid-1980s. Traveling is becoming more cumbersome. Learning new ways of managing, and new ways to increase productivity takes its toll. *Merely living* in America today and participating as a functioning member of society guarantees that your day, week, month, year and life, and your physical, emotional, and spiritual energy will easily be depleted without the proper vantage point from which to approach each day and conduct your life.

Do you personally know *anyone* who works for a living who consistently has unscheduled, free stretches? Five factors, or "mega-realities," are simultaneously contributing to the perceptual and actual erosion of leisure time among Americans, including:

–Population growth;
–An expanding volume of knowledge;
–Mass media growth and electronic addiction;
–The paper trail culture; and
–An over-abundance of choices.

Population

From the beginning of creation to 1850 A.D. world population grew to one billion. It grew to two billion by 1930, three billion by 1960, four billion by 1979, and five billion by 1987, with six billion en route. Every 33 months, the current population of America, 257,000,000 people, is added to the planet.

The world of your childhood is gone, forever. The present is crowded and becoming more so. Each day, world population (births minus deaths) increases by *more than 260,000 people.* Regardless of your political, religious, or economic views, the fact remains that geometric growth in human population permeates and dominates every aspect of the planet and its resources, the environment and all living things. This is the most compelling aspect of our existence, and will be linked momentarily to the four other mega-realities.

When JFK was elected President, domestic population was 180 million. It grew by 70 million in one generation. Our growing population has *not* dispersed over the nation's 5.4 million square miles. About 97 percent of the U.S. population resides on 3 percent of the land mass. Half of our population resides within 50 miles of the Atlantic or the Pacific Ocean, and 75 percent of the U.S. population live in urban areas, with 80 percent predicted by the end of the nineties.

More densely packed urban areas have resulted predictably in a gridlock of the nation's transportation systems. It *is* taking you longer merely to drive a few blocks; it's not your imagination, it's not the day of the week or the season, and it's not going to subside soon. Our population and road use grow faster than our ability to repair highways, bridges and arteries. In fact, vehicles (primarily cars) are multiplying twice as fast as people, currently approaching 400,000,000 vehicles, compared to 165,000,000 registered motorists.

From *Vital Speeches of the Day,* May 15, 1993, pp. 470-473. © 1993 by Jeff Davidson, M.B.A., C.M.C. Based on the book *Breathing Space: Living and Working at a Comfortable Pace in a Sped-Up Society* by Jeff Davidson, M.B.A., C.M.C., Chapel Hill, NC. Reprinted by permission.

Some 86 percent of American commuters still get to work by automobile, and 84 percent of inner city travel is by automobile. The average American now commutes 157,600 miles to work during his working life, equal to six times around the earth. Commuting snarls are increasing.

City planners report there will be no clear solution to gridlock for decades, and all population studies reveal that our nation's metropolitan areas will become home *to an even greater percentage of the population.* Even less populated urban areas will face unending traffic dilemmas. If only the gridlock were confined to commuter arteries. However, shoppers, air travelers, vacationers, even campers – everyone in motion is or will be feeling its effects.

Knowledge

Everybody in America fears that he/she is under-informed. This moment, you, and everyone you know, are being bombarded on all sides. *Over-information* wreaks havoc on the receptive capacities of the unwary. The volume of new knowledge broadcast and published in every field is enormous and exceeds anyone's ability to keep pace. All told, more words are published or broadcast *in a day* than you could comfortably ingest in the rest of your life. By far, America leads the world in the sheer volume of information generated and disseminated.

Increasingly, there is no body of knowledge that everyone can be expected to know. In its 140th year, for example, the Smithsonian Museum in Washington D.C. added 942,000 items to its collections. Even our language keeps expanding. Since 1966, more than 60,000 words have been added to the English language – equal to half or more of the words in some languages. Harvard Library subscribes to 160,000 journals and periodicals.

With more information comes more misinformation. Annually, more than 40,000 scientific journals publish over one million new articles. "The number of scientific articles and journals published worldwide is starting to confuse research, overwhelm the quality-control systems of science, encourage fraud, and distort the dissemination of important findings," says *New York Times* science journalist William J. Broad.

In America, too many legislators, regulators and others *entrusted* to devise the rules which guide the course of society *take shelter in the information overglut by intentionally adding to it.* We are saddled with 26-page laws that could be stated in two pages, and regulations that contradict themselves every fourth page. And, this phenomenon is not confined to Capital Hill. Impossible VCR manuals, insurance policies, sweepstakes instructions, and frequent flyer bonus plans all contribute to our immobility.

Media Growth

The effect of the mass media on our lives continues unchecked. Worldwide media coverage certainly yields benefits. Democracy springs forth when oppressed people have a chance to see or learn about how other people in free societies live. As we spend more hours tuned to electronic media, we are exposed to tens of thousands of messages and images.

In America, more than three out of five television households own VCRs, while the number of movie tickets sold and videos rented in the U.S. each exceeded one billion annually starting in 1988. More than 575 motion pictures are produced each year compared to an average of 175 twelve years ago. In 1972, three major television networks dominated television – ABC, NBC and CBS. There are now 339 full-power independent television stations and many cable TV subscribers receive up to 140 channels offering more than 72,000 shows per month.

> All told, the average American spends more than eight solid years watching electronically how other people supposedly live.

To capture overstimulated, distracted viewers, American television and other news media increasingly rely on sensationalism. LIke too much food at once, too much data, in any form, isn't easily ingested. You can't afford to pay homage to everyone else's 15 minutes of fame. As Neil Postman observed, in *Amusing Ourselves to Death: Public Discourse in the Age of Television,* with the three words, "and now this . . ." television news anchors are able to hold your attention while shifting gears 180 degrees.

Radio power – Radio listenership does not lag either. From 5:00 a.m. to 5:00 p.m. each weekday in America, listenership far surpasses that of television viewership. Unknown to most people, since television was first introduced, the number of radio stations has increased tenfold, and 97 percent of all households own an average of five radios, not counting their car radios. On weekdays, 95.2 percent of Americans listen to radio for three hours and fourteen minutes. Shock-talk disc jockeys make $300,000 to $600,000 per year and more, plus bonuses.

With a planet of more than five billion people, American media are easily furnished with an endless supply of turmoil for mass transmission. At any given moment somebody is fomenting revolution somewhere. Such turmoil is packaged daily for the evening news, whose credo has become, "If it bleeds, it leads." We are lured with images of crashes, hostages, and natural disasters. We offer our time and rapt attention to each new hostility, scandal or disaster. Far more people die annually from choking on food than in plane crashes or by guns, but crashes and shootings make for great footage, and play into people's fears.

With its sensationalized trivia, the mass medias overglut obscures fundamental issues that *do* merit concern, such as preserving the environment.

Meanwhile, broadcasts themselves regularly imply that it is uncivil or immoral not to tune into the daily news – "all the news you need to know," and "we won't keep you waiting for the latest. . . ." It is *not* immoral to not "keep up" with the news that is offered. However to "tune out" – turn your back on the world – is not appropriate either. Being more selective in what you give your attention to, and to how long you give it, makes more sense.

Tomorrow, while dressing, rather than plugging in to the mass media, quietly envision how you would like your day to be. Include everything that's important to you. Envision talking with others, making major decisions, having lunch, attending meetings, finishing projects, and walking out in the evening. You'll experience a greater sense of control over aspects of your position that you may have considered uncontrollable.

There is only one party who controls the volume and frequency of information that you're exposed to. That person is you. As yet, few people are wise information consumers. Each of us needs to vigilantly guard against being deluded with excess data. The notion of "keeping up" with everything is illusory, frustrating, and self-defeating. The sooner you give it up the better you'll feel and function.

Keen focus on a handful of priorities has never been more important. Yes, some compelling issues must be given short shrift. Otherwise you run the risk of being overwhelmed by more demanding issues, and *feeling overwhelmed always exacerbates feeling overworked.*

Paper Trails

Paper, paper everywhere but not a thought to think. Imagine staring out the window from the fifth floor of a building and seeing a stack of reports from the ground up to your eye level. This 55-foot high stack would weigh some 659 pounds. *Pulp & Paper* reports that Americans annually consume 659 pounds of paper per person. In Japan, it's only 400 pounds per person; in Europe, Russia, Africa, Australia, and South America, far less.

Similar to too much information, or too many eyewitness reports, having too much paper to deal with is going to make you feel overwhelmed, and over-worked. Americans today are consuming at least three times as much paper as 10 years ago. The long held prediction of paperless offices, for now, is a laugher.

There are two basic reasons why our society spews so much paper:

– We have the lowest postal rates in the world, and
– We have the broadest distribution of paper-generating technology.

Last year, Congress received more than 300,000,000 pieces of mail, up from 15,000,000 in 1970. Nationwide more than 55,000,000 printers are plugged into at least 55,000,000 computers, and annually kick out billions of reams. Are 18,000 sheets enough? Your four-drawer file cabinet, when full, holds 18,000 pages.

The Thoreau Society reports that last year, Henry David Thoreau, who personally has been unable to make any purchases since 1862, received 90 direct mail solicitations at Walden Pond. Under our existing postal rates, catalog publishers and junk mail producers can miss the target 98 percent of the attempts and still make a profit – *only 2 percent of recipients* need to place an order for a direct mailer to score big.

Direct mailers, attempting to sell more, send you record amounts of unsolicted mail. In 1988, 12 billion catalogs were mailed in the U.S., up from five billion in 1980 – equal to 50 catalogs for every man, woman, and child in America. In the last decade, growth in the total volume of regular, third-class bulk mail (junk mail) was 13 times faster than growth in the population. The typical (over-worked? or overwhelmed?) executive receives more than 225 pieces of unsolicited mail each month, or about 12 pieces daily. Even Greenpeace, stalwart protector of the environment, annually sends out 25,000,000 pieces of direct mail.

Attempting to contain what seems unmanageable, our institutions create paper accounting systems which provide temporary relief and some sense of order, while usually becoming ingrained and immovable, and creating more muddle. Certainly accounting is necessary, but why so complicated? Because in our over-information society reams of data are regarded as a form of protection.

Why is documentation, such as circulating a copy to your boss, so critical to this culture? Because everyone is afraid of getting his derriere roasted! We live in a culture of fear, not like a marshall law dictatorship, but a form of fear nonetheless. "If I cannot document or account, I cannot prove, or defend myself." Attempting to contain what seems unmanageable, organizations and institutions, public and private, create paper accounting systems. These systems provide temporary relief and some sense of order. Usually they become ingrained and immovable, while creating more muddle. These accounting systems go by names such as federal income taxes, deed of trust, car loan, etc. Sure, accounting is necesssary, but why so complicated? Because in the era of over-information, over-information is used as a form of protection.

Of the five mega-realities, only paper flow promises to diminish some day as virtual reality, the electronic book, and the gigabyte highway are perfected. For the foreseeable future, you're likely to be up to your eyeballs in paper.

Start where you are – It is essential to clear the in-bins of your mind and your desk. Regard each piece of paper entering your personal domain as a potential mutineer or rebel. Each sheet has to earn its keep and remain worthy of your retention.

An Over-abundance of Choices

In 1969, Alvin Toffler predicted that we would be overwhelmed by too many choices. He said that this would inhibit action, result in greater anxiety, and trigger the perception of less freedom and less time. Having choices is a blessing of a free market economy. Like too much of everything else, however, having too many choices leaves the feeling of being ovewhelmed and results not only in increased time expenditure but also in a mounting form of exhaustion.

Consider the supermarket glut: Gorman's *New Product News* reports that in 1978 the typical supermarket carried 11,767 items. By 1987, that figure had risen to an astounding 24,531 items – more than double in nine years. More than 45,000 other products were introduced during those years, but failed. Elsewhere in the supermarket, Hallmark Cards now offers cards for *105* familial relationships. Currently more than 1260 varieties of shampoo are on the market. More than 2000 skin care products are currently selling. Seventy-five different types of exercise shoes are now available, each with scores of variations in style, and features. A *New York Times* article reported that even buying leisure time goods has become a stressful, overwhelming experience.

Periodically, the sweetest choice is choosing from what you already have, choosing to actually have what you've already chosen. More important is to avoid engaging in low level decisions. If a tennis racquet comes with either a black, or brown handle, and it's no concern to you, take the one the clerk hands you.

Whenever you catch yourself about to make a low level decision, consider: does this really make a difference? Get in the habit of making *fewer* decisions each day – the ones that count.

A Combined Effect

In a *Time Magazine* cover story entitled, "Drowsy America," the director of Stanford University's sleep center concluded that, "Most Americans no longer know what it feels like to be fully alert." Lacking a balance between work and play, responsibility and respite, "getting things done" can become an end-all. We function like human doings instead of human beings. We begin to link executing the items on our growing "to do" lists with feelings of self-worth. As the list keeps growing longer, the lingering sense of more to do infiltrates our sense of self-acceptance. What's worse, our entire society seems to be irrevocably headed toward a new epoch of human existence. Is frantic, however, any way to exist as a nation? Is it any way to run your life?

John Kenneth Galbraith studied poverty stricken societies on four continents. In *The Nature of Mass Poverty,* he concluded that some societies remain poor (often for centuries) because they *accommodate* poverty. Although it's difficult to live in abject poverty, Galbraith found that many poor societies are not willing to accept the difficulty of making things better.

As Americans, we appear poised to accommodate a frenzied, time-pressured existence, as if this is the way it has to be and always has been. *This is not how it has to be.* As an author, I have a vision. I see Americans leading balanced lives, with rewarding careers, happy home lives, and the ability to enjoy themselves. Our ticket to living and working at a comfortable pace is to not accommodate a way of being that doesn't support us, and addressing the true nature of the problem head-on:

The combined effect of the five mega-realities will continue to accelerate the feeling of pressure. Meanwhile, there will continue to be well-intentioned but misdirected voices who choose to condemn "employers" or "Washington DC" or what have you for the lack of true leisure in our lives.

A Complete Self

We are, however, forging our own frenetic society. Nevetheless, the very good news is that the key to forging a more palatable existence can occur one by one. *You,* for example, are whole and complete right now, and you can achieve balance in your life. You *are not* your position. You are not your tasks; they don't define you and they don't constrain you. You have the capacity to acknowledge that your life is finite; you cannot indiscriminately take in the daily deluge that our culture heaps on each of us and expect to feel anything but overwhelmed.

Viewed from 2002, 1992 will appear as a period of relative calm and stability when life moved at a manageable pace. When your days on earth are over and the big auditor in the sky examines the ledger of your life, she'll be upset if you *didn't* take enough breaks, and if you don't enjoy yourself.

On a deeply felt personal level, recognize that from now on, you will face an *ever-increasing* array of items competing for your attention. Each of the five mega-realities will proliferate in the '90's. You *cannot handle everything,* nor is making the attempt desirable. It is time to make compassionate though difficult choices about what is best ignored, versus what does merit your attention and action.

A De-Moralized Society: The British/American Experience

Gertrude Himmelfarb

Gertrude Himmelfarb is professor emeritus of history at the City University of New York. She is the author of numerous books, including The Idea of Poverty, The New History and the Old, *and* On Looking Into the Abyss.

"THE PAST is a foreign country," it has been said. But it is not an unrecognizable country. Indeed, we sometimes experience a "shock of recognition" as we confront some aspect of the past in the present. One does not need to have had a Victorian grandmother, as did Margaret Thatcher, to be reminded of "Victorian values." One does not even have to be English; "Victorian America," as it has been called, was not all that different, at least in terms of values, from Victorian England. Vestigial remains of that Victorianism are everywhere around us. And memories of them persist, even when the realities are gone, rather like an amputated limb that still seems to throb when the weather is bad.

How can we not think of our present condition when we read Thomas Carlyle on the "Condition of England" one hundred and fifty years ago? While his contemporaries were debating "the standard of living question"—the "pessimists" arguing that the standard of living of the working classes had declined in that early period of industrialism, and the "optimists" that it had improved—Carlyle reformulated the issue to read, "the condition of England question." That question, he insisted, could not be resolved by citing "figures of arithmetic" about wages and prices. What was important was the "condition" and "disposition" of the people: their beliefs and feelings, their sense of right and wrong, the attitudes and habits that would dispose them either to a "wholesome composure, frugality, and prosperity," or to an "acrid unrest, recklessness, gin-drinking and gradual ruin."

In fact, the Victorians did have "figures of arithmetic" dealing with the condition and disposition of the people as well as their economic state. These "moral statistics" or "social statistics," as they called them, dealt with crime, illiteracy, illegitimacy, drunkenness, pauperism, vagrancy. If they did not have, as we do, statistics on drugs, divorce, or teenage suicide, it is because these problems were then so negligible as not to constitute "social problems."

It is in this historical context that we may address our own "condition of the people question." And it is by comparison with the Victorians that we may find even more cause for alarm. For the current moral statistics are not only more troubling than those a century ago; they constitute a trend that bodes even worse for the future than for the present. Where the Victorians had the satisfaction of witnessing a significant improvement in their moral and social condition, we are confronting a considerable deterioration in ours.

The 'Moral Statistics': Illegitimacy

In nineteenth-century England, the illegitimacy ratio—the proportion of out-of-wedlock births to total births—rose from a little over 5 percent at the beginning of the century to a peak of 7 percent in 1845. It then fell steadily until it was less than 4 percent at the turn of the century. In East London, the poorest section of the city, the figures are even more dramatic, for illegitimacy was consistently well below the average: 4.5 percent in mid-century and 3 percent by the end of the century. Apart from a temporary increase during both world wars, the ratio continued to hover around 5 percent until 1960. It then began to rise: to over 8 percent in 1970, 12 percent in 1980, and then precipitously, to more than 32 percent by the end of 1992—a two-and-one-half-times increase in the last decade alone and a sixfold rise in three decades. In 1981, a married woman was half as likely to have a child as she was in 1901, while an unmarried woman was three times as likely.

In the United States, the figures are no less dramatic. Starting at 3 percent in 1920 (the first year for which there are national statistics), the illegitimacy ratio rose gradually to slightly over 5 percent by 1960, after which it grew rapidly: to almost 11 percent in 1970, over 18 percent in 1980, and 30 percent by 1991—a tenfold increase from 1920 and a sixfold increase from 1960. . . .

There are no national crime statistics for the United States for the nineteenth century and only partial ones (for homicides) for the early twentieth century. Local statistics, however, suggest that, as in England, the decrease in crime started in the latter part of the nineteenth century (except for a few years following the Civil War) and continued into the early twentieth century. There was even a decline of homicides in the larger cities, where they were most common; in Philadelphia, the rate fell from 3.3 per 100,000 population in mid-century to 2.1 by the end of the century.

National crime statistics became available only in 1960, when the rate was under 1,900 per 100,000 population.

From *American Educator,* Winter 1994/95, pp. 14-21, 40-43. Originally from *The Public Interest,* Fall 1994. Adapted from *The De-Moralization of Society: From Victorian Virtues to Modern Values* by Gertrude Himmelfarb.

That figure doubled within the decade and tripled by 1980. . . . In 1987, the Department of Justice estimated that eight of every ten Americans would be a victim of violent crime at least once in their lives.

Homicide statistics go back to the beginning of the century, when the national rate was 1.2 per 100,000 population. That figure skyrocketed during prohibition, reaching as high as 9.7 by one account (6.5 by another) in 1933, when prohibition was repealed. The rate dropped to between five and six during the 1940s and to under five in the fifties and early sixties. In the mid-sixties, it started to climb rapidly, more than doubling between 1965 and 1980. . . .

More Moral Statistics

There are brave souls, inveterate optimists, who try to put the best gloss on the statistics. But it is not much consolation to be told that the overall crime rate in the United States has declined slightly from its peak in the early 1980s if the violent crime rate has risen in the same period—and increased still more among juveniles and girls (an ominous trend, since the teenage population is also growing). Nor that the divorce rate has fallen somewhat

Violent crime has become so endemic that we have practically become inured to it.

in the past decade, if it had doubled in the previous two decades; if more parents are co-habitating without benefit of marriage (the rate in the United States has increased sixfold since 1970); and if more children are born out of wedlock and living with single parents. (In 1970, one out of ten families was headed by a single parent; in 1990, three out of ten were). Nor that the white illegitimacy ratio is considerably lower than the black, if the white ratio is rapidly approaching the black ratio of a few decades ago, when Daniel Patrick Moynihan wrote his percipient report about the breakdown of the black family. (The black ratio in 1964, when that report was issued, was 24.5 percent; the white ratio now is 22 percent. In 1964, 50 percent of black teenage mothers were single; in 1991, 55 percent of white teenage mothers were single.)

Nor is it reassuring to be told that two-thirds of new welfare recipients are off the rolls within two years, if half of those soon return, and a quarter of all recipients are on for more than eight years. Nor that divorced mothers leave the welfare rolls after an average of five years, if never-married mothers remain for more than nine years, and unmarried mothers who bore their children as teenagers stay on for ten or more years. (Forty-three percent of the longest-term welfare recipients started their families as unwed teenagers.)

Nor is the cause of racial equality promoted by the news of an emerging "white underclass," smaller and less conspicuous than the black (partly because it is more dispersed) but rapidly increasing. If, as has been conclusively demonstrated, the single-parent family is the most important factor associated with the "pathology of poverty"—welfare dependency, crime, drugs, illiteracy, homelessness—a white illegitimacy ratio of 22 percent, and twice that for white women below the poverty line, signifies a new and dangerous trend. In England, Charles Murray has shown, a similar underclass is developing with twice the illegitimacy of the rest of the population; there it is a purely class rather than racial phenomenon.

Redefining Deviancy

The English sociologist Christie Davies has described a "U-curve model of deviance," which applies to both Britain and the United States. The curve shows the drop in crime, violence, illegitimacy, and alcoholism in the last half of the nineteenth century, reaching a low at the turn of the century, and a sharp rise in the latter part of the twentieth century. The curve is actually more skewed than this image suggests. It might more accurately be described as a "J-curve," for the height of deviancy in the nineteenth century was considerably lower than it is today—an illegitimacy ratio of 7 percent in England in the mid-nineteenth century, compared with over 32 percent toward the end of the twentieth; or a crime rate of about 500 per 100,000 population then compared with 10,000 now.

In his *American Scholar* essay, "Defining Deviancy Down," Senator Moynihan has taken the idea of deviancy a step further by describing the downward curve of the concept of deviancy. What was once regarded as deviant behavior is no longer so regarded; what was once deemed abnormal has been normalized. As deviancy is defined downward, so the threshold of deviancy rises: Behavior once stigmatized as deviant is now tolerated and even sanctioned. Mental patients, no longer institutionalized, are now treated, and appear in the statistics, not as mentally incapacitated but as "homeless." Divorce and illegitimacy, once seen as betokening the breakdown of the family, are now viewed more benignly; illegitimacy has been officially rebaptized as "nonmarital childbearing," and divorced and unmarried mothers are lumped together in the category of "single-parent families." And violent crime has become so endemic that we have practically become inured to it. The St. Valentine's Day Massacre in Chicago in 1929, when four gangsters killed seven other gangsters, shocked the nation and became legendary, immortalized in encyclopedias and history books; in Los Angeles today, James Q. Wilson observes, as many people are killed every weekend.

It is ironic to recall that only a short while ago criminologists were accounting for the rise of the crime rates in terms of our "sensitization to violence." As a result of the century-long decline of violence, they reasoned, we had become more sensitive to "residual violence"; thus, more crimes were being reported and apprehended. This "residual violence" has by now become so overwhelming that, as Moynihan points out, we are being desensitized to it.

Charles Krauthammer has proposed a complementary concept in his *New Republic* essay, "Defining Deviancy Up." As deviancy is normalized, so the normal becomes deviant. The kind of family that has been regarded for centuries as natural and moral—the "bourgeois" family, as it is invidiously called—is now seen as pathological, concealing behind the façade of respectability the new "original sin," child abuse. While crime is underreported because we have become desensitized to it, child abuse is overreported, including fantasies imagined (often inspired by therapists and social workers) long after the supposed events. Similarly, rape has been "defined up" as "date rape," to include sexual relations that the participants themselves may not at the time have perceived as rape.

The combined effect of defining deviancy up and defining it down has been to normalize and legitimate what was once regarded as abnormal and illegitimate, and, conversely, to stigmatize and discredit what was once normal and respectable. This process too, has occurred with startling rapidity. One might expect that attitudes and values would lag behind the reality, that people would continue to pay lip service to the moral principles they were brought up with, even while violating those principles in practice. What is startling about the 1960s "sexual revolution," as it has properly been called, is how revolutionary it was, in sensibility as well as reality. In 1965, 69 percent of American women and 65 percent of men under the age of thirty said that premarital sex was always or almost always wrong; in 1972, those figures plummeted to 24 percent and 21 percent. For women over the age of thirty, the figures dropped from 91 percent to 62 percent, and for men from 62 percent to 47 percent—this in seven short years. Thus language, sensibility, and social policy conspire together to redefine deviancy.

. . . There is an occasional, but not consistent, relation between crime and economic depression and poverty. In England in the 1890s, in a period of severe unemployment, crime (including property crime) fell. Indeed, the inverse relationship between crime and poverty at the end of the nineteenth century suggests, as one study put it, that "poverty-based crime" had given way to "prosperity-based crime."

In the twentieth century, the correlation between crime and unemployment has been no less erratic. While crime did increase in England during the depression of the 1930s, that increase had started some years earlier. A graph of unemployment and crime between 1950 and 1980 shows no significant correlation in the first fifteen years and only a rough correlation thereafter. The crime figures, a Home Office bulletin concludes, would correspond equally well, or even better, with other kinds of data. "Indeed, the consumption of alcohol, the consumption of ice cream, the number of cars on the road, and the Gross National Product are highly correlated with rising crime over 1950-1980."

The situation is similar in the United States. In the high-unemployment years of 1949, 1958 and 1961, when unemployment was 6 or 7 percent, crime was less than 2 percent; in the low-unemployment years of 1966 to 1969, with unemployment between 3 and 4 percent, crime was almost 4 percent. Today in the inner cities there is a correlation between unemployment and crime, but it may be argued that it is not so much unemployment that causes crime as a culture that denigrates or discourages employment, making crime seem more normal, natural, and desirable than employment. The "culture of criminality," it is evident, is very different from the "culture of poverty" as we once understood that concept.

Nor can the decline of the two-parent family be attributed, as is sometimes suggested, to the economic recession of recent times. Neither illegitimacy nor divorce increased during the far more serious depression of the 1930s—or, for that matter, in previous depressions, either in England or in the United States. In England in the 1980s, illegitimacy actually increased more in areas where the employment situation improved than in those where it got worse. Nor is there a correlation between illegitimacy and poverty; in the latter part of the nineteenth century, illegitimacy was significantly lower in the East End of London than in the rest of the country. Today there is a correlation between illegitimacy and poverty, but not a causal one; just as crime has become part of the culture of poverty, so has the single-parent family.

The Language of Morality

These realities have been difficult to confront because they violate the dominant ethos, which assumes that moral progress is a necessary byproduct of material progress. It seems incomprehensible that in this age of free, compulsory education, illiteracy should be a problem, not among immigrants but among native-born Americans; or illegitimacy, at a time when sex education, birth control, and abortion are widely available. Even more important is the suspicion of the very idea of morality. Moral principles, still more moral judgments, are thought to be at best an intellectual embarrassment, at worst evidence of an illiberal and repressive disposition. It is this reluctance to speak the language of morality, far more than any specific values, that separates us from the Victorians.

We are constantly beseeched to be "nonjudgmental," to be wary of crediting our beliefs with any greater validity than anyone else's.

Most of us are uncomfortable with the idea of making moral judgments even in our private lives, let alone with the "intrusion," as we say, of moral judgments into public affairs. We are uncomfortable not only because we have come to feel that we have no right to make such judgments and impose them upon others, but because we have no confidence in the judgments themselves, no assurance that our principles are true and right for us, let alone for others. We are constantly beseeched to be "nonjudgmental," to be wary of crediting our beliefs with any greater validity than anyone else's, to be conscious of how "Eurocentric" and "culture bound" we are. *Chacun à son goût,* we say of morals, as of taste; indeed, morals have become a matter of taste.

Public officials in particular shy away from the word "immoral," lest they be accused of racism, sexism, or elitism. When members of the president's Cabinet were asked if it is immoral for people to have children out of wedlock, they drew back from that distasteful phrase. The Secretary of Health and Human Services replied, "I don't like to put this in moral terms, but I do believe that having children out of wedlock is just wrong." The Surgeon General was more forthright: "No. Everyone has different moral standards . . . . You can't impose your standards on someone else."

It is not only our political and cultural leaders who are prone to this failure of moral nerve. Everyone has been infected by it, to one degree or another. A moving testimonial to this comes from an unlikely source: Richard Hoggart, the British literary critic and very much a man of the left, not given to celebrating Victorian values. It was in the course of criticizing a book espousing traditional virtues that Hoggart observed about his own hometown:

> In Hunslet, a working-class district of Leeds, within which I was brought up, old people will still enunciate, as guides to living, the moral rules they learned at Sunday School and Chapel. Then they almost always add, these days: "But it's only my opinion, of course." A late-twentieth century insurance clause, a recognition that times have changed towards the always shiftingly relativist. In that same council estate, any idea of parental guidance has in many homes been lost. Most of the children there live in, take for granted, a violent, jungle world.

De-moralizing Social Policy

In Victorian England, moral principles and judgments were as much a part of social discourse as of private discourse, and as much a part of public policy as of personal life. They were not only deeply ingrained in tradition, they were also imbedded in two powerful strains of Victorian thought: Utilitarianism on the one hand, Evangelicalism and Methodism on the other. These may not have been philosophically compatible, but in practice they complemented and reinforced each other, the Benthamite calculus of pleasure and pain, rewards and punishments, being the secular equivalent of the virtues and vices that Evangelicalism and Methodism derived from religion.

It was this alliance of a secular ethos and a religious one that determined social policy, so that every measure of poor relief or philanthropy, for example, had to justify itself by showing that it would promote the moral as well as the material well-being of the poor. The distinction between pauper and poor, the stigma attached to the "abled-bodied pauper," indeed, the word "pauper" itself, today seem invidious and inhumane. At the time, however, they were the result of a conscious moral decision: an effort to discourage dependency and preserve the respectability of the independent poor, while providing at least minimal sustenance for the indigent.

In recent decades, we have so completely rejected any kind of moral calculus that we have deliberately, systematically divorced welfare from moral sanctions or incentives. This reflects in part the theory that society is responsible for all social problems and should therefore assume the task of solving them; and in part the prevailing spirit of relativism, which makes it difficult to pass any moral judgments or impose any moral conditions upon the recipients of relief. We are now confronting the consequences of this policy of moral neutrality. Having made the most valiant attempt to "objectify" the problem of poverty, to see it as the product of impersonal economic and social forces, we are discovering that the economic and social aspects of that problem are inseparable from the moral and personal ones. And having made the most determined effort to devise social policies that are "value free," we find that these policies imperil both the moral and the material well-being of their intended beneficiaries.

In de-moralizing social policy—divorcing it from any moral criteria, requirements, even expectations—we have demoralized, in the more familiar sense, both the individuals receiving relief and society as a whole. Our welfare system is counterproductive not only because it aggravates the problem of welfare, creating more incentives to enter and remain within it than to try to avoid or escape from it. It also has the effect of exacerbating other, more serious, social problems, so that chronic dependency has become an integral part of the larger phenomenon of "social pathology."

The Supplemental Security Income program is a case in point. Introduced in 1972 to provide a minimum income for the blind, the elderly, and the disabled poor, the program has been extended to drug addicts and alcoholics as the result of an earlier ruling defining "substance abusers" as "disabled" and therefore eligible for public assistance. Apart from encouraging these "disabilities" ("vices," the Victorians would have called them), the program has the effect of rewarding those who remain addicts or alcoholics while penalizing (by cutting off funds) those who try to overcome their addiction. This is the reverse of the principle of "less eligibility" that was the keystone of Victorian social policy: the principle that the dependent poor be in a less "eligible," less desirable, condition than the independent poor. One might say that we are now operating under a principle of "more eligibility," the recipient of relief being in a more favorable position than the self-supporting person.

Just as many intellectuals, social critics, and policy makers were reluctant for so long to credit the unpalatable facts about crime, illegitimacy, or dependency, so they find it difficult to appreciate the extent to which these facts themselves are a function of values—the extent to which "social pathology" is a function of "moral pathology" and social policy a function of moral principle.

Victims of the Upperclass

The moral divide has become a class divide. The same people who have long resisted the realities of social life also find it difficult to sympathize with those, among the working classes especially, who feel acutely threatened by a social order that they perceive to be in an acute state of disorder. (The very word "order" now sounds archaic.) The "new class," as it has been called, is not, in fact, all that new; it is by now firmly established in the media, the academy, the professions, and the government. In its denigration of "bourgeois values" and the "Puritan ethic," the new class has legitimized, as it were, the values of the underclass and illegitimized those of the working class, who are still committed to bourgeois values, the Puritan ethic, and other such benighted ideas.

In a powerfully argued book, Myron Magnet has analyzed the dual revolution that led to this strange alliance between what he calls the "Haves" and the "Have-Nots." The first was a social revolution, intended to liberate the poor from the political, economic, and racial oppression that kept them in bondage. The second was a cultural revolution, liberating them (as the Haves themselves were being liberated) from the moral restraints of bourgeois values. The first created the welfare programs of the Great Society, which provided counter-incentives to leaving poverty. And the second disparaged the behavior and attitudes that traditionally made for economic improvement—"deferral of gratification, sobriety, thrift, dogged industry, and so on through the whole catalogue of antique-sounding bourgeois virtues." Together these revolutions had the unintended effect of miring the poor in their poverty—a poverty even more demoralizing and self-perpetuating than the old poverty.

The underclass is not only the victim of its own culture, the "culture of poverty." It is also the victim of the upperclass culture around it. The kind of "delinquency" that a white suburban teenager can absorb with relative (only relative) impunity may be literally fatal to a black inner-city teenager. Similarly, the child in a single-parent family headed by an affluent professional woman is obviously in a very different condition from the child (more often, children) of a woman on welfare. The effects of the culture, however, are felt at all levels. It was only a matter of time before there should have emerged a white underclass with much the same pathology as the black. And not only a white underclass but a white upper class; the most affluent suburbs are beginning to exhibit the same pathological symptoms: teenage alcoholism, drug addiction, crime, and illegitimacy.

By now this "liberated," anti-bourgeois ethic no longer seems so liberating. The social realities have become so egregious that it is now finally permissible to speak of the need for "family values." President Clinton himself has put the official seal of approval on family values, even going so far as to concede—a year after the event—that there were "a lot of very good things" in Quayle's famous speech about family values (although he was quick to add that the "Murphy Brown thing" was a mistake). . . .

The Use and Abuse of History

One of the most effective weapons in the arsenal of the "counter-counterculture" is history—the memory not only of a time before the counterculture but also of the evolution of the counterculture itself. In 1968, the English playwright and member of Parliament A.P. Herbert had the satisfaction of witnessing the passage of the act he had sponsored abolishing censorship on the stage. Only two years later, he complained that what had started as a "worthy struggle for reasonable liberty for honest writers" had ended as the "right to represent copulation, veraciously, on the public stage." About the same time, a leading American civil liberties lawyer, Morris Ernst, was moved to protest that he had meant to ensure the publication of Joyce's *Ulysses,* not the public performance of sodomy.

In the last two decades, the movements for cultural and sexual liberation in both countries have progressed far beyond their original intentions. Yet, few people are able to resist their momentum or to recall their initial principles. In an unhistorical age such as ours, even the immediate past seems so remote as to be antediluvian; anything short of the present state of "liberation" is regarded as illiberal. And in a thoroughly relativistic age such as ours, any assertion of value—any distinction between the publication of *Ulysses* and the public performance of sodomy—is thought to be arbitrary and authoritarian.

It is in this situation that history may be instructive, to remind us of a time, not so long ago, when all societies, liberal as well as conservative, affirmed values very different from our own. (One need not go back to the Victorian age; several decades will suffice.) To say that history is instructive is not to suggest that it provides us with models for emulation. One could not, even if one so desired, emulate a society—Victorian society, for example—at a different stage of economic, technological, social, political, and cultural development. Moreover, if there is much in the ethos of our own times that one may deplore, there is no less in Victorian times. Late-Victorian society was more open, liberal, and humane than early-Victorian society, but it was less open, liberal, and humane than most people today would think desirable. Social, ethnic, and sexual discriminations, class rigidities and political inequalities, autocratic men, submissive women, and overly disciplined children, constraints, restrictions, and abuses of all kinds—there is enough to give pause to the most ardent Victoriaphile. Yet there is also much that might appeal to even a modern, liberated spirit.

Victorian Lessons

The main thing the Victorians can teach us is the importance of values—or, as they would have said, "virtues"—in our public as well as private lives. The Victorians were, candidly and proudly, "moralists." In recent decades, that has almost become a term of derision. Yet, contemplating our own society, we may be prepared to take a more appreciative view of Victorian moralism—of the "Puritan ethic" of work, thrift, temperance, cleanliness; of the idea of "respectability" that was as powerful among the working classes as among the middle classes; of the reverence for "home and hearth"; of the stigma attached to the "able-bodied pauper," as a deterrent to the "independent" worker; of the spirit of philanthropy that made it a moral duty on the part of the donors to give not only money but their own time and effort to the charitable cause, and a moral duty on the part of the recipients to try to "better themselves."

We may even be on the verge of assimilating some of that moralism into our own thinking. It is not only "values" that are being rediscovered but "virtues" as well. That long-neglected word is appearing in the most unlikely places: in books, newspaper columns, journal articles, and scholarly discourse. An article in the *Times Literary Supplement,* reporting on a spate of books and articles from "virtue revivalists" on both the right and the left of the political spectrum, observes that "even if the news that Virtue is back is not in itself particularly exciting to American pragmatism, the news that Virtue is good for you most emphatically is." The philosopher Martha Nussbaum, reviewing the state of Anglo-American philosophy, focuses upon the subject of "Virtue Revived," and her account suggests a return not to classical ethics but to something very like Victorian ethics: an ethics based on "virtue" rather than "principle," on "tradition and particularity" rather than "universality," on "local wisdom" rather than "theory," on the "concreteness of history" rather than an "ahistorical detached ethics." . . .

A Society's Ethos

The historical perspective is also useful in reminding us of our gains and losses—our considerable gains in material goods, political liberty, social mobility, racial and sexual equality—and our no-less-considerable losses in moral well-being. There are those who say that it is all of a piece, that what we have lost is the necessary price of what we have gained. ("No pain, no gain," as the motto has it.) In this view, liberal democracy, capitalism, affluence, and modernity are thought to carry with them the "contradictions" that are their undoing. The very qualities that encourage economic and social progress—individuality, boldness, the spirit of enterprise and innovation—are said to undermine conventional manners and morals, traditions, and authorities. This echoes a famous passage in *The Communist Manifesto:*

> The bourgeoisie, wherever it has got the upper hand, has put an end to all feudal, patriarchal, idyllic relations. It has pitilessly torn asunder the motley feudal ties that bound man to his "natural superior," and has left no other bond between man and man then naked self-interest, than callous "cash payment." ... The bourgeoisie has torn away from the family its sentimental veil and has reduced the family relation to a mere money relation.

Marx was as wrong about this as he was about so many things. Victorian England was a crucial test case for him because it was the first country to experience the industrial-capitalist-bourgeois revolution in its most highly developed form. Yet, that revolution did not have the effects he attributed to it. It did not destroy all social relations, tear asunder the ties that bound man to man, strip from the family its sentimental veil, and reduce everything to "cash payment" (the "cash nexus," in other translations). It did not do this, in part because the free market was never as free or as pervasive as Marx thought (*laissez-faire,* historians now agree, was less rigorous, both in theory and in practice, that was once supposed); and in part because traditional values and institutions continued to play an important role in society, even in those industrial and urban areas most affected by the economic and social revolution.

Industrialism and urbanism—"modernism," as it is now known—so far from contributing to the de-moralization of the poor, seem to have had the opposite effect. At the end of the nineteenth century, England was a more civil, more pacific, more humane society than it had been in the beginning. "Middle-class" manners and morals had penetrated into large sections of the working classes. The traditional family was as firmly established as ever, even as women began to be liberated from their "separate sphere." And religion continued to thrive, in spite of the premature reports of its death.

If Victorian England did not succumb to the moral and cultural anarchy that are said to be the inevitable consequences of economic individualism, it is because of a powerful ethos that kept that individualism in check. For the Victorians, the individual, or "self," was the ally rather than the adversary of society. Self-help was seen in the context of the community as well as the family; among the working classes, this was reflected in the virtue of "neighbourliness," among the middle classes, of philanthropy. Self-interest stood not in opposition to the general interest but, as Adam Smith had it, as the instrument of the general interest. Self-discipline and self-control were thought of as the source of self-respect and self-betterment; and self-respect as the precondition for the respect and approbation of others. The individual, in short, was assumed to have responsibilities as well as rights, duties as well as privileges.

That Victorian "self" was very different from the "self" that is celebrated today. Unlike "self-help," "self-esteem" does not depend upon the individual's actions or achievements; it is presumed to adhere to the individual regardless of how he behaves or what he accomplishes. Moreover, it adheres to him regardless of the esteem in which he is held by others, unlike the Victorian's self-respect, which always entailed the respect of others. The current notions of self-fulfillment, self-expression, and self-realization derive from a self that does not have to prove itself by reference to any values, purposes, or persons outside itself—that simply is, and by reason of that alone deserves to be fulfilled and realized. This is truly a self divorced from others, narcissistic and solipsistic.

This is the final lesson we may learn from the Victorians: that the ethos of society, its moral and spiritual character, cannot be reduced to economic, material, political, or other factors, that values—or, better yet, virtues—are a determining factor in their own right; so far from being a "reflection," as the Marxist says, of the economic realities, they are themselves, as often as not, the crucial agent in shaping those realities. If in a period of rapid economic and social change, the Victorians showed a substantial improvement in their "condition" and "disposition," it may be that economic and social change do not necessarily result in personal and public disarray. If they could retain and even strengthen an ethos that had its roots in religion and tradition, it may be that we are not as constrained by the material conditions of our time as we have thought. A post-industrial economy, we may conclude, does not necessarily entail a postmodernist society or culture, still less a de-moralized society or culture.

The Decline of Bourgeois America

Stanley Rothman

According to the latest national polls, Americans are still more concerned about the state of the economy today than they are about the decline of what have come to be called "family values." However, the family value issue and economic concerns are, in reality, related to each other in very practical ways, even though many fail to make the necessary connections. Despite the renewed economic growth and growth in productivity during the past three years, the United States no longer epitomizes an efficient industrial capitalist country to the rest of the world or even to its own citizens. Japan, Korea, and Taiwan are clearly outpacing the United States in the growth of productivity and quality of product. Japan's current problems should not blind one to that fact.

Alfred Chandler, in his magisterial study of the growth of modern business enterprise in the United States, England, and Germany, argues that the key variable was innovative management that took advantage of the economies of scope and size. More recently, an MIT study of American industry, *Made in America: Regaining the Productive Edge,* laid some of the blame for America's relative economic decline at the feet of management. U.S. managers, the authors argue, seek quick profits and their horizons are narrow. They have little sense of commitment to the larger community and little concern for the future of the firm itself. On the other hand, "executives in other countries tend to be less preoccupied with earnings, dividends and share prices. Japanese and European firms hold that managers are not only responsible to shareholders, but also have a commitment to the larger community of employees, customers, neighbors and suppliers and to the continuity and growth of the firm itself."

Is the MIT study correct? The authors are not the only ones to criticize the behavior of America's business class. Indeed, this notion has become part of the conventional wisdom of the informed public, as has the notion that the U.S. work ethic is generally in decline. In a recent public opinion poll, only 42 percent of respondents characterized Americans as hard working and productive, as against 70 percent of those who described the Japanese in those terms. I believe that, in this case, the conventional wisdom has some merit. Americans, in general, have lost their competitive edge. Though many other factors are at work, this loss is at least partly a function of changing values, including changes in the nature and structure of family life. I plan to deal with this question by drawing upon the tradition of culture and personality in anthropology and sociology that flourished from World War II until the mid-1960s, when it was replaced by rational choice models on the right and a rather crude Marxism on the left. However, the approach has reappeared in recent years, despite the continuing widespread critique of the psychoanalytic categories that have generally been its foundation.

Scholarship in the field of culture and personality ranges from the work of Margaret Mead and Melford Spiro to that Geoffrey Gorer, David Reisman, and Eric

Erikson. However, the most influential attempts to examine the cultural and personality sources of the rise, and in their minds probable decay, of capitalism are those of the German "Frankfurt school" of intellectuals whose goal was to mutually enrich psychoanalysis and Marxism by integrating the two. Many social analysts were involved in this enterprise, including Theodor Adorno and Walter Benjamin, but two names stand out: Erich Fromm and Herbert Marcuse. Marcuse and Fromm were sharply critical of each other. However, despite their differences, they both accepted the notion that Christianity had played a large role in the emergence of individualistic, acquisitive capitalism. They also both agreed that capitalism had produced individuals who are psychologically impaired; that is, in the words of Marcuse, capitalism produces "one-dimensional" men and women. The analysis that follows also attempts to integrate psychoanalysis with sociology, but it is based upon a sociological tradition that finds its sources in Max Weber rather than in Karl Marx.

The Rise and Decay of Liberal Capitalism

David Gutmann argues that the essential differences between the traditional and the modern psyche stem from the development of a powerful superego (conscience). To primitive man, power lies outside the self, although the boundaries of self and nonself are not clear. One seeks power by following rules set by the gods, through the charismatic leader or tightly knit community structures. The problem is that unless the gods continually speak, their message dims. Social organization of any kind, therefore, requires rigid and continuous community controls. Communities adapt to their environment. They rarely attempt to master it beyond what is necessary for survival. The stranger and the different are suspect. Forbidden impulses are projected upon them. The superego of primitive man is largely located outside himself. While he longs to identify with shamans or gods, he has little sense of a consistent self as an actor who can exercise power over others to serve internal needs.

Modern Western man has been different. While psychoanalysis tends to emphasize the punitive qualities of the superego, it is not without positive aspects. By providing an *inner* mechanism of control, the superego permits the individual to develop a stronger sense of self. Primitive man believes that the gods help the one who does their bidding. Modern men and women attempt to overcome obstacles by living up to their inner ideals. The source of power is within, not without. Furthermore, the sublimated energy derived from their internal control of both erotic and aggressive drives fosters the development of the ego. This enables them to examine and to manipulate nature, to adapt to new circumstances, and to bring creative energies to bear upon them. It also enables them to overcome fear of the unknown, and of other groups, as they develop rational models for understanding nature. In these ways the development of superego and ego strength permit the emergence of the principled, nonauthoritarian, relatively democratic, flexible person.

Drawing upon Max Weber and others, we can trace this development to Christian doctrines and, more particularly, to the Reformation. Weber demonstrates that the unintended consequences of Christian and, more specifically, Calvinist doctrines included the emergence of the modern world in the form of "liberal capitalism," that is, a capitalist economic system associated with individualism and gradual, if imperfect, democratization. Weber emphasizes the capitalist side of the equation, but the contribution of this cultural complex to democratization is well documented.

Obviously, none of the distinctions I have drawn is absolute. Some internalization occurs in all societies, even the most primitive. It is also clear that every great civilization has been accompanied by a heightening of superego development, usually based on the emergence of a new, more universalist religious system. It is possible, moreover, that in some of these civilizations, superego development was comparable to that of the West. As Weber argues, however, cultural developments in the West were unusual from the outset. First, the emergence of a prophetic religion gave a peculiar intensity to the superego. Second, the emphasis was on an individual rather than a communal relationship with God. Third, religious-cultural imperatives stressed general, universal, moral rules. Fourth, God was conceived as standing apart from nature, and his laws could be comprehended through reason. Finally, great emphasis was placed upon repressing the passions in the service of worldly asceticism, that is, fulfilling one's obligations through activity in this world.

To be sure, some of these themes have been present individually in other civilizations. Historically and comparatively though, this was a unique combination. It is undoubtedly true that Confucianism, with its emphasis on the control of the passions, produced a similar result in China and Japan via a shame culture. Confucian doctrine and the quality of popular religion in both countries was such that neither the Japanese nor Chinese could bring the modern world into existence; however, once that world came into existence, they could use the energy derived from the repression of sexual and aggressive drives in the service of science and industry. Thus, given an appropriate response by elites in Japan and later Hong Kong, Singapore, Taiwan, Korea, and, finally, mainland China, these areas could adapt to the requirements of an industrial society fairly easily, as compared to other Asian, African, or even Latin American nations.

Restraint and Capitalism

The United States can be examined as a paradigmatic case of the relationship between Calvinism, capitalism, and democracy, for it has epitomized the Western version of modernity. Perceived as an "empty land" by the English Protestants who initially shaped its culture, it came to represent the ideal type of liberal capitalism for observers as diverse as Alexis de Tocqueville, G.W.F. Hegel, and Marx. To be sure, there were myriad other influences on the development of U.S. capitalism, among them the Amerindians of the ever moving frontier and vast numbers of immigrants, including 400,000 Africans who were brought, as slaves and against their wills, to what is now the United States and Canada. Nevertheless, American liberalism, with its highly individualistic and relatively egalitarian and "pragmatic" social ideology and reality, was essentially derived from European culture, though it crossed the Atlantic without the baggage of the earlier communal European traditions with which European liberals had to struggle.

American liberal individualism emphasized freedom, but within a framework that set stringent limits upon its expression. Individuals were free to act but were held responsible for their actions. However, their behavior was mediated by an emphasis upon the restraint of impulse, including sexual impulse, and their energies were concentrated on hard work and the accumulation of material wealth. For example, contrary to popular folklore, the Puritans were not antisex. Rather, they emphasized fidelity and measure in marriage and constancy of emotion in sexual as well as other aspects of behavior. As Edmund Leites points out:

> There were ascetic tendencies in Puritanism, but the Puritans also stressed constancy and self control for a major purpose that was not ascetic.... Puritan preachers and theologians urged spouses to maintain a steady and reliable delight in their mates, a pleasure both sensuous and spiritual. They did not call for ascetic denial of impulse but for a fusion of self-denial and worldly desire, for a style of feeling and action which was at once self-controlled and free.

Originally, then, the justification for restraint, the acceptance of discipline and hierarchy, and the accumulation of wealth all rested on religious foundations. These foundations, however, assumed a sense of community that also set boundaries for individualism. Those who were fortunate enough to achieve wealth had a duty to exhibit "mercy, gentleness, [and] temperance." As late as the end of the nineteenth century, this creed, in somewhat different form, played an important role in shaping the ethos of the United States. And, of course, it evolved in tandem with patterns of child rearing.

Anthony Wallace summarizes some of the elements of the dominant U.S. creed of the late nineteenth century in his perceptive discussion of the development of a small U.S. industrial community:

> It was in Rockdale, and in dozens of other industrial communities like Rockdale, that an American world view developed which pervades the present—or did so until recently—with a sense of superior Christian virtue, a sense of global mission, a sense of responsibility and capability of bringing enlightenment to a dark and superstitious world, for overthrowing ancient and new tyrannies, and for making backward infidels into Christian men of enterprise.

He might have added that religious justification was reinforced by the deference with which businessmen were treated by ordinary folk.

Twentieth-century intellectuals have not been kind to nineteenth-century businessmen. The vast majority of nineteenth-century entrepreneurs were not ruthless "robber barons." Certainly these entrepreneurs believed in the morality of unrestrained market competition, but that was not the whole story. Most of them combined the desire for wealth and power with the desire to create something important for the community. The vigor with which they fought the nascent trade union movement was counterbalanced by the large amount of money they donated to charity. Indeed, philanthropic giving has been far more extensive in the United States than in any other country in the world, and that, too, has been an integral part of the liberal tradition.

Daniel Bell has pointed out that the religious values that underlay U.S. culture began to erode during the late nineteenth century, partly as the result of rationalizing tendencies inherent in liberal capitalism itself. As these values decayed, religious justifications for the goals and limitations imposed by the culture were replaced by a belief in material progress as an end in itself. Hard work and self-restraint in a liberal capitalist system would lead to secular progress and ever better tomorrows for all. And, because they were contributing to that end, businessmen could still count on the respect of ordinary citizens to persuade them that their work constituted a genuine achievement.

Religion and Capitalism

Of course many businessmen retained their religious beliefs, and others continued to follow patterns of behavior that were, at their source, derived from a Calvinist perspective even though they no longer considered themselves to be particularly religious. By the 1940s, however, as Leo Lowenthal discovered, the "idols of production" had been partially replaced by the "idols of consumption," and by the late 1950s segments of an affluent middle class were adding the consumption of experience ("self-realization") to the consumption of material goods as a goal. Lacking a religious base, the requirements of work and self-discipline had been further undermined by affluence. The Beats of the 1950s were probably an expression of this, but there were many others, as the popularity of the writings of Erich Fromm and Abraham Maslow attest.

The shift in cultural values was accompanied and encouraged by the growth of a stratum of new strategic elites in the society, themselves products of economic and social as well as ideological changes. Many of these people were highly educated professionals working either in the public sector or, if in the private sector, in areas associated with the production and transmission of knowledge. These elites include psychiatrists, social workers, college professors, journalists, and, more recently, producers, writers, and directors of motion picture and television entertainment.

These new "Metro Americans," as Erich Goldman called them, tended to be rather skeptical of traditional values, of the economics of liberal capitalism, and of U.S. foreign policy. By the mid-1960s, traditional bourgeois liberalism had been replaced in large segments of the professional middle class by new strands of liberalism, among them what Robert Bellah has called "expressive individualism." As a cultural system, expressive individualism is characterized by a shift from the perception of an individual as a "being" to the perception of the individual as a "self"and a shift from restraint of impulse to its free expression, or, more generally, a rejection of the old for the new. The culture of expressive individualism is centered around the exploration of experience and sensation—unfettered, impulsive, and nonrational. In the arts, humanities, and letters, it rejects traditional standards in favor of the avant-garde. It emphasizes the search for new and what it claims to be superior modes of artistic and literary representation.

The rejection of the old is inevitable, given the expressive individual's trust in self-generated truths. In classical, Catholic, and even traditional Calvinist thought, all human beings possess a common nature related to their humanity and their place in the cosmos. Their humanity and their place tie them to, even as they separate them from, other species. In contrast, as unique individuals realizing themselves through a set of freely chosen meaningful experiences, men and women lack a fixed nature that defines them. They are merely the creations of themselves, of the societies of which they are part, or both. The contrast between expressive individualism and prior bourgeois culture could not be starker. The Protestant ethic stressed sobriety over playfulness, restraint over expression, reason over emotion. The expressive individualistic ethic turns the Protestant world upside down.

Despite the conservative veneer of the 1950s, expressive individualism had clearly won the cultural battle in the United States by the end of that decade, and the antibourgeois aesthetic triumphed especially within institutions in which intellectuals played a significant role (such as the universities, mass media outlets, television, motion pictures, or book publishing).

Another strand of the new ideology is "collectivist liberalism," a phrase coined by John Dewey. The threads of this strand have been extensively analyzed. Suffice it to say that a major component is the value placed on equality: All individuals are alike in some fundamental way, and therefore no person is of greater worth than any other, although some income differentiation is permitted—that is, professionals can be permitted to earn larger salaries than ordinary workers. Tied closely to this is the belief that the political order should play a major role in maximizing social, political, and economic equality. In contrast, the central value for the more traditional view is economic freedom, rather than economic equality. Traditional free-market capitalism rests on the assumption that the greatest good arises from each individual freely pursuing his or her own chosen activities. The ideology of free-market capitalism assumes that each individual knows best what is in his or her economic interest. Individuals ought then to be free to pursue their own economic ends. The collective good is the aggregation of all individuals acting in pursuit of their individual private goods within the rules set by the community to regulate human intercourse.

In the 1920s, there were few government programs protecting workers from the harshness of the private economic sector—unemployment, inadequate wages, dangerous and unsanitary working conditions, industrial accidents—and almost no regulations dealing with work in the private factory. Needless to say, there was no assumption of federal intervention in education, civil rights for minorities, women, and the disabled, health and medical care for the poor and elderly, poverty programs, and so on. In contrast, collectivist liberalism, in its contemporary form, demands that the regulatory state create among various groups of Americans an equality of both opportunity and (at least to some extent) result in wealth, housing, education, civil rights, health and medical care, and culture and the arts.

In many ways, the new liberalism is an inversion of the old. Traditional liberalism called for economic freedom within a framework of emotional and expressive restraint. The new liberalism discards expressive restraints but adds bureaucratic controls. The crisis of the 1960s integrated both collectivist liberalism and expressive individualism into a new pattern, adding the dimension of system alienation. In the view of the New Left, reason and objectivity represent and legitimate power in bourgeois society. Because the social order of bourgeois–liberal society is inherently dehumanizing and repressive, rejecting authority and order require also rejecting reason and objectivity. The New Left died in the 1970s, but a significant segment of the population remains alienated from the culture.

The description of the new liberalism I have offered is not simply a creative construct. Its content was suggested to my colleagues and to me by a factor analysis of questions asked of the random samples of various U.S. elite groups that we surveyed during the mid-1980s. Naturally, we can not measure the changes that have taken place since the turn of the century, but the new cultural elites and traditional elites continue to differ fairly sharply from each other on all three factors in expected directions. Relatively, journalists are liberal collectivists and expressive individualists; businessmen and women are neither, though we strongly suspect that they are more supportive of both perspectives than was a previous generation of businessmen. On the other hand, the creators of motion pictures are only moderately liberal on economic issues, but they are very strong supporters of expressive individualism.

In the shift to expressive individualism, the mass media, especially television, has undoubtedly played a key role. By its very nature, television adds new dimensions to the communication of information and radically changes the rules of the game. The consequences for certain aspects of U.S. life are clear. Far more than newspapers, radio, or movies, television provides its audience with a sense that what it sees is true and real. The audience sees events taking place in its living room. Stories, documentaries, even drama take on a reality with which other media cannot compete. The written word and even the spoken word remain somewhat abstract to most readers and listeners, but moving pictures seen in the privacy of one's home are extremely compelling. Even if one knows that footage may have been spliced together and may therefore conceivably present a somewhat distorted perspective of the events being portrayed (and few are aware of that fact), it is hard to escape the perception that one is viewing reality.

Television and the New Sensibility

Television has broken down class and regional boundaries to a far greater extent than have other media. Books and newspapers are segregated by area and readership. Only the well-educated can read serious books, and the style of the *New York Times* only appeals to those with a certain level of education and affluence. Thus, to some extent, newspapers and books encourage the segregation of knowledge. Radio began to break down that segregation. Television goes much further. There are programs that cater to more elite audiences and are watched only by them, but insofar as television seeks the lowest common denominator and finds it, Americans as a group are introduced to the same themes in the same way. *Roots* and other "docudramas," as well as the six o'clock news, are watched by millions of Americans of all educational and social backgrounds, and they see the same pictures and receive the same information.

The process begins early in childhood. As Joshua Meyrowitz points out, cultures in which knowledge is dependent on the ability to read require substantial preparation before one can penetrate many of the secrets of adult life. Television has broken that barrier. Children can and do watch television programs that tell them about the off-stage behavior of parents and introduce them to themes that, in the past, they would not have encountered until much later in life. Young children are exposed to the news almost every day along with their parents. Most so-called family programs deal with concerns with which children would not have been familiar even twenty-five years ago, and millions of children are still awake at hours when television programs for "mature audiences" are shown.

It is impossible to understand the revolution that took place in U.S. values and attitudes from the 1960s to the 1990s without taking into account the influence of television on the fabric of our life, including its breaking down of old barriers and its weakening of old ties. For the first time, the metropolitan United States was becoming all of the United States. In the 1920s, the new therapeutic ethic of self-realization had only permeated a small section of America's metropolitan upper middle class. By the 1970s, as the authors of *Habits of the Heart* point out, it had spread far more widely. Not surprisingly, few realize how rapid the pace of change has been. The events of the 1960s, including the rapid loss of faith in U.S. institutions, and the legitimation of lifestyles once considered to be deviant, could not have occurred in a pre-television age. The Reagan administration conceivably slowed change down a little. It clearly did not reverse direction or even stop it.

The United States has become, as Richard Merelman points out in *Making Something of Ourselves,* a "loose bounded culture." Americans' primordial ties to family, locality, church, and what is considered appropriate behavior have eroded, and Americans have lost their sense of place. They are not alone in this, of course. Their experience is increasingly shared by Europeans, Japanese, and, perhaps, even Russians. Certainly mass television is not the only factor at work. The revolution is real, however, and the epoch we live in is quite new.

Working-class parochials may continue to identify with those they know and with whom they work and live, but public reality is now such that we also know and develop ersatz-intimate and intense relationships with public figures of all kinds, from anchormen to rock performers to politicians. As my colleagues and I, as well as others, have demonstrated, the liberal cosmopolitanism of television writers and producers is reflected to at least some extent in the way they describe the world, with significant consequences for the larger culture.

To be sure, television entertainment does seek the lowest common denominator and stresses sensationalism in order to sell. However, that factor alone can not explain the changes that have taken place in the stories told on television.

Television's United States once looked like California's Orange County writ large—Waspish, businesslike, religious, patriotic, and middle-American. Today it resembles more the upper-middle-class "liberated" world of California's Marin County—trendy, self-expressive, culturally diverse, and cosmopolitan. *Pace* neo-Marxist and deconstructionist arguments, this limited sector of U.S. society has accelerated the acceptance of a liberal cosmopolitan perspective by other segments of the population.

Decay and the Erosion of Restraint

Liberal capitalism, then, is in a state of decay. The unconscious restraints that underlie the rationality of action in liberal capitalist societies are eroded by both affluence and rationality itself. Rationality undermines the religious foundations of that restraint, and affluence undermines the need to discipline one's behavior in the marketplace. Thus rational self-interest, restrained by unconscious assumptions about the legitimate parameters of behavior, is replaced by the pursuit of any sensation or experience that gives satisfaction without directly harming others or by the immediate satisfactions of wealth or power. Rationality itself finally comes under attack, because it is perceived as derived partly from the superego, thus limiting possibilities for choosing one's own lifestyle.

Marriage, childbearing, and the family have become experiences in self-realization rather than duties and obligations, and this fundamental change in emphasis has had a far-reaching impact. For example, children are no longer disciplined in the traditional manner, both because parents (married and single) want to allow them to "realize themselves" and because it is too much trouble and takes too much time. Added to this are an escalating divorce rate that leaves an increasingly large number of children to be raised by one parent, usually the mother; the development of widespread drug dependency; and a popular culture that stresses both sexuality and violence.

As Alice Rossi notes:

> western societies are undergoing a second fundamental transformation of basic values, a shift from a child centered to an individual or self-centered orientation. The second transformation is a further development in the ongoing erosion of traditional beliefs under the impact of increasing secularism, the loosened hold of religion over private behavior and the further application of the concept of individual freedom of choice to the private world of the family.

The consequences for personality structure are not difficult to discern. These fit in with the requirements of the new ideologies. Power replaces achievement as an ideal goal, and with that comes heightened mutual suspiciousness, even as the number of narcissistic personality types increases. More significantly, an ever larger number of children are initiated into sexual activity before they are ready for it; even as increasing numbers of children are born with AIDS or brain damaged because of their mother's sexual activity, drug abuse, or both.

Some intellectuals have attempted to deny the negative impact on children of both divorce and other forms of single parenthood, and the new styles of child rearing. However, the evidence is now overwhelming. Whatever the virtues and faults of traditional family patterns, the new ones are damaging to young people, especially young males, since children of both sexes tend to remain with their mothers and the lack of a paternal model is harder upon the male. In a worst-case scenario, in the black community, large numbers of children are being born and growing up without the capacity to hold down even minimal jobs in the work-a-day world. Some are growing up complete sociopaths, killing promiscuously at ever earlier ages without remorse. More and more children, especially of the poor, are abused by stepparents or by their mothers' boyfriends, who have no biological investment in the children of the woman with whom they are associating or whom they have married. It is clear that these consequences are more than a function of poverty.

The evidence of sociopathy is all around us, as is reduced capacity for achievement.

Freud would have predicted both the emergence of such sociopathy and lessened capacity for achievement for males raised without fathers or by parents unwilling to invest energy into caring for them at an early age. The evidence of sociopathy is all around us, as is reduced capacity for achievement. For example, scores on the SAT, whatever the limitations of these tests, have dropped overall, despite a partial recovery. Even more significant, fewer and fewer young people are achieving high scores.

Traditional elites (businessmen, corporate lawyers, and so on) are being won over by the new ideologies. Those who enter business today are less likely to regard it as a calling than were their nineteenth-century counterparts. In addition to personality factors, both the religious sources of and social respect for that drive have withered. Large businesses today are often anonymous bureaucratic enterprises whose leaders may inspire envy but rarely the respect that their forbears did. They are managers rather than genuine entrepreneurs. They may become celebrities, but they have no personal ties with other members of a viable local community. Further, why should they seek public esteem for their work when under the best of circumstances it probably will not be forthcoming?

John DeLorean is, in some ways, representative of the new breed of businessmen and women.

His criminal drug activity was the logical outcome of the ideology of the age and the personal qualities which drove him.

The character of contemporary culture is exemplified by the popularity of such stars as Madonna and by the growing popularity of explicitly violent "rap" music. It is not accidental that a few years ago a Florida jury easily decided that the following lyrics of the rap group 2 Live Crew are not obscene: "Grabbed her by the hair, threw her on the floor / opened her thighs and guess what I saw.... He'll tear the cunt open 'cause it's satisfaction.... Bust your cunt then break your backbone.... I wanna see you bleed."

There is good reason, then, to believe that the patterns of ego control, so laboriously constructed in Europe and the United States, are breaking down, even as are conventional superego restraints. The result is an erosion of the capacity to sublimate both aggressive and erotic drives in the service of civilization, and the replacement of bourgeois commitments to achievement and constancy by an increase in defensive projection, "acting out," and drives for power and control. There is at least some evidence that the number of persons in the United States who no longer evince these patterns of ego control is on the increase, despite countertrends.

Liberal democracy and capitalism in Europe emerged out of a particular cultural and personality matrix. The pattern was unique. Their limited success has depended both on particular public policies and upon the character and underlying cultural commitments of their citizens. The mere belief that a society should be democratic and rely upon markets is not enough. If both are to work in the relatively successful manner in which they have in the West in the past, a particular personality structure is required. Of course, liberal capitalism, whatever its limitations, historically has encouraged rationality, emotional complexity, growth, and the capacity for a democratic polity.

The liberal democrat can be an individualist precisely because his individualism is supported by an internal psychic structure that defines limits and moderates desire by a sense of social responsibility. When this internal psychic structure collapses, no system of institutions will sustain the kind of capitalist order that has characterized parts of Europe and America.

Halfway between a narrowly defined religious outlook and the loss of a religious worldview, policymakers like Averill Harriman, Robert Lovett, John McCloy, Charles (Chip) Bohlen, Dean Acheson, and George Kennan represented the best of U.S. bourgeois culture. Cosmopolitan in outlook, they nevertheless still lived on the borrowed capital of an older moral worldview and psychological balance.

The question is, in the place of the Protestant, bourgeois ideology that once characterized so much more of U.S. society and that appears to have been associated with certain personality types, what do we now see? The evidence around us in the culture suggests that many of those in the middle and upper middle classes, having lost the internal gyroscope (and metaphors) that gave the lives of previous generations structure and meaning, feel torn between the desire for power and gratification on the one hand and the fear of losing control on the other. They lurch between longings for complete autonomy (that is, the destruction of an already weakened ego and superego) and the wish to totally lose themselves in something that will give their lives order; hence the proliferation of cults—from New Age channeling to Scientology to belief in satanism, witchcraft, and UFOs—that appeal to so many in this social stratum in our era. I am not persuaded that the change bodes well for the future of a liberal democratic order in the United States.

The assumption of permanent moral progress is an illusion.

Events of the past several years have not diminished my skepticism. Marxism may for the time being have been discredited by the failures of Soviet-type regimes. Some social critics are making pronouncements about the end of history. Such speculations are not new, but new ideologies always emerge that promise to save humanity and create a world free of violence, poverty, and all its other evils. It is still my expectation (or fear) that the future does lie with authoritarian bureaucratic societies, even if these are partly associated with the market and private property.

The assumption of permanent moral progress is an illusion. The Western bourgeois order has been built on one method of channeling sexuality and aggression into socially useful activities. It never eliminated aggression, and, indeed, violence was always breaking through the patina of civilization. The notion that we can have a society in which we can fully express ourselves in every way as long as we do not harm others is a chimera. There is every evidence that the current decay of bourgeois standards is translating itself into escalating physical violence.

I am skeptical of attempts by intellectuals to restore or create a sense of personal responsibility. A renewal may occur in the society, or there may even be a shift in orientation on the basis of new cultural understandings, but such a shift will not be initiated by secular intellectuals. Max Weber had some rather sharp comments for those who were responding to the crisis of the Weimar Republic by seeking answers in secular argument or prophecy: "To the person who can not bear the fate of the times...one must say...the arms of the old churches are opened widely and compassionately for him.... In my eyes such religious return stands higher than...academic prophecy."

Unlike Freud and Marx, Weber was not persuaded that human beings could maintain a civilization or achieve happiness without some religious orientation, though he was personally a religious skeptic.

This then is where we are. We are a much more tolerant society in many ways than we once were, even as signs of pathology increase around us. Paradoxically, the two may be interrelated. Our willingness to accept all kinds of lifestyles may go hand in hand with our inability to maintain a civilized community.

This surely is not the whole story. The world has changed in many ways. The difficulties of overcoming our racist past, dealing with new international issues, absorbing a huge new influx of immigrants, responding to environmental problems, and reexamining relations between men and women are not merely a function of the contradictions of liberal capitalism nor are they a function of the role of cultural elites. Nevertheless, it seems reasonably clear that a society sufficiently sure of its own values would be handling many of these factors rather differently than is the United States.

Stanley Rothman is Mary Huggins Gamble Professor of Government at Smith College and Director of the Center for the Study of Social and Political Change. He is the author or co-author of Roots of Radicalism; The Media Elite; The IQ Controversy: The Media and Public Policy; Prime Time; *and* The Mass Media in Liberal Democratic Societies.

Socialization and Social Control

- Childhood and Influences on Personality and Behavior (Articles 6 and 7)
- Crime, Law Enforcement, and Social Control (Articles 8–11)

UNIT 2

Why do we behave the way we do? Three forces are at work: the shaping influences of both biology and socialization, and the human will or internal decision maker. The focus in sociology is on socialization, which is the conscious and unconscious process whereby we learn the norms and behavior patterns that enable us to function appropriately in our social environment. Socialization is based on the need to belong, because the desire for acceptance is the major motivation for internalizing the socially approved attitudes and behaviors. Fear of punishment is another motivation. It is utilized by parents and institutionalized in the law enforcement system. The language we use, the concepts we use in thinking, the image we have of ourselves, our gender roles, and our masculine and feminine ideals are all learned through socialization. Socialization may take place in many contexts. The most basic socialization takes place in the family, but churches, schools, communities, the media, and workplaces also play major roles in the process.

The first section deals with basic influences on the development of our character and behavior patterns. Laura Shapiro analyzes biological and social influences on the development of children, focusing on gender roles. Her review of the evidence concludes that social influences are preeminent. "Children of the Universe" by Amitai Etzioni looks at the other side of the picture—the results of widespread inadequate socialization due to the "parenting deficit." His point is simple: well-brought-up children are a great benefit not only to parents but also to the community and to the larger society, while poorly brought-up children are very costly.

The second section of the unit deals with crime, law enforcement, and social control—major concerns today because crime and violence seem to be out of control. The first article in this section assesses the crime situation in America and comes to some surprising conclusions. Crime rates have declined slightly over the past two decades, and crime rates in America are comparable to crime rates in European countries, except for America's extremely high murder rate. America also "imprisons seven times as many people (proportionately) as does the average European country, largely as a result of get-tough-on-crime laws." Then, America's crime policies are examined and judged to be irrational, ineffective, expensive, and often unjust.

Next, Paul Robinson analyzes how crime can be reduced by finding out why people obey the law and strengthening those forces. Social disapproval by one's own group and one's own personal morality have far greater deterrent effects than the threat of legal punishment. Legal forces have been greatly weakened by the low level of moral credibility of the criminal justice system, but Robinson proposes many ways to strengthen its moral authority. Then, the focus is on crimes of domestic abuse that are often unreported and that are only recently being recognized by the legal system. This problem was thrust into the public eye by the two O. J. Simpson criminal and civil trials. The last article in the section deals with the question of legalization of some drugs. This is a tough issue, and much can be said on both sides, but given here are many sound reasons why legalization may cause far more harm than the present system. These arguments, however, do not seem to apply well to marijuana.

Looking Ahead: Challenge Questions

How can the ways in which children are socialized in America be improved?

Why is socialization a lifelong process?

What are the principal factors that make people what they are?

How are girls socialized differently from boys?

What traditional American norms are no longer widely honored?

What are some reasons for the increase of crime in the United States?

What are the advantages and disadvantages of legalizing some of the illegal drugs?

Guns and Dolls

Alas, our children don't exemplify equality any more than we did. Is biology to blame? Scientists say maybe—but parents can do better, too.

LAURA SHAPIRO

Meet Rebecca. She's 3 years old, and both her parents have full-time jobs. Every evening Rebecca's father makes dinner for the family—Rebecca's mother rarely cooks. But when it's dinner time in Rebecca's dollhouse, she invariably chooses the Mommy doll and puts her to work in the kitchen.

Now meet George. He's 4, and his parents are still loyal to the values of the '60s. He was never taught the word "gun," much less given a war toy of any sort. On his own, however, he picked up the word "shoot." Thereafter he would grab a stick from the park, brandish it about and call it his "shooter."

Are boys and girls *born* different? Does every infant really come into the world programmed for caretaking or war making? Or does culture get to work on our children earlier and more inexorably than even parents are aware? Today these questions have new urgency for a generation that once made sexual equality its cause and now finds itself shopping for Barbie clothes and G.I. Joe paraphernalia. Parents may wonder if gender roles are immutable after all, give or take a Supreme Court justice. But burgeoning research indicates otherwise. No matter how stubborn the stereotype, individuals can challenge it; and they will if they're encouraged to try. Fathers and mothers should be relieved to hear that they do make a difference.

Biologists, psychologists, anthropologists and sociologists have been seeking the origin of gender differences for more than a century, debating the possibilities with increasing rancor ever since researchers were forced to question their favorite theory back in 1902. At that time many scientists believed that intelligence was a function of brain size and that males uniformly had larger brains than women—a fact that would nicely explain men's pre-eminence in art, science and letters. This treasured hypothesis began to disintegrate when a woman graduate student compared the cranial capacities of a group of male scientists with those of female college students; several women came out ahead of the men, and one of the smallest skulls belonged to a famous male anthropologist.

GIRLS

Girls' cribs have pink tags and boys' cribs have blue tags; mothers and . . .

NEWBORNS

BOYS

. . . fathers should be on the alert, for the gender-role juggernaut has begun

Gender research has become a lot more sophisticated in the ensuing decades, and a lot more controversial. The touchiest question concerns sex hormones, especially testosterone, which circulates in both sexes but is more abundant in males and is a likely, though unproven, source of aggression. To postulate a biological determinant for behavior in an ostensibly egalitarian society like ours requires a thick skin. "For a while I didn't dare talk about hormones, because women would get up and leave the room," says Beatrice Whiting, professor emeritus of education and anthropology at Harvard. "Now they seem to have more self-confidence. But they're skeptical. The data's not in yet."

Some feminist social scientists are staying away from gender research entirely—"They're saying the results will be used against women," says Jean Berko Gleason, a professor of psychology at Boston University who works on gender differences in the acquisition of language. Others see no reason to shy away from the subject. "Let's say it were proven that there were biological foundations for the division of labor," says Cynthia Fuchs Epstein, professor of sociology at the City University of New York, who doesn't, in fact, believe in such a likelihood. "It doesn't mean we couldn't do anything about it. People can make from scientific findings whatever they want." But a glance at the way society treats those gender differences already on record is not very encouraging. Boys learn to read more slowly than girls, for instance, and suffer more reading disabilities such as dyslexia, while girls fall behind in math when they get to high school. "Society can amplify differences like these or cover them up," says Gleason. "We rush in reading teachers to do remedial reading, and their classes are almost all boys. We don't talk about it, we just scurry around getting them to catch up to the girls. But where are the remedial math teachers? Girls are *supposed* to be less good at math, so that difference is incorporated into the way we live."

No matter where they stand on the question of biology versus culture, social scientists agree that the sexes are much more alike than they are different, and that variations within each sex are far greater than variations between the sexes. Even differences long taken for granted have begun to disappear. Janet Shibley Hyde, a professor of psychology at the University of Wisconsin, analyzed hundreds of studies on verbal and math ability and found boys and girls alike in verbal ability. In math, boys have a moderate edge; but only among highly precocious math students is the disparity large. Most important, Hyde found that verbal and math studies dating from the '60s and '70s showed greater differences than more recent research. "Parents may be making more efforts to tone down the stereotypes," she says. There's also what academics call "the file-drawer effect." "If you do a study that shows no differences, you assume it won't be published," says Claire Etaugh, professor of psychology at Bradley University in Peoria, Ill. "And until recently, you'd be right. So you just file it away."

The most famous gender differences in academics show up in the annual SAT results, which do continue to favor boys. Traditionally they have excelled on the math portion, and since 1972 they have slightly outperformed girls on the verbal side as well. Possible explanations range from bias to biology, but the socioeconomic profile of those taking the test may also play a role. "The SAT gets a lot of publicity every year, but nobody points out that there are more women taking it than men, and the women come from less advantaged backgrounds," says Hyde. "The men are a more highly selected sample: they're better off in terms of parental income, father's education and attendance at private school."

GIRLS

Girls are encouraged to think about how their actions affect others . . .

2–3 YEARS

BOYS

. . . boys often misbehave, get punished and then misbehave again

Another longstanding assumption does hold true: boys tend to be somewhat more active, according to a recent study, and the difference may even start prenatally. But the most vivid distinctions between the sexes don't surface until well into the preschool years. "If I showed you a hundred kids aged 2, and you couldn't tell the sex by the haircuts, you couldn't tell if they were boys or girls," says Harvard professor of psychology Jerome Kagan. Staff members at the Children's Museum in Boston say that the boys and girls racing through the exhibits are similarly active, similarly rambunctious and similarly interested in model cars and model kitchens, until they reach first grade or so. And at New York's Bank Street preschool, most of the 3-year-olds clustered around the cooking table to make banana bread one recent morning were boys. (It was a girl who gathered up three briefcases from the costume box and announced, "Let's go to work.")

By the age of 4 or 5, however, children start to embrace gender stereotypes with a determination that makes liberal-minded parents groan in despair. No matter how careful they may have been to correct the disparities in "Pat the Bunny" ("Paul isn't the *only* one who can play peekaboo, *Judy* can play peekaboo"), their children will delight in the traditional male/female distinctions preserved everywhere else: on television, in books, at day care and preschool, in the park and with friends. "One of the things that is very helpful to children is to learn what their identity is," says Kyle Pruett, a psychiatrist at the Yale Child Study Center. "There are rules about being feminine and there are rules about being masculine. You can argue until the cows come home about whether those are good or bad societal influences, but when you look at the children, they love to know the differences. It solidifies who they are."

Water pistols: So girls play dolls, boys play Ghostbusters. Girls take turns at hopscotch, boys compete at football. Girls help Mommy, boys aim their water pistols at guests and shout, "You're dead!" For boys, notes Pruett, guns are an inevitable part of this developmental process, at least in a television-driven culture like our own. "It can be a cardboard paper towelholder, it doesn't have to be a miniature Uzi, but it serves as the focus for fantasies about the way he is going to make himself powerful in the world," he says. "Little girls have their aggressive side, too, but by the time they're socialized it takes a different form. The kinds of things boys work out with guns, girls work out in terms of relationships—with put-downs and social cruelty." As if to underscore his point, a 4-year-old at a recent Manhattan party turned to her young hostess as a small stranger toddled up to them. "Tell her we don't want to play with her," she commanded. "Tell her we don't like her."

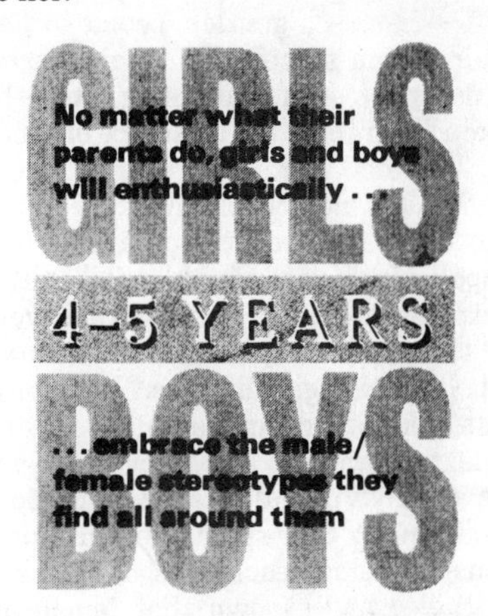

Once the girls know they're female and the boys know they're male, the powerful stereotypes that guided them don't just disappear. Whether they're bred into our chromosomes or ingested with our cornflakes, images of the aggressive male and the nurturant female are with us for the rest of our lives. "When we see a man with a child, we say, 'They're playing'," says Epstein. "We never say, 'He's nurturant'."

The case for biologically based gender differences is building up slowly, amid a great deal of academic dispute. The theory is that male and female brains, as well as bodies, develop differently according to the amount of testosterone circulating around the time of birth. Much of the evidence rests on animal studies showing, for instance, that brain cells from newborn mice change their shape when treated with testosterone. The male sex hormone may also account for the different reactions of male and female rhesus monkeys, raised in isolation, when an infant monkey is placed in the cage. The males are more likely to strike at the infant, the females to nurture it. Scientists disagree—vehemently—on whether animal behavior has human parallels. The most convincing human evidence comes from anthropology, where cross-cultural studies consistently find that while societies differ in their predilection toward violence, the males in any given society will act more aggressively than the females. "But it's very important to emphasize that by aggression we mean only physical violence," says Melvin Konner, a physician and anthropologist at Emory University in Atlanta. "With competitive, verbal or any other form of aggression, the evidence for gender differences doesn't hold." Empirical findings (i.e., look around you) indicate that women in positions of corporate, academic or political power can learn to wield it as aggressively as any man.

Apart from the fact that women everywhere give birth and care for children, there is surprisingly little evidence to support the notion that their biology makes women kinder, gentler people or even equips them specifically for motherhood. Philosophers—and mothers, too—have taken for granted the existence of a maternal "instinct" that research in female hormones has not conclusively proven. At most there may be a temporary hormonal response associated with childbirth that prompts females to nurture their young, but that doesn't explain women's near monopoly on changing diapers. Nor is it likely that a similar hormonal surge is responsible for women's tendency to organize the family's social life or take up the traditionally underpaid "helping" professions—nursing, teaching, social work.

Studies have shown that female newborns cry more readily than males in response to the cry of another infant, and that small girls try more often than boys to comfort or help their mothers when they appear distressed. But in general the results of most research into such traits as empathy and altruism do not consistently favor one sex or the other. There is one major exception: females of all ages seem better able to "read" people, to discern their emotions, without the help of verbal cues. (Typically researchers will display a picture of someone expressing a strong reaction and ask test-takers to identify the emotion.) Perhaps this skill—which in evolutionary terms would have helped females survive and protect their young—is the sole biological foundation for our unshakable faith in female selflessness.

Infant ties: Those who explore the unconscious have had more success than other researchers in trying to account for male aggression and female nurturance, perhaps because their theories cannot be tested in a laboratory but are deemed "true" if they suit our intuitions. According to Nancy J. Chodorow, professor of sociology at Berkeley and the author of the influential book "The Reproduction of Mothering," the fact that both boys and girls are primarily raised by women has crucial effects on gender roles. Girls, who start out as infants identifying with their mothers and continue to do so, grow up defining themselves in relation to other people. Maintaining human connections remains vital to them. Boys eventually turn to their fathers for self-definition, but in order to do so must repress those powerful infant ties to mother and womanhood. Human connections thus become more problematic for them than for women. Chodorow's book, published in 1978, received national attention despite a dense, academic prose style; clearly, her perspective rang true to many.

Harvard's Kagan, who has been studying young children for 35 years, sees a different constellation of influences at work. He speculates that women's propensity for caretaking can be traced back to an early awareness of their role in nature. "Every girl knows, somewhere between the ages of 5 and 10, that she is different from boys and that she will have a child—something that everyone, including children, understands as quintessentially natural," he says. "If, in our society, nature stands for the giving of life, nurturance, help, affection, then the girl will conclude unconsciously that those are the qualities she should strive to attain. And the boy won't. And that's exactly what happens."

Kagan calls such gender differences "inevitable but not genetic," and he emphasizes—as does Chodorow—that they need have no implications for women's status, legally or occupationally. In the real world, of course, they have enormous implications. Even feminists who see gender differences as cultural artifacts agree that, if not inevitable, they're hard to shake. "The most emancipated families, who really feel they want to engage in gender-free behavior toward their kids, will still encourage boys to be boys and girls to be girls," says Epstein of CUNY. "Cultural constraints are acting on you all the time. If I go to buy a toy for a friend's little girl, I think to myself, why don't I buy her a truck? Well, I'm afraid the parents wouldn't like it. A makeup set would really go against my ideology, but maybe I'll buy some blocks. It's very hard. You have to be on the alert every second."

In fact, emancipated parents have to be on

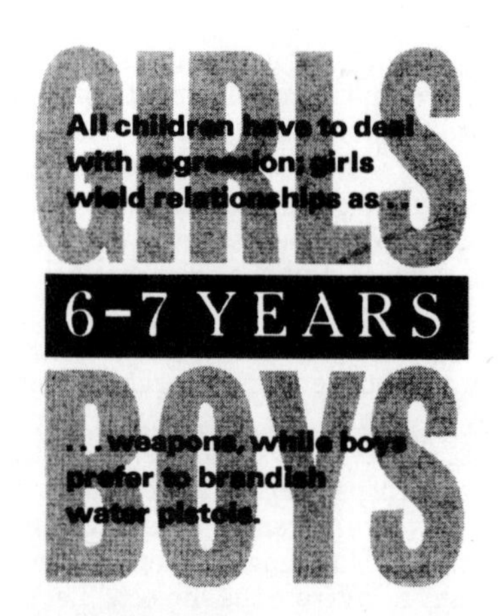

the alert from the moment their child is born. Beginning with the pink and blue name tags for newborns in the hospital nursery—I'M A GIRL/I'M A BOY—the gender-role juggernaut is overwhelming. Carol Z. Malatesta, associate professor of psychology at Long Island University in New York, notes that baby girls' eyebrows are higher above their eyes and that girls raise their eyebrows more than boys do, giving the girls "a more appealing, socially responsive look." Malatesta and her colleagues, who videotaped and coded the facial expressions on mothers and infants as they played, found that mothers displayed a wider range of emotional responses to girls than to boys. When the baby girls displayed anger, however, they met what seemed to be greater disapproval from their mothers than the boys did. These patterns, Malatesta suggests, may be among the reasons why baby girls grow up to smile more, to seem more sociable than males, and to possess the skill noted earlier in "reading" emotions.

The way parents discipline their toddlers also has an effect on social behavior later on. Judith G. Smetana, associate professor of education, psychology and pediatrics at the University of Rochester, found that mothers were more likely to deal differently with similar kinds of misbehavior depending on the sex of the child. If a little girl bit her friend and snatched a toy, for instance, the mother would explain why biting and snatching were unacceptable. If a boy did the same thing, his mother would be more likely to stop him, punish him and leave it at that. Misbehavior such as hitting in both sexes peaks around the age of 2; after that, little boys go on to misbehave more than girls.

Psychologists have known for years that boys are punished more than girls. Some have conjectured that boys simply drive their parents to distraction more quickly; but as Carolyn Zahn-Waxler, a psychologist at the National Institute of Mental Health, points out, the difference in parental treatment starts even before the difference in behavior shows up. "Girls receive very different messages than boys," she says. "Girls are encouraged to care about the problems of others, beginning very early. By elementary school, they're showing more caregiver behavior, and they have a wider social network."

Children also pick up gender cues in the process of learning to talk. "We compared fathers and mothers reading books to children," says Boston University's Gleason. "Both parents used more inner-state words, words about feelings and emotions, to girls than to boys. And by the age of 2, girls are using more emotion words than boys." According to Gleason, fathers tend to use more directives ("Bring that over here") and more threatening language with their sons than their daughters, while mothers' directives take more polite forms ("Could you bring that to me, please?"). The 4-year-old boys and girls in one study were duly imitating their fathers and mothers in that very conversational pattern. Studies of slightly older children found that boys talking among themselves use more threatening, commanding, dominating language than girls, while girls emphasize agreement and mutuality. Polite or not, however, girls get interrupted by their parents more often than boys, according to language studies—and women get interrupted more often than men.

Despite the ever-increasing complexity and detail of research on gender differences, the not-so-secret agenda governing the discussion hasn't changed in a century: how to understand women. Whether the question is brain size, activity levels or modes of punishing children, the traditional implication is that the standard of life is male, while the entity that needs explaining is female. (Or as an editor put it, suggesting possible titles for this article: "Why Girls Are Different.") Perhaps the time has finally come for a new agenda. Women, after all, are not a big problem. Our society does not suffer from burdensome amounts of empathy and altruism, or a plague of nurturance. The problem is men—or more accurately, maleness.

"There's one set of sex differences that's ineluctable, and that's the death statistics," says Gleason. "Men are killing themselves doing all the things that our society wants them to do. At every age they're dying in accidents, they're being shot, they drive cars badly, they ride the tops of elevators, they're two-fisted hard drinkers. And violence against women is incredibly pervasive. Maybe it's men's raging hormones, but I think it's because they're trying to be a *man*. If I were the mother of a boy, I would be very concerned about societal pressures that idolize behaviors like that."

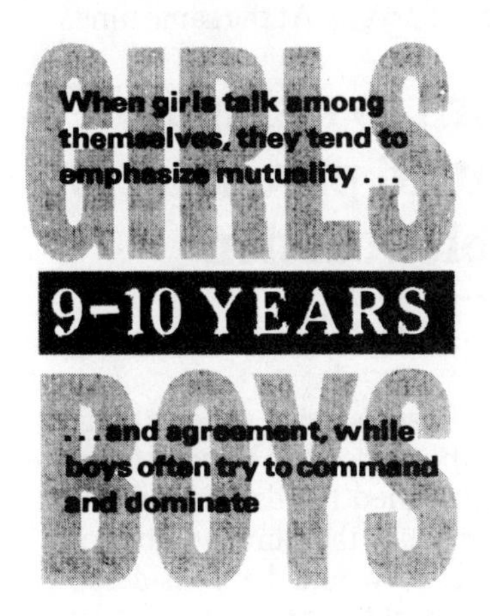

Studies of other cultures show that male behavior, while characteristically aggressive, need not be characteristically deadly. Harvard's Whiting, who has been analyzing children cross-culturally for half a century, found that in societies where boys as well as girls take care of younger siblings, boys as well as girls show nurturant, sociable behavior. "I'm convinced that infants elicit positive behavior from people," says Whiting. "If you have to take care of somebody who can't talk, you have to learn empathy. Of course there can be all kinds of experiences that make you extinguish that eliciting power, so that you no longer respond positively. But on the basis of our data, boys make very good baby tenders."

In our own society, evidence is emerging that fathers who actively participate in raising their children will be steering both sons and daughters toward healthier gender roles. For the last eight years Yale's Pruett has been conducting a groundbreaking longitudinal study of 16 families, representing a range of socioeconomic circumstances, in which the fathers take primary responsibility for child care while the mothers work full time. The children are now between 8 and 10 years old, and Pruett has watched subtle but important differences develop between them and their peers. "It's not that they have conflicts about their gender identity—the boys are masculine and the girls are feminine, they're all interested in the same things their friends are," he says. "But when they were 4 or 5, for instance, the stage at preschool when the boys leave the doll corner and the girls leave the block corner, these children didn't give up one or the other. The boys spent time playing with the girls in the doll corner, and the girls were building things with blocks, taking pride in their accomplishments."

Little footballs: Traditionally, Pruett notes, fathers have enforced sex stereotypes more strongly than mothers, engaging the boys in active play and complimenting the girls on their pretty dresses. "Not these fathers," says Pruett. "That went by the boards. They weren't interested in bringing home little footballs for their sons or little tutus for the girls. They dealt with the kids according to the individual. I even saw a couple of the mothers begin to take over those issues—one of them brought home a Dallas Cowboys sleeper for her 18-month-old. Her husband said, 'Honey, I thought we weren't going to do this, remember?' She said, 'Do what?' So that may be more a function of being in the second tier of parenting rather than the first."

As a result of this loosening up of stereotypes, the children are more relaxed about gender roles. "I saw the boys really enjoy their nurturing skills," says Pruett. "They knew what to do with a baby, they didn't see that as a girl's job, they saw it as a human job. I saw the girls have very active images of the outside world and what their mothers were doing in the workplace—things that become interesting to most girls when they're 8 or 10, but these girls were interested when they were 4 or 5."

Pruett doesn't argue that fathers are better at mothering than mothers, simply that two involved parents are better than "one and a lump." And it's hardly necessary for fathers to quit their jobs in order to become more involved. A 1965-66 study showed that working mothers spent 50 minutes a day engaged primarily with their children, while the fathers spent 12 minutes. Later studies have found fathers in two-career households spending only about a third as much time with their children as mothers. What's more, Pruett predicts that fathers would benefit as much as children from the increased responsibility. "The more involved father tends to feel differently about his own life," he says. "A lot of men, if they're on the fast track, know a lot about competitive relationships, but they don't know much about intimate relationships. Children are experts in intimacy. After a while the wives in my study would say, 'He's just a nicer guy'."

Pruett's study is too small in scope to support major claims for personality development; he emphasizes that his findings are chiefly theoretical until more research can be undertaken. But right now he's watching a motif that fascinates him. "Every single one of these kids is growing something," he says. "They don't just plant a watermelon seed and let it die. They're really propagating things, they're doing salad-bowl starts in the backyard, they're breeding guinea pigs. That says worlds about what they think matters. Generativity is valued a great deal, when both your mother and your father say it's OK." Scientists may never agree on what divides the sexes; but someday, perhaps, our children will learn to relish what unites them.

Children of the universe

Good parenting benefits the community as well as kids

Amitai Etzioni

Making a child is a moral act. Obviously, it obligates the parents to the child. But it also obligates the parents to the community. We must all live with the consequences of children who are not brought up properly, whether bad economic conditions or self-centered parents are to blame. Juvenile delinquents do more than break their parents' hearts, and drug abusers do more than give their parents grief. They mug the elderly, hold up stores and gas stations, and prey on innocent children returning from school. They grow up to be problems, draining society's resources and patience. In contrast, well-brought-up children are more than a joy to their families; they are (oddly, it is necessary to reiterate this) a foundation of proud and successful communities. Therefore, parents have a moral responsibility to the community to invest themselves in the proper upbringing of their children, and communities have a similar responsibility to enable parents to so dedicate themselves.

A word about proper upbringing: I do not mean merely feeding children, cleaning their rear ends, and making sure that they do not roam the streets. These custodial responsibilities are obvious and quite well reflected in our laws. As psychology professor Urie Bronfenbrenner writes, "Basic medical services and adequate diet, while essential, are not enough by themselves to insure normal physical and psychological development....Beyond health care and nutrition, certain other essential requirements must also be met."

If all that children receive is custodial care and morally careless education, their bodies will mature but their souls will not. If the moral representatives of society do not reach children, television and the streets will. We are all too familiar with, and frequently bemoan, the results of this type of "education." Now I will examine one of the root causes: Like charity, education—or the lack thereof—begins at home. In order for education to start at home, there must be a home.

I rarely discuss this matter in public or with friends without someone exclaiming, "Hey, you're dumping on women!" or "You believe that women must stay at home and attend to the family's children! Women have the same right as men to work outside the home!" As I see it, this is not the issue; the issue is the dearth of involvement of both mothers and fathers.

Consider for a moment parenting as an industry. As farming declined, most fathers left to work away from home generations ago. Over the past 20 years, millions of American mothers have sharply curtailed their work in the "parenting industry" by moving to work outside the home. By 1991 two-thirds (66.7 percent) of all mothers with children under 18 were in the labor force, and more than half (55.4 percent) of women with children under the age of 3 were. At the same time, a much smaller number of child-care personnel moved into the parenting industry.

We need to return to a situation in which committed parenting is an honorable vocation.

If this were any other business, say, shoemaking, and more than half of the labor force had been lost and replaced with fewer, less-qualified hands and still we asked the shoemakers to produce the same number of shoes of the same quality, we would be considered crazy. But this is what happened to parenting. As first men and then women left to work outside the home, they were replaced by some child-care services, a rela-

From *Utne Reader,* May/June 1993, pp. 52-61. Excerpt from *The Spirit of Community: Rights, Responsibilities and the Communitarian Agenda* by Amitai Etzioni. © 1993 by Amitai Etzioni. Reprinted by permission of Crown Publishers, Inc.

tively small increase in babysitters and nannies, and some additional service by grandparents—leaving parenting woefully shorthanded. The millions of latch-key children, who are left alone for long stretches of time, are but the most visible result of the parenting deficit.

Is this the "fault" of the women's movement, feminism, or mothers per se? Obviously not. All women did was demand for themselves what men had long possessed, working outside the home not only for their own personal satisfaction but because of what they often perceived as the economic necessity of a second income. Whatever the cause, the result is an empty nest.

While parenting is the responsibility of both parents—and may well be most effectively discharged in two-parent families immersed in a community context of kin and neighbors—most important is the scope of commitment. Single parents may do better than two-career absentee parents. Children require attention and a commitment of time, energy, and, above all, self.

Parenting cannot be carried out over the phone, however well meaning and loving the calls may be. It requires physical presence. The notion of "quality time" is a lame excuse for parental absence; it presupposes that bonding and education can take place in brief time bursts, on the run. Quality time occurs within quantity time. As you spend time with children—fishing, gardening, camping, or "just" eating a meal—there are unpredictable moments when an opening occurs, and education takes hold. As family expert Barbara Dafoe Whitehead puts it: "Maybe there is indeed such a thing as a one-minute manager, but there is no such thing as a one-minute parent."

Is the answer to the parenting deficit building more child-care centers? After all, other societies have delegated the upbringing of their children—to black nannies in the antebellum South, to Greek slaves in ancient Rome. True enough. But in these historical situations, the person who attended to the children was an adjunct to the parents, rather than a replacement for them, and an accessory reserved mostly for upper-class families with leisure. A caregiver remained with the family throughout the children's formative years, and often beyond; she was, to varying degrees, integrated into the family. The caregiver, in turn, reflected, at least in part, the family's values and educational posture. Some children may have been isolated from their parents, but, as a rule, there was a warm, committed figure dedicated to them, one who bonded and stayed with them.

Today most child-care centers are woefully understaffed with poorly paid and underqualified personnel. Child-care workers are in the lowest 10th of all wage earners (with an average salary of $5.35 per hour in 1988), well below janitors. They frequently receive no health insurance or other benefits, which makes child care an even less attractive job. As Edward Zigler, a professor of child development at Yale, put it: "We pay these people less than we do zookeepers—and then we expect them to do wonders." The personnel come and go, at a rate of 41 percent per year at an average day-care center.

Bonding between children and caregivers under these circumstances is very difficult to achieve. Moreover, children suffer a loss every time their surrogate parents leave the day-care center for another job. It would be far from inaccurate to call the worst of these facilities kennels for kids. Sure, there are exceptions, but these exceptions should not distract us from the basically dismal picture: substandard care and all-too-frequent warehousing of children, with overworked parents frantically trying to make up the deficit in their free time.

Certainly many low-income couples and single parents have little or no choice except to use the minimal, if expensive, care that such centers provide. All we can offer here is to urge that before parents put their children in such institutions, they should check them out as extensively as possible (including surprise visits in the middle of the day). Moreover, we should all support these parents' quest for additional support from corporations and government if they cannot themselves spend more on child care.

Particularly effective are cooperative arrangements that require each parent to contribute some time—four hours a week?—to serve at his or her child's center. Not only do such arrangements reduce the center's costs, they also allow parents to see firsthand what actually goes on, ensuring some measure of built-in accountability. It provides for continuity—while staff may come and go, parents stay. And parents come to know other parents of children in the same stage of development. They form social bonds, which can be drawn upon to work together to make these centers more responsive to children's needs.

Above all, age matters. Young infants, under 2 years old, are particularly vulnerable to separation anxiety. Several bodies of data strongly indicate that infants who are institutionalized at a young age will not mature into well-adjusted adults. As Edward Zigler puts it, "We are cannibalizing children. Children are dying in the system, never mind achieving optimum development."

Unless the parents are absent or abusive, children are better off at home. Older children, between 2 and 4, may be able to handle some measure of institutionalization in child-care centers, but their personalities often seem too unformed to be able to cope well with a nine-to-five separation from a parent.

As a person who grew up in Israel, I am sometimes asked whether it is true that kibbutzim succeed in bringing up toddlers in child-care centers. I need to note first that, unlike the personnel in most American child-care centers, the people who care for children on a kibbutz are some of the most dedicated members of the work force, because these communities consider child care to be a very high priority. As a result, child-care positions are highly sought after and there is little turnover, which allows for essential bonding to take place. In addition, both parents are intimately involved in bringing up their children, and they frequently visit the child-care centers, which are located very close to where they live and work.

Even so, Israeli kibbutzim are rapidly dismantling their collective child-care centers and returning children to live with their families—because both the families and the community established that even a limited disassociation of children from their parents at a tender age is unacceptable.

There is no sense looking back and beating our breasts over how we got ourselves into the present situation. But we must acknowledge that as a matter of social policy (as distinct from some individual situations) we have made a mistake in assuming that strangers can be entrusted with the effective personality formation of infants and toddlers. Over the last 25 years, we have seen the future, and it is not a wholesome one. If we fervently wish for our children to grow up in a civilized society, and if we seek to live in one, let's face facts: It will not happen unless we dedicate more of ourselves to our children.

When communitarians [communitarianism is a new political movement emphasizing community needs] discuss parental responsibilities we are often asked, "How can we have more time for the kids if we need to work full time just to make ends meet?" Our response requires an examination of the value of children as compared to other "priorities."

Few people who advocated equal rights for women favored a society in which sexual equality would mean a society in which all adults would act like men, who in the past were relatively inattentive to children. The new gender-equalized world was supposed to be a combination of all that was sound and ennobling in the traditional roles of women and men. Women were to be free to work anyplace they wanted, and men would be free to show emotion, care, and domestic commitment. For children, this was not supposed to mean, as it too often has, that they would be lacking dedicated parenting. Now that we have seen the result of decades of widespread neglect of children, the time has come for both parents to revalue children and for the community to support and recognize their efforts. Fathers and mothers are not just entitled to equal pay for equal work, equal credit and housing opportunities, and the right to choose a last name; they also must bear equal responsibilities—above all, for their children.

A major 1991 report by the National Commission on Children in effect is a national call for revaluation of children. Joseph Duffey, president of The American University, and Anne Wexler, a leading liberal, have also expressed the renewed commitment: "Perhaps, in the end," they wrote, "the great test for American society will be this: Whether we are capable of caring and sacrificing for the future of children. For the future of children other than our own, and for children of future generations. Whether we are capable of caring and sacrificing that they might have a future of opportunity."

In the 1950s, mothers who worked outside the home were often made to feel guilty by questions such as "Doesn't Jenny mind eating lunch in school?" By the 1980s, the moral voice had swung the other way. Now, women, not to mention men, who chose to be homemakers were put down by such comments as "Oh, you're not working...," the implication being that if you did not pursue a career outside the house, there was nothing to talk to you about. We need to return to a situation in which committed parenting is an honorable vocation.

In 1987, poor families spent 25 percent of their income on child care.
—Mother Jones
May/June 1991

In France, new mothers get four months of paid leave at 84 percent of earnings. About 30 percent of children attend public day-care programs, for which parents pay on a sliding scale according to income.
—In These Times
Feb. 22, 1993

An estimated 8 to 10 million families nationwide have childen who are left alone each weekday before or after school. A 1990 study by the Child Welfare League of America, a Washington, D.C.-based organization that lobbies for affordable child-care programs, found that 42 percent of children age 9 and under are left home alone occasionally, if not regularly.
—Pacific Sun
Jan. 29, 1993

One major way that commitment may be assessed is by the number of hours that are dedicated to a task over the span of a day. According to a 1985 study by a University of Maryland sociologist, parents spent an average of only 17 hours per week with their children, compared to 30 in 1965. Even this paltry amount of time is almost certainly an overstatement of the case because it is based on self-reporting.

This revaluation of the importance of children has two major ramifications. For potential parents, it means that they must consider what is important to them: more income or better relationships with their children. Most people cannot "have it all." They must face the possibility that they will have to curtail their gainful employment in order to invest more time and energy into their offspring. This may hurt their chances of making money directly (by working fewer hours) or indirectly (by advancing more slowly in their careers).

Many parents, especially those with lower incomes, argue that they both desire gainful employment not because they enjoy it or seek self-expression, but because they "cannot make ends meet" otherwise. They feel that both parents have no choice but to work full time outside the home if they are to pay for rent, food, clothing, and other basics. A 1990 Gallup Poll found that half of those households with working mothers would want the mother to stay home if "money were not an issue." (The same question should have been asked about fathers.)

This sense of economic pressure certainly has a

strong element of reality. Many couples in the '90s need two paychecks to buy little more than what a couple in the early '70s could acquire with a single income. There are millions of people in America these days, especially the poor and the near poor, who are barely surviving, even when both parents do work long and hard outside the home. And surely many single parents must work to support themselves and their children. But at a certain level of income, which is lower than the conventional wisdom would have us believe, parents do begin to have a choice between enhanced earnings and attending to their children.

There is considerable disagreement as to what that level might be. Several social scientists have shown that most of what many people consider "essentials" are actually purchases that their cultures and communities tell them are essential, rather than what is actually required. They point out that people need rather little: shelter, liquids, a certain amount of calories and vitamins a day, and a few other such things that can be bought quite cheaply. Most of what people feel that they "must have"—from VCRs to $150 Nike sneakers—is socially conditioned.

A colleague who read an earlier version of these pages suggested that the preceding line of argument sounds as if social scientists wish to cement the barriers between the classes and not allow lower-class people to aspire to higher status. Hardly so. They are not arguing that people should lead a life of denial and poverty, but that they have made, and do make, choices all the time, whether or not they are aware of this fact. They choose between a more rapid climb up the social ladder and spending more time with their children. Communitarians would add that in the long run parents will find more satisfaction and will contribute more to their community if they heed their children more and their social status less. But even if they choose to order their priorities the other way around, let it not be said that they did not make a choice. Careerism is not a law of nature.

We return then to the value we as a community put on having and bringing up children. In a society that places more value on Armani suits and winter vacations than on education, parents are under pressure to earn more, whatever their income. They feel that it is important to work overtime and to dedicate themselves to enhancing their incomes and advancing their careers. We must recognize now, after two decades of celebrating greed and in the face of a generation of neglected children, the importance of educating our children.

While the shift from consumerism and careerism to an emphasis on children is largely one of values, it has some rewarding payoffs. Corporations keep complaining, correctly, that the young workers who present themselves on their doorsteps are undertrained. A large part of what they mean is a deficiency of character and an inability to control impulses, defer gratification, and commit to the task at hand. If businesses would cooperate with parents to make it easier for them to earn a living and attend to their children, the corporate payoffs would be more than social approbation: They would gain a labor force that is much better able to perform. The community, too, would benefit, by having members who are not merely more sensitive to one another and more caring but also more likely to contribute to the commonweal. Last but not least, parents would discover that while there are some failures despite the best intentions and strongest dedication, and while there are no guarantees or refunds in bringing up children, by and large you reap what you sow. If people dedicate a part of their lives to their kids, they are likely to have sons and daughters who will make them proud and fill their old age with love.

The community—that is, all of us—suffers the ill effects of absentee parenting. According to a study by social scientist Jean Richardson and her colleagues, for example, eighth-grade students who took care of themselves for 11 or more hours a week were twice as likely to be abusers of controlled substances (that is, to smoke marijuana or tobacco or to drink alcohol) as those who were actively cared for by adults. "The increased risk appeared no matter what the sex, race, or socioeconomic status of the children," Richardson and her associates noted. And students who took care of themselves for 11 or more hours per week were one and a half to two times more likely "to score high on risk-taking, anger, family conflict, and stress" than those who did not care for themselves, a later study by Richardson and her colleagues found.

Travis Hirschi reports in *Causes of Delinquency* that the number of delinquent acts, as reported by the children themselves, was powerfully influenced by the children's attachment to the parents. The closer the mother's supervision of the child, the more intimate the child's communication with the father, and the greater the affection between child and parents, the less the delinquency. Even when the father held a low-status job, the stronger the child's attachment to him the less the delinquency. Other factors also contributed to delinquency, such as whether the child did well in and liked school, but these factors were themselves affected by family conditions.

Other studies point to the same conclusions.

Gang warfare in the streets, massive drug abuse, a poorly committed work force, and a strong sense of entitlement and a weak sense of responsibility are, to a large extent, the product of poor parenting. True, economic and social factors also play a role. But a lack of effective parenting is a major cause, and the other factors could be handled more readily if we remained committed to the importance of the upbringing of the young. The fact is, in poor neighborhoods one finds decent and hardworking youngsters right next to antisocial ones. Likewise, in affluent suburbs one finds antisocial youngsters right next to decent, hardworking ones. The difference is often a reflection of the homes they come from.

What we need now, first of all, is to return more hands and, above all, more voices to the "parenting industry." This can be accom-

The family leave bill only covers 5 percent of all firms and roughly 60 percent of all workers. According to a study by 9 to 5, the national association of working women, fewer than 40 percent of working women can take advantage of the full unpaid leave without severe financial hardship. *—In These Times* Feb. 22, 1993	In 1989 just 3 percent of U.S. firms offered paid maternity leave and only 37 percent offered unpaid leave. (When the U.S. family leave bill takes effect this year, employers with 50 or more workers must allow employees—male and female—to take up to 12 weeks of unpaid family leave per year.) *—Dollars & Sense* Jan./Feb. 1993

plished in several ways, all of which are quite familiar but are not being dealt with nearly often enough.

Given the acceptance of labor unions and employers, it is possible for millions of parents to work at home. Computers, modems, up- and downlinks, satellites, and other modern means of communication can allow you to trade commodities worldwide without ever leaving your den or to provide answers on a medical hotline from a corner of the living room or to process insurance claims and edit books from a desk placed anywhere in the house.

If both parents must work outside the household, it is preferable if they can arrange to work different shifts, to increase the all-important parental presence. Some couples manage with only one working full time and the other part time. In some instances, two parents can share one job and the parenting duties. Some find flextime work that allows them to come in late or leave early (or make some other adjustments in their schedule) if the other parent is detained at work, a child is sick, and so on.

These are not pie-in-the-sky, futuristic ideas. Several of the largest corporations already provide one or more of these family-friendly features. Du Pont in 1992 had 2,000 employees working part time and between 10,000 and 15,000 working flextime. IBM has a "flexible work leave of absence" plan that allows employees to work up to three years part time and still collect full-employment benefits. Avon Products and a subsidiary of Knight-Ridder newspapers have their own versions of these programs, and the list goes on.

Public policies could further sustain the family. Child allowances, which are common in Europe, could provide each family with some additional funds when a child is born. Others suggest a program, modeled after the GI Bill, that would give parents who stay home "points" toward future educational or retraining expenses.

These measures require a commitment on the part of parents to work things out so that they can discharge more of their parenting responsibilities, and on the side of corporations and the government to make effective parenting possible.

The recent national debate over whether parents should be allowed three months of unpaid leave is ridiculous, a sign of how much we have lost our sense of the importance of parenting. A bill finally passed by Congress in 1993 and signed by President Clinton mandated only 12 weeks of unpaid leave and only for companies with more than 50 employees. (In Canada, employees receive 15 weeks at 60 percent pay; in Sweden, they receive 90 percent pay for 36 weeks and prorated paid leave for the next 18 months.)

No one can form the minimal bonding a newborn child requires in 12 weeks, a woefully brief period of time. A typical finding is that infants who were subject to 20 hours a week of non-parental care are insecure in their relationship with their parents at the end of the first year, and more likely to be aggressive between the ages of 3 and 8. If children age 2 or younger are too young to be institutionalized in child-care centers, a bare minimum of two years of intensive parenting is essential.

The fact that this recommendation is considered utopian is troubling, not merely for parents and children but also for all who care about the future of this society. Let's state it here unabashedly: Corporations should provide six months of paid leave, and another year and a half of unpaid leave. (The costs should be shared by the employers of the father and the mother.) The government should cover six months of the unpaid leave from public funds (many European countries do at least this much), and the rest would be absorbed by the family.

Given increased governmental support and corporate flexibility, each couple must work out its own division of labor. In one family I know, the mother is a nurse and the father a day laborer. She is earning much more, and he found it attractive to work occasionally outside the home while making the care of their two young daughters his prime responsibility. He responds to calls from people who need a tow truck; if the calls come while his wife is not at home, he takes his daughters with him. I met them when he towed my car. They seemed a happy lot, but he was a bit defensive about the fact that he was the home parent. The community's moral voice should fully approve of this arrangement, rather than expect that the woman be the parent who stays at home. At the same time, there should be no social stigma attached to women who prefer to make raising children their vocation or career. We need more fathers and mothers who make these choices; stigmatizing any of them is hardly a way to encourage parenting. Re-elevating the value of children will help bring about the needed change of heart.

CRIME IN AMERICA

Violent and irrational—and that's just the policy

Recent falls in America's crime rate have led policy-makers round the world to look admiringly at the country's get-tough policies. They should not. American criminal-justice policy is misconceived and dangerous

REMEMBER serial killers? A few years ago, these twisted creatures haunted not just the American imagination but, it seemed, America's real streets and parks: an official of the Justice Department was widely reported as saying that 4,000 of America's annual 24,000-or-so murders were attributable to serial killers.

America loves its myths—and that was pretty much what the "wave of serial killings" turned out to be: 4,000 people are not victims of serial murderers; 4,000 murders remain unsolved each year. According to cool-headed academic research, maybe 50 people a year are victims of serial murderers; the figure has been stable for 20 years.

Serial murderers obviously form a bizarre and special category of criminal. People might well believe extraordinary things about them. But about crime in general, surely ordinary folk have a better understanding—don't they? Well, consider two widely-held beliefs:

"America has experienced a crime wave in the past 20 years." No. According to the National Crime Victimisation Survey, violent crime fell in the first half of the 1980s, rose in the second half, and has been falling in the 1990s. Over the past two decades, it has fallen slightly. Non-violent property crimes (theft, larceny and burglary) have followed similar patterns. So has murder: its peak was in 1980 (see chart on next page).

"America is more criminal than other countries." Again, no. According to an International Crime Survey, carried out by the Ministry of Justice in the Netherlands in 1992, America is not obviously more criminal than anywhere else. You are more likely to be burgled in Australia or New Zealand. You are more likely to be robbed with violence in Spain; you are more likely to be robbed without violence in Spain, Canada, Australia and New Zealand. You are more likely to be raped or indecently assaulted in Canada, Australia or western Germany. And so on.

American misconceptions raise two questions. First, why are Americans so afraid of crime? (As according to Gallup polls, they are: in recent years Americans have put crime either first or second in their list of problems facing the country; in Britain, crime limps along between second and sixth in people's priorities.) Second, why should Americans be so punitive in their attitude to criminals? (As they also seem to be: when asked by the International Crime Survey what should happen to a young burglar who has committed more than one offence, 53% of Americans reckoned he should go to prison, compared with 37% of English and Welsh, 22% of Italians, and 13% of Germans and French.)

One possible explanation is that Americans are irrational in their attitudes to crime. But that cannot be right: crime imposes huge costs on the country and has helped turn parts of American inner cities into nightmares of violence. Given that, it is hardly surprising that Americans should fear the spread of crime. But it remains surprising that American public attitudes should be so different from those in other countries which also have dangerous inner cities. No, there seems to be something else feeding Americans' fear and loathing of criminals. More probably, two things: the violence of American crime, and its irrationality. And it is with these that America's real crime-policy problems begin.

Murder as public choice

America tops the developed-country crime league only in one category: murder. While you are more likely to be burgled in Sydney than in Los Angeles, you are 20 times more likely to be murdered in Los Angeles than you are in Sydney.

American crime is not only more violent; it is also irrational in its violence. Think about a person held up at gunpoint who fails to co-operate with a robber. "Since both the risk of apprehension and the potential punishment escalate when the victim is killed," says Franklin Zimring, a criminologist at the University of California, Berkeley, "the rational robber would be well advised to meet flight or refusal by avoiding conflict and seeking another victim." Yet Americans commonly get killed in these circumstances, and it is the irrationality of such violence that terrifies.

There is nothing odd or surprising in the observation that America is more violent than other countries, that Americans are more afraid of crime, and they are therefore more punitive. But the problem with America's criminal-justice policy lies in that sequence of thought. By eliding violence and crime, Americans fail to identify the problem that sets them apart from the rest of the rich world, which is violence, rather than crime generally. Americans are right to think they have a special problem of violence. They are wrong to think their country is being overwhelmed by crime of every sort. Yet because many people do think that, they are throwing their weight behind indiscriminate policies which, at huge cost, bludgeon crime as a whole but fail to tackle the problem of violence.

America now imprisons seven times as many people (proportionately) as does the average European country, largely as a result of get-tough-on-crime laws. These are the laws other countries are now studying with admiration.

First came mandatory sentencing laws, requiring courts to impose minimum sentences on offenders for particular crimes. Michigan, for instance, has a mandatory life sentence for an offender caught with 650 grams of cocaine. A federal law condemns anybody convicted of possession of more than five grams of crack to a minimum of five years in prison.

Then came "three-strike laws", supported by Bill Clinton and adopted by 20-odd states and the federal government. These impose a mandatory life sentence on anybody convicted of a third felony. The seriousness of the felony, and therefore the impact of the law, varies from state to state. In California, in the most celebrated case, a man who stole a pizza as his third felony got life. His case was extreme, but not unique: another man got life after stealing three steaks.

Three steaks and you're out

Now, the fashion is for "truth-in-sentencing". Such laws require the criminal to spend most of his sentence (usually 85%) in prison, rather than making him eligible for parole after, say, four to six years of a ten-year sentence. There is much to be said for a system that does not leave the public feeling cheated about what sentences actually amount to. But, by imposing the 85% average on all offenders, "truth in sentencing" makes it impossible to discriminate between people who seem genuinely remorseful and might be let out early and the more dangerous types who should serve the whole of their sentence.

Since the early 1970s, when the first tough-sentencing laws were introduced, the prison population has risen from 200,000 to 1.1m. If that increase were made up mostly of the violent people that have engendered America's crime panic, that could be counted as a blow against violent crime. But it is not: the biggest increase is in non-violent drug offenders.

Between 1980 and now, the proportion of those sentenced to prison for non-violent property crimes has remained about the same (two-fifths). The number of those sentenced for drugs has soared (from one-tenth to over one-third). The share sentenced for violent crimes has fallen from half to under one-third.

And so what, you might ask? Non-violent crime still matters. Even if America's crime panic is related to violence, it is right and proper that the system should be seeking to minimise all crime. The prison population is going up. The crime figures are going down. Let 'em rot. As the right says: "Prison works."

Or does it? That depends on what you mean by "works". To many people, prison can strongly influence the trend in the crime rate: putting a lot of people in prison, they believe, can achieve a long-term reversal of rising crime. This must be doubtful. Yes, crime is falling now. But it also fell in the early 1980s, rose in the late 1980s and fell again in the early 1990s. The prison population rose through the whole period.

If there is any single explanation for these changes, it would seem to lie in demographics. Young men commit by far and away the largest number of crimes, so when there are more of them around, proportionately, the crime rate goes up. That was what happened in the 1960s, the period of the big, sustained post-war rise in the crime rate. Demography also tells you that there will be more young men around in ten years' time to commit more crimes.

But demographics cannot be the only explanation. If it were, crime would have fallen in the second half of the 1980s, when there were fewer teenagers. In fact, it rose.

Why? The answer is probably drugs. What seems to have happened is that the appearance of crack in late 1985 shook up the drugs-distribution business. The number of dealers increased, kids with no capital got into the business and gangs competed murderously for market share.

This theory would account for the decline in homicides in the 1990s. Crack consumption seems to be falling—possibly just because drugs go in and out of fashion, possibly because teenagers have seen how bad the stuff is. And the market has matured as well as declined. Policemen and researchers say territories have been carved out, boundaries set. With competition less rife, murders have declined.

The significance of all this is that it loosens the connection between the rise in the prison population and the fall in the crime rate. Crime might have fallen anyway. A combination of demographic and social explanations, rather than changes in the prison population, seems to account for much of the changing pattern of crime.

Vox populi, vox dei, vox dementiae

That said, there might still be a justification for putting more people in prison: if by doing so you lowered the overall level of crime by taking criminals out of circulation. Indeed, if a small number of young men commit a disproportionately large number of crimes, then locking up this particular group might depress crime a lot.

Liberal criminologists sometimes appear to doubt this. "It seems," says John DiIulio, the right-wing's favourite thinker on crime, "that you need a PhD in criminology to doubt the proposition that putting criminals in prison will keep down crime." Of course, the proposition is self-evidently true. If you banged up for life anyone who had ever committed a crime, however trivial, crime would plummet. But the question is: is this sensible, even if it does work?

To many ordinary Americans, it is and politicians are happy to oblige the voters by promising to get ever tougher on crime. But what is the evidence about whether prison is an effective way of reducing crime?

Looking across the states' different crime rates and imprisonment rates, there is no correlation between the two. True, you would not necessarily expect one: states are different and tough-sentencing laws might be a reaction to a high crime rate as much as a way of bringing it down. But more sophisticated analyses confirm there is no link. Mr Zimring took the adult and juvenile crime rates in California and studied what happened over the period when tough laws were being introduced for adults, but not for juveniles. No relationship is detectable: for most crimes, offences committed by juveniles either fell or rose significantly less than did those committed by adults.

And, just as there is no convincing argument that prison effectively reduces the level of crime, nor does there seem to be a convincing cost-benefit argument in favour of prison. The problem lies in costing crime. One often-used estimate, which monetises intangibles like pain and suffering, calculates the annual costs of crime at $450 billion. This makes prison look a bargain: its annual bill is $35 billion, while the criminal-justice system, including police

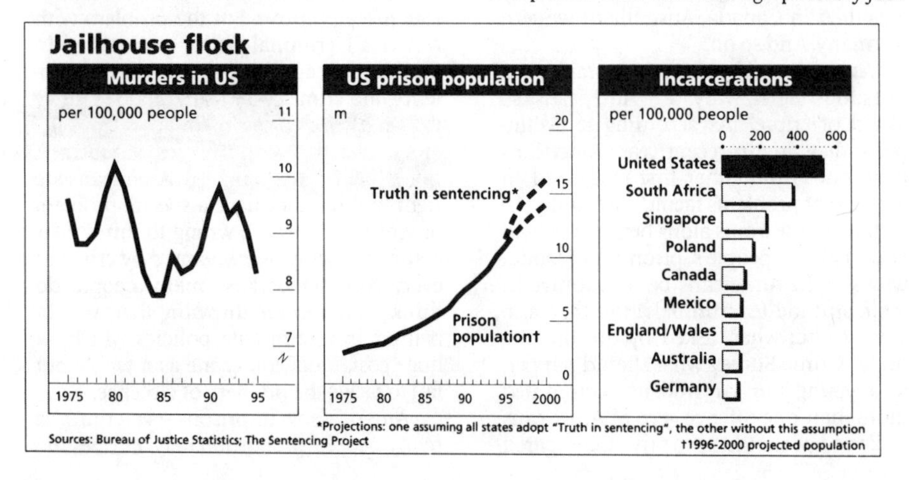

and courts, costs $100 billion. But if you calculate the costs of crime on the basis of physical damage—hospital bills or the cost of replacing stolen goods—the figure comes out at a mere $18 billion a year. The moral is that, while the cost of crime must be high, no one has any real idea what it is.

What you can say is that, out of the range of options for dealing with criminals, prison is among the most expensive. One currently popular alternative is the "drugs court". Under this system, people charged with possession or small dealing may opt to go through a drugs-treatment programme rather than stand trial. Treatment costs $3,500–15,000 a year, depending on whether it is residential or not; prison costs $22,000. There is also some evidence that these courts are better than prisons at discouraging reoffending, though, since they are relatively new, the evidence is not conclusive.

Of course, get-tough policies raise questions other than that of efficacy. One is moral. Is it right to lock somebody up for life for stealing a pizza? Another is racial (see box). These concerns have not, it seems, made much of an impact on public opinion. According to Mr DiIulio, "Americans have lost interest in the Anglo-Saxon, innocent-until-proven-guilty model of justice. They want to get the bad guys."

Yet even by this measure, the get-tough policies are misfiring. Around 100,000 people go to prison for the 6m-odd violent crimes committed a year. The system is not getting the bad guys. What it is getting is a great many drug-taking, drug-dealing, small-time thieves. Conservatives argue that most people in prison are either violent or repeat offenders. True, but many of the repeat offenders are addicts financing their habit through drug dealing or burglary. Nobody suggests that they are unfortunates for whom one should merely be sorry; but it is not clear that sending a crack-user to prison for five years is a rational solution to America's violent-crime problem.

America is awash with academics, judges, commissioners and policemen who know and study crime. The Justice Department's research arm, the National Institute of Justice, spent $53m last year on research of a higher standard, and in a larger quantity, than goes on anywhere else in the world.

Almost all of this stuff doubts the efficacy of what is going on in criminal justice, and fears for the consequences. Almost all the professionals agree that America's problem is violence, and that the way to reduce violence is to restrict access to guns. And on this—though the point is rarely noticed—the public agrees: 62%, according to a recent Gallup poll, favour stricter gun control.

Yet none of it makes much difference to public policy. The administration promotes a three-strike policy even though it knows that the main effect of three-strike laws is to bung up the prison system with people long past crime-committing age.

American crime policy seems to have become an area where the arguments—admittedly often complex and finely balanced—take second place to the lobbying power of special-interest groups. The effectiveness of one, the National Rifle Association, has been well-documented. A less familiar one is the prison-building lobby.

Prisons have been likened to the defence industry as a government subsidy to the white working class. For areas hit by the end of the cold war, and by the ups and downs of agriculture, prisons provide attractively recession-proof employment. As the flier for the American Jail Association last year said, "Jails are BIG BUSINESS." Towns compete to get them.

The prison guards' union has also become a powerful voice. According to a study of campaign contributions in California in 1991-92, the local version, the California Correctional Peace Officers' Association, was the second-largest donor in the state. It spends around $1m on political contributions for the governorship and the legislature in each electoral cycle.

But more important than the lobbying, and more worrying, is the failure of public debate on prison, its costs, and the alternatives. According to Bobby Scott, a Democratic congressman opposed to tough-sentencing laws, "When you call for more incarceration, you do not have to explain yourself; when you argue for effective alternatives, you do. And in politics, when you start explaining, you've lost." If that is true—and it sounds painfully accurate—something has gone badly wrong not just with American crime policy, but with America's capacity for reasoned public debate.

One-third and rising

BLACKS are more likely to commit crimes than are whites. Around 45% of those arrested for serious crimes are black. But they are also more harshly treated. Numberless studies have shown that the criminal-justice system is not colour blind. There are more unfounded arrests of blacks. Blacks pay on average twice as much bail as whites. They are more likely to be jailed before trial and get heavier sentences for the same crime.

Since the "war on drugs", the bias seems to have got worse. Blacks make up 12% of the American population, and, according to government surveys, 13% of those who say they have used drugs in the past month. But they account for 35% of arrests for drug possession, 55% of convictions and 74% of prison sentences.

Partly, that is because drug laws implicitly target blacks. Crack is a drug favoured by blacks. The mandatory federal penalty for possessing five grams of crack (a couple of days' supply for an addict) is five years in jail. Cocaine is principally a white person's drug. To get the same sentence a cocaine user has to have half a kilo in his possession.

The implementation of anti-drug laws also affects blacks disproportionately. Partly that is because the police raid black areas, not nice white suburbs, but that is not the full explanation. A study of sentencing in the 1980s, which divided blacks between "underclass" and "non-underclass", concluded that the biggest increase in the prison population was among "non-underclass" blacks convicted for drug offences.

The figures on blacks in the criminal-justice system are shocking. According to the Sentencing Project, a Washington-based penal-reform group, one-third of 20-29-year-old black men are on probation, on parole or in prison. As the prison population rises, that share will increase yet further. Think about that.

People who have been in prison have a slim chance of regular employment on release. Their families are therefore poorer than others. Their children are fatherless while they are inside. Prison becomes the norm; "normal" life abnormal. America is on a dangerous course.

Moral Credibility and Crime

Moral authority, rather than rehabilitation or deterrence, may be the key ingredient in a criminal-justice system that can reduce violent crime. That authority has been deeply eroded by the system's own rules and procedures, the author contends, and he offers ten major cases in point

PAUL H. ROBINSON

Paul H. Robinson is a professor at Northwestern University School of Law. He is a co-author, with John Darley, of Justice, Liability, and Blame: Community Views and the Criminal Law, *which was published [February 1995].*

WE are in a panic over crime. Legislators compete with one another to propose the toughest anti-crime legislation. The $30 billion federal anti-crime bill got strong support in the midst of a deficit-reduction drive. In the states "three strikes and you're out" proposals are trumped by "two strikes" proposals.

Clearheaded commentators point out that the panic is unjustified: crime rates have not in fact increased recently, they correctly note. But a complete description of our situation includes two other important pieces of information. First, one reason that crime has not increased recently is that we have altered the way we live in order to avoid it. We no longer go out at night. We no longer let our children walk to a friend's house to play. We install locks, carry Mace, and readily pay more for apartment buildings with security. In 1980 private expenditures on security were $20 billion—considerably more than the $14 billion of public spending in the same category. By 1990 annual private expenditures for security had risen to $52 billion. The rate of crime has stabilized because we increasingly diminish the quality of our lives to avoid it. The injury of crime escalates even where its incidence does not.

The other important fact is that crime *has* dramatically increased over the longer term, albeit in increments sufficiently small that no single reported increase has justified panic. In 1955 forty-six robberies occurred annually per 100,000 in the population; today the rate is more than 270—a sixfold increase. Rape rates have more than tripled. Murder per capita has more than doubled. The aggravated-assault rate has increased more than sixfold. Overall, taking into account both urban and rural areas, the major-crime rate is more than four times what it was four decades ago.

Perhaps the current panic is born not of new data but of realized frustration—reminding us of the monkey who works methodically to free himself from a trap and goes berserk only when he realizes he can't escape. Frustration, not crime, has boiled over.

To many, our situation is intolerable. Five years ago, before the current panic, 82 percent of those polled in one study believed that crime was getting worse; 34 percent felt "truly desperate" about rising crime.

From *The Atlantic Monthly,* March 1995, pp. 72-78.

What We Have Tried

THE steady worsening of the crime problem is not the result of inattention; rather, it has occurred despite our best efforts to halt it. In the 1950s we thought we could best stop crime by not just imprisoning criminals but rehabilitating them. Today we would call it attacking the root cause of the problem: offenders have a disease; we will treat it. The logic of the rehabilitative model dictated that sentences for all felonies be indeterminate—from one day to life, depending on the treatment needed and how the offender responded. But Robert Martinson, a sociologist, concluded in his 1974 survey article "What Works?," in *The Public Interest,* "I am bound to say that these data, involving over two hundred studies and hundreds of thousands of individuals as they do, are the best available and give us very little reason to hope that we have in fact found a sure way of reducing recidivism through rehabilitation."

Deterrence became popular as an alternative. Potential offenders would be dissuaded from committing offenses by the threat of serious penalties. The greater the threatened penalty (the longer the prison term), the greater the disincentive. The high cost of imprisonment would normally put a natural limit on the severity of the deterrent threat, but the threat could be made dramatic without courting fiscal crisis if longer sentences than would actually be served were publicly imposed. The idea was that the deterrent benefit would be fully realized at the moment of public imposition, and that offenders could later be quietly released by parole boards (conveniently left over from the rehabilitation approach). Typical was the federal system, in which offenders could be eligible for release immediately, and which generally required release after serving a third of the sentence.

But in our open society the shell game was soon seen for what it was. The public came to understand that a twenty-year sentence really meant a maximum of seven years and often much less. To counteract this discounting of sentences, judges imposed ever greater sentences, sometimes of hundreds of years. But this only increased public skepticism, because the potential discount also increased; even a sentence many times as long as the human life-span could end in release after a few years if the parole board so chose. Many states are following the lead of the federal system, which recently shifted to "real-time" sentencing, under which offenders must serve at least 85 percent of the term imposed. With time, sentencing will regain credibility.

But does imprisonment deter crime? Deterrence requires that potential offenders think about the consequences of their actions, as many fail to do. More important, deterrence requires that those who do think about the consequences see some real risk that they will be caught and punished—a risk that must outweigh the benefits they expect from the crime. Unfortunately for deterrence, potential offenders think that the threat of capture and punishment applies to others but not to them. Unlike the other guy, *they* will avoid detection by taking the necessary precautions. Thus even those who might think about the consequences of their actions do not think that the threat of punishment applies to them.

But even if most offenders thought the threat did apply to them, what would be the nature of the threat they faced? An astounding number of serious offenses are never reported to the police (for example, 21 percent of rapes and 40 percent of burglaries), out of embarrassment, out of fear of reprisal, or in the belief that the police are impotent to do anything. For the offenses reported, clearance rates (the rates at which the police identify and arrest suspects) have been dropping steadily for decades. The nationwide clearance rate for homicide, which was 93 percent in 1955, has steadily declined to 67 percent. That for rape has declined from 79 to 52 percent, and that for burglary from 32 to 14 percent.

And, as our realistic potential offender knows, being arrested is a far cry from being punished. The overall conviction rate among those arrested for the most serious offenses is 30 percent. Fewer than half of those convicted of a felony are sentenced to prison. Finally, the median time served by those actually sentenced to state prison ranges from 5.5 years for murder to 2.2 years for kidnapping to 1.4 years for arson.

The cumulative effect of the many escape hatches from punishment leaves a deterrent threat that looks like this: Homicide offers a less than 45 percent chance of being caught, convicted, and imprisoned, rape a 12 percent chance, robbery a four percent chance. Assault, burglary, larceny, and motor-vehicle theft are each a hundred-to-one shot. Our potential offender may not be cowed by these threats. These statistics also explain why longer prison terms can have only a limited effect in deterring crime: if a robber faces a mere four percent chance of going to prison, why should it matter to him whether the likely sentence is two years or ten years?

Our fallback crime-control policy has increasingly been to keep in prison those offenders we think may commit another crime. We know this works (at least to protect the public—victimization of other prisoners is another matter). Unfortunately, we are no better at predicting future dangerousness than we are at rehabilitating. In a monograph published by the National Institute of Mental Health, John Monahan, a psychologist, summarizes studies indicating that we are wrong about two out of every three people predicted to commit a serious offense. In other words, as many as two thirds of the prisoners detained for dangerousness are detained needlessly. Besides its obvious unfairness, this approach makes the incapacitation strategy wildly expensive. And even at its best, the value of incapacitating dangerous offenders is limited to avoiding further offenses while the person is in prison. It does nothing to avoid offenses after imprisonment.

This is a truly depressing picture. Is this the future of mankind: ever increasing numbers of people in prison? Can't we do something to prevent people from committing crimes in the first place?

Moral Credibility as Crime Control

WE tend to think of criminals as a distinct class of people. This image is reinforced by reports that one group of offenders is responsible for a disproportionately large share of crimes. If we could only do something with these criminal types in our society, the logic goes, we could solve our crime problem. And then the standard punishment-versus-prevention debate is off and running: should we keep more of these folks in prison, or should we target this group for more social services, or should we strike some balance between the two?

Some offenders truly are career criminals. But the greater truth is that no offenders, except for a small group of the mentally ill, are irrevocably driven to crime. For nearly any criminal, whatever the person's age, race, social background, or economic status, one can identify hundreds of thousands of people with essentially the same characteristics who have chosen to act differently and to remain law-abiding.

People who commit crimes are people like us who have chosen to do bad things. The study *Ordinary People and Extraordinary Evil* (1993), by the sociologist Fred Katz, shows how ordinary people in the course of everyday choices can come step by step to undertake serious wrongdoing—even horrendous acts like those of the Nazi Holocaust and the My Lai massacre.

Asking "How should we deal with our criminal class?" distracts us from the positive inquiry "Why do people obey the law?" Why, even in difficult situations of need and temptation, and even when they are unlikely to be caught and punished, do the vast majority of people remain law-abiding? Perhaps if we better understood what makes a person choose not to commit a crime, even when temptation and opportunity present themselves, we could develop and enlarge that influence.

Here is what preliminary social-science research hints at: beyond the threat of legal punishment, people obey the law because they fear the disapproval of their social group and because they generally see themselves as moral beings who want to do the right thing as they perceive it. In a 1980 study the sociologists Harold Grasmick and Donald Green concluded, "Each of the three independent variables [threat of legal punishment, social disapproval, and personal moral commitment] makes a significant, independent contribution to the explained variance [the rate of criminal behavior]."

As to social disapproval specifically, the sociologists Robert Meier and Weldon Johnson found in a 1977 study that "despite contemporary predisposition toward the importance of legal sanctions, our findings are . . . consistent with the accumulated literature concerning the primacy of interpersonal influence [that is, social disapproval]" over legal sanction.

As to moral commitment specifically, the social psychologist Tom Tyler concluded in *Why People Obey the Law* (1990) that "the most important incremental contribution is made by personal morality."

> This high level of normative commitment to obeying the law offers an important basis for the effective exercise of authority by legal officials. People clearly have a strong predisposition toward following the law. If authorities can tap into such feelings, their decisions will be more widely followed.

Can legal authorities tap into these powerful forces for compliance? If they can, the potential benefits are enormous. First, as the studies quoted above suggest, both the fear of social disapproval by one's group and one's own moral commitment have strong effects on compliance—stronger than the present deterrent threat of legal punishment.

Second, unlike the threat of legal punishment, these sources of compliance do not require the likelihood of being arrested, convicted, and imprisoned to be high. A person's family or friends may suspect that he is committing crimes even if the authorities do not, or cannot prove it. In any case, *the person* always knows about his crimes. Thus reinforcing the compliance powers of the social group and personal morality could reduce crime even if policing and prosecuting functions cannot be made more effective.

Finally, neither of these sources of compliance is as staggeringly expensive as large-scale incarceration is. Nor do they require the increased intrusions on privacy that more-effective crime investigation would require, or the increased errors in adjudication that easier rules of prosecution might cause. In other words, they offer the best of both worlds—significantly better compliance at lower cost.

But one key condition must exist if personal moral commitment and the power of social disapproval are to be harnessed: criminal law must be seen by the potential offender and by the potential offender's social group as an authoritative source of what is moral, of what is right—much as, within a functional family, a parent may be seen as such an authoritative source. More specifically, the social-science studies suggest, the extent of the law's power to gain compliance depends upon the extent of the law's moral credibility.

By "moral credibility" and "moral authority" I mean criminal law's reputation for punishing those who deserve it, under rules perceived as just; protecting from punishment those who do not deserve it; and where punishment is deserved, imposing the amount deserved—no more and no less. I do not underestimate how complex a matter it is to determine liability rules that will be perceived as just. But, as John Darley and I show in our new book, *Justice, Liability, and Blame: Community*

Views and the Criminal Law, shared community intuitions on morally just principles of punishment can be determined and articulated.

I would argue that people in our present society do not see the criminal-justice system as having a moral authority even loosely comparable to that which operates within, say, a typical functional family, or within a voluntary association. The criminal-justice system has been shaped over the years by forces that have little to do with moral desert or, sometimes, common sense, and everything to do with legal abstractions, procedural expediency, and the criminological theory of the day. The result is a system in which the dynamic of moral authority that works in successful families and small groups is sadly lacking.

What can we do to increase the moral authority of the criminal-justice system? What current rules and practices undercut the system's moral credibility? Full answers would require a detailed review, but in this brief space I can at least touch upon the highlights.

First a caveat. Some failures of the system are inevitable. Less-than-perfect clearance rates by the police, for example, admittedly limit the system's ability to do justice. But most people understand that not every offender can be caught and punished, that policing and prosecution have practical limits. Failures of the system due to these limits may be frustrating in the individual case, but over time they are not likely to hurt the system's credibility. What can hurt is failures of justice that are avoidable.

Avoidable Failures of Justice

DOES the system regularly fail to impose deserved punishment when it has the power to do so? Does it sometimes appear to have *chosen* a course that frustrates justice? Let me suggest five instances in which the answer is yes.

First, the American criminal-justice system routinely excludes reliable evidence under what is called the exclusionary rule. This suggests a system less strongly committed to doing justice than to discouraging overreaching by the police. But one may ask, If that discouraging hurts the system's moral credibility and therefore its power to encourage compliance, can't police overreaching be discouraged by other means? Why not discourage police misconduct by making officers directly liable or by helping victims to get compensation from municipalities? If limiting overreaching by the police is such an important goal, why not attack it head-on rather than "punishing" the offending officers by letting the criminal go free—which punishes society rather than the officers?

Plea bargaining is a second practice seen by many as illustrating the system's moral poverty: an offender who gets a "bargain" does not get justice. Some plea bargains reflect genuine disputes over the facts, but in many cases—the vast majority in some jurisdictions—plea bargains are struck for reasons of expediency. If we are willing to spend an additional $30 billion to fight crime, why not spend a small portion of the money to fight crime by doing justice?

Arbitrary limits on police power are a third reason why the system lacks credibility. Police power is properly limited in the name of individual freedoms. But many will argue that in a democracy the majority ought to be free to choose, for example, less privacy in exchange for less crime. If a majority of residents in a public-housing project want periodic gun sweeps of their building, should their preference be frustrated by the courts? Why not at least allow those who prefer gun sweeps to live together in a building where sweeps are permitted, leaving those who oppose sweeps to live together and to bear the burden of their choice?

Fourth, the law recognizes non-exculpatory defenses. Diplomatic immunity, the statute of limitations, and the entrapment defense, for example, allow blameworthy offenders to remain exempt from criminal liability. Efforts to reduce the scope of diplomatic immunity face practical obstacles. For one thing, diplomatic immunity is mandated by international treaty. But no such obstacles prevent our lengthening the periods of limitation or restricting the entrapment defense to instances of duress (for which a separate defense already exists) or to instances of unconstitutional police conduct (which, like exclusionary-rule violations, might better be deterred by methods other than letting criminals go free).

One more feature of the system which undermines its moral credibility is sentences that are perceived as soft or as no punishment at all—probation, for example. When properly and selectively applied, non-incarcerative sanctions can be a source of real punishment that costs much less than prison. But to many reformers, the intermediate-sanctions movement, as it is called, is just another opportunity to avoid imposing earned punishment. If non-incarcerative sanctions are to succeed, the total punitive bite of the sanctions—as the community perceives it—must match the amount of punishment the offender deserves.

Avoidable Injustice

BUT doing justice is only half of earning moral authority. As important, if not more so, is that the system does not do injustice. Does the system sometimes seem to have chosen to permit injustice? Here are five instances.

First, American jurisdictions have increasingly defined as criminal actions that do not require proof of a defendant's culpable state of mind. Criminal liability can be imposed for an honest and reasonable mistake or an unavoidable accident. A person can be criminally liable for killing migratory birds even if the person has done everything possible to prevent their death. One can be criminally liable for statutory rape even if any reasonable person would similarly have thought the partner was of legal age. In the same spirit, almost no American jurisdiction recognizes as a defense one's reasonable mistake in understanding the law. Liability without culpability may ease the burden for prosecutors, but it also dilutes the moral significance of a successful prosecution. Every criminal conviction without a showing of blameworthiness increases the likelihood that subsequent criminal convictions will fail to inspire moral condemnation.

Does the justice system regularly fail to impose deserved punishment when it has the power to do so? Does it sometimes appear to have chosen a course that frustrates justice?

Second, and similarly, some states have abolished their insanity defense, others are moving in that direction, and still others have achieved de facto abolition by allowing a verdict of "guilty but mentally ill," which encourages juries to convict even when an offender is insane. Abolition of the insanity defense is touted as necessary to keep dangerous mentally ill people incarcerated, given the limitations imposed by courts upon their civil commitment. But if limitations on civil commitment are the problem, those limitations ought to be attacked directly. Abolition is also fueled by high-profile insanity acquittals in controversial cases—for example, the acquittal of John Hinckley for the attempted assassination of President Ronald Reagan. But such acquittals could be avoided simply by narrowing the scope of the defense or by shifting the burden of proof. Instead of giving the prosecution the burden of proving sanity beyond a reasonable doubt, as it had in the Hinckley case, the defendant could be given the burden of proving insanity by a preponderance of the evidence. The number of insane offenders is too small for abolition of the defense to lead to large-scale injustice, but the effect of the abolition trend is to create the impression that the system does not care whether an offender is blameworthy or not.

Third, criminal law is increasingly used against purely regulatory offenses, such as those involving the activities permitted in public parks, the maintenance procedures at warehouses, and the foodstuffs that may be imported into a state. The move is understandable: reformers seek to enlist the moral force implicit in criminal conviction for the sake of deterrence—a force that civil liability does not carry. But the use of criminal conviction in the absence of serious criminal harm that deserves moral condemnation weakens that very force. As the label "criminal" is increasingly applied to minor violations of a merely civil nature, criminal liability will increasingly become indistinct from civil and will lose its particular stigma.

A similar effect occurs when purely legal entities are criminally "convicted." Legal fictions like corporations cannot make immoral choices; only the human beings within them can. To criminally convict a legal fiction is to undercut the claim that criminal conviction ought to bring moral condemnation.

A fourth source of damage to the system's moral credibility is the current state of correctional facilities. When made a prisoner, a person is stripped of the ability to defend himself and to avoid places and situations of danger. Prison authorities take complete control and, with it, responsibility for protection. Given this responsibility, and given the considerable authority granted to officials to meet it, a single assault on a prisoner by another prisoner is objectionable. Currently more than 15,000 prisoners are assaulted each year in our prisons.

The criminologist Robert Johnson, in *Hard Time* (1987), wrote, "From the mid-60's to the present, a new prison type has emerged. It is defined by the climate of violence and predation on the part of the prisoners that often marks its yards and other public areas." The trend is borne out by statistics. In California in 1973, for example, 289 assaults on inmates were reported. By 1983 the prison population had approximately doubled, and the number of assaults had nearly quintupled, to 1,438. In Texas in 1973 there were 130 prison assaults. In 1982 the prison population was a bit more than twice as large, and the number had grown to 887—a tripling of the assault rate in less than ten years. The next year in Texas 3,411 assaults were reported: one in ten prisoners in Texas was the victim of an officially reported assault. Imagine the daily fear that these statistics suggest. Imagine those prisoners' views of the system's moral authority.

One final damaging practice, and one of the most pervasive, is a point touched on earlier—setting punishment according to the perceived dangerousness of an offender rather than according to the offender's deserts. Under the assumption that a prior offense proves long-term dangerousness, prior offenses are widely used to increase the term of imprisonment under so-called recidivist statutes, including the "three strikes and you're out" rules of current reforms. The same rationale is used by parole boards in setting release policy and by sentencing commissions in setting guidelines. The United States Supreme Court has approved such practices as constitutional. In *Rummel* v. *Estelle*, for example, it approved a life term for a third fraud conviction, for passing a bad check for $120.75 (the two previous offenses involved an $80 credit-card fraud and a forged check for $28.36).

Even assuming that past offenses are a good predictor of future dangerousness, one simply cannot deserve punishment for an offense not yet committed—for an act that others only think will be done. This is why the law requires, for example, that a person must perform some act before he or she can be held liable for even an offense like attempt. Thinking about committing a crime, we have said since criminal law existed, is not enough to establish criminal liability, because moral lapse occurs only when one chooses to act upon the intention. Yet under recidivist statutes we routinely punish people for offenses they have not yet even thought of.

Certainly society must be able to protect itself from dangerous people, but to criminally commit a person because of a predicted future offense is to undercut the law's moral authority. If we feel we must incarcerate people to protect ourselves against crimes not yet committed, civil rather than criminal commitment ought to be used, just as we civilly commit the mentally ill and persons with contagious diseases when they pose a threat. Civil commitment has requirements that criminal commitment does not. First, because commitment is based on present dangerousness rather than a past offense, periodic reviews test for continuing dangerousness. Second, because commitment is for our protection rather than for deserved punishment, its conditions are nonpunitive in nature. The detainee's liberty is restricted only to the extent required for our protection.

Last year a *New York Times* op-ed piece criticized a jury's death verdict that jurors justified on the grounds that a life sentence would have allowed for the possibility of release. The writer was much offended that the jury could return a death verdict without concluding that the offender actually deserved the death penalty. But the logic of the verdict is entirely consistent with the increasingly common practice of American criminal justice. If we are offended because the death penalty is imposed on the basis of dangerousness rather than deserts, then we ought also to be offended by the hundreds of cases resolved each day in which prison terms are set according to dangerousness rather than deserts. Indeed, the argument against basing sentences on dangerousness is stronger with regard to imprisonment. Incarceration under civil commitment provides society with protection identical to that available through criminal commitment, whereas the death penalty provides a level of protection that civil law cannot match.

WE have lost much ground in the past forty years in fighting crime. Ironically, that same period has been one of significant efforts to revolutionize the fight against crime. Our recent insights into the law's moral authority as a force for compliance may help explain our past failures. By setting sentences that would best rehabilitate offenders or would best deter other potential offenders or would best incapacitate dangerous offenders, each of our past programs distributed punishment in a way that could be seriously disproportionate to an offender's blameworthiness. The cumulative effect of these policies has been to divert the criminal-justice system from doing justice, and public perception of that shift has undercut the system's moral authority. Our past crime-control reforms may well have increased rather than decreased crime.

The research is incomplete. We do not yet fully understand the interaction between the system's moral credibility and crime. How much decrease in credibility causes how much increase in crime? Is the relationship a continuous one, or can we find credibility trigger points below which crime increases dramatically? Will research confirm my speculations about the practices that most undercut the system's moral authority?

We do know that human beings share a desire for justice, and giving it to them seems to carry little risk. If further research supports the conclusions of the preliminary studies, the search for effective crime control as well as for justice will demand that we re-examine every practice that contributes to the moral poverty of the American criminal-justice system.

WHEN VIOLENCE HITS HOME

Suddenly, domestic abuse, once perniciously silent, is exposed for its brutality in the wake of a highly public scandal

JILL SMOLOWE

DANA USED TO HIDE THE BRUISES ON HER neck with her long red hair. On June 18, her husband made sure she could not afford even that strand of camouflage. Ted ambushed Dana (not their real names) as she walked from her car to a crafts store in Denver. Slashing with a knife, Ted, a pharmaceutical scientist, lopped off Dana's ponytail, then grabbed her throat, adding a fresh layer of bruises to her neck.

Dana got off easy that time. Last year she lost most of her hearing after Ted slammed her against the living-room wall of their home and kicked her repeatedly in the head, then stuffed her unconscious body into the fireplace. Later, he was tearfully despondent, and Dana, a former social worker, believed his apologies, believed he needed her, believed him when he whispered, "I love you more than anything in the world." She kept on believing, even when more assaults followed.

Last Tuesday, however, Dana finally came to believe her life was in danger. Her change of mind came as she nursed her latest wounds, mesmerized by the reports about Nicole Simpson's tempestuous marriage to ex-football star O.J. "I grew up idolizing him," she says. "I didn't want to believe it was O.J. It was just like with my husband." Then, she says, "the reality hit me. Her story is the same as mine—except she's dead."

THE HORROR HAS ALWAYS BEEN WITH US, A PERSISTENT secret, silent and pernicious, intimate and brutal. Now, however, as a result of the Simpson drama, Americans are confronting the ferocious violence that may erupt when love runs awry. Women who have clung to destructive relationships for years are realizing, like Dana, that they may be in dire jeopardy. Last week phone calls to domestic-violence hot lines surged to record numbers; many battered women suddenly found the strength to quit their homes and seek sanctuary in shelters. Although it has been two years since the American Medical As-

"Women are at more risk of being killed by their

sociation reported that as many as 1 in 3 women will be assaulted by a domestic partner in her lifetime—4 million in any given year—it has taken the murder of Nicole Simpson to give national resonance to those numbers.

"Everyone is acting as if this is so shocking," says Debbie Tucker, chairman of the national Domestic Violence Coalition on Public Policy. "This happens all the time." In Los Angeles, where calls to abuse hot lines were up 80% overall last week, experts sense a sort of awakening as women relate personally to Simpson's tragedy. "Often a woman who's been battered thinks it's happening only to her. But with this story, women are saying, 'Oh, my God, this is what's happening to me,'" says Lynn Moriarty, director of the Family Violence Project of Jewish Family Services in Los Angeles. "Something as dramatic as this cracks through a lot of the denial."

Time and again, Health and Human Services Secretary Donna Shalala has warned, "Domestic violence is an unacknowledged epidemic in our society." Now, finally, lawmakers are not only listening—they are acting. In New York last week, the state legislature unanimously passed a sweeping bill that mandates arrest for any person who commits a domestic assault. Members of the California legislature are pressing for a computerized registry of restraining orders and the confiscation of guns from men arrested for domestic violence. This week Colorado's package of anti-domestic-violence laws, one of the nation's toughest, will go into effect. It not only compels police to take abusers into custody at the scene of violence but also requires arrest for a first violation of a restraining order. Subsequent violations bring mandatory jail time.

Just as women's groups used the Anita Hill–Clarence Thomas hearings as a springboard to educate the public about sexual harassment, they are now capitalizing on the Simpson controversy to further their campaign against domestic violence. Advocates for women are pressing for passage of the Violence Against Women Act, which is appended to the anticrime bill that legislators hope to have on President Clinton's desk by July 4. Modeled on the Civil Rights Act of 1964, it stipulates that gender-biased crimes violate a woman's civil rights. The victims of such crimes would therefore be eligible for compensatory relief and punitive damages.

Heightened awareness may also help add bite to laws that are on the books but are often underenforced. At present, 25 states require arrest when a reported domestic dispute turns violent. But police often walk away if the victim refuses to press charges. Though they act quickly to separate strangers, law-enforcement officials remain wary of interfering in domestic altercations, convinced that such battles are more private and less serious.

Yet, of the 5,745 women murdered in 1991, 6 out of 10 were killed by someone they knew. Half were murdered by a spouse or someone with whom they had been intimate. And that does not even hint at the level of violence against women by loved ones: while only a tiny percentage of all assaults on women result in death, the violence often involves severe physical or psychological damage. Says psychologist Angela Browne, a pioneering researcher in partner violence: "Women are at more risk of being killed by their current or former male partners than by any other kind of assault."

AFTER DANA DECIDED TO LEAVE TED IN MAY, SHE used all the legal weapons at her disposal to protect herself. She got a restraining order, filed for a divorce and found a new place to live. But none of that gave her a new life. Ted phoned repeatedly and stalked her. The restraining order seemed only to provoke his rage. On Memorial Day, he trailed her to a shopping-mall parking garage and looped a rope around her neck. He dragged her along the cement floor and growled, "If I can't have you, no one will." Bystanders watched in shock. But no one intervened.

After Ted broke into her home while she was away, Dana called the police. When she produced her protective order, she was told, "We don't put people in jail for breaking a restraining order." Dana expected little better after Ted came at her with the knife on June 18. But this time a female cop, herself a battering victim, encouraged Dana to seek shelter. On Tuesday, Dana checked herself into a shelter for battered women. There, she sleeps on a floor with her two closest friends, Sam and Odie—two cats. Odie is a survivor too. Two months ago, Ted tried to flush him down a toilet.

THOUGH DOMESTIC VIOLENCE USUALLY GOES UNDEtected by neighbors, there is a predictable progression to relationships that end in murder. Typically it begins either with a steady diet of battery or isolated incidents of violence that can go on for years. Often the drama is fueled by both parties. A man wages an assault. The woman retaliates by deliberately trying to provoke his jealousy or anger. He strikes again. And the cycle repeats, with the two locked in a sick battle that binds—and reassures—even as it divides.

When the relationship is in risk of permanent rupture, the violence escalates. At that point the abused female may seek help outside the home, but frequently the man will refuse counseling, convinced that she, not he, is at fault. Instead he will reassert his authority by stepping up the assaults. "Battering is about maintaining power and dominance in a relationship," says Dick Bathrick, an instructor at the Atlanta-based Men Stopping Violence, a domestic-violence intervention group. "Men who batter believe that they have the right to do whatever it takes to regain control."

When the woman decides she has had enough, she may move out or demand that her partner leave. But "the men sometimes panic about losing [their women] and will do anything to prevent it from happening," says Deborah Burk, an Atlanta prosecutor.

male partners than by any other kind of assault.”

“The men who batter believe that they have the

To combat feelings of helplessness and powerlessness, the man may stalk the woman or harass her by phone.

Women are most in danger when they seek to put a firm end to an abusive relationship. Experts warn that the two actions most likely to trigger deadly assault are moving out of a shared residence and beginning a relationship with another man. “There aren't many issues that arouse greater passion than infidelity and abandonment,” says Dr. Park Dietz, a forensic psychiatrist who is a leading expert on homicide.

Disturbingly, the very pieces of paper designed to protect women—divorce decrees, arrest warrants, court orders of protection—are often read by enraged men as a license to kill. “A restraining order is a way of getting killed faster,” warns Dietz. “Someone who is truly dangerous will see this as an extreme denial of what he's entitled to, his God-given right.” That slip of paper, which documents his loss, may be interpreted by the man as a threat to his own life. “In a last-ditch, nihilistic act,” says Roland Maiuro, director of Seattle's Harborview Anger Management and Domestic Violence Program, “he will engage in behavior that destroys the source of that threat.” And in the expanding range of rage, victims can include children, a woman's lawyer, the judge who issues the restraining order, the cop who comes between. Anyone in the way.

For that reason, not all battered women's organizations support the proliferating mandatory arrest laws. That puts them into an unlikely alliance with the police organizations that were critical of New York's tough new bill. “There are cases,” argues Francis Looney, counsel to the New York State Association of Chiefs of Police, “where discretion may be used to the better interest of the family.”

Proponents of mandatory-arrest laws counter that education, not discretion, is required. “I'd like to see better implementation of the laws we have,” says Vickie Smith, executive director of the Illinois Coalition Against Domestic Violence. “We work to train police officers, judges and prosecutors about why they need to enforce them.”

“I TOOK IT VERY SERIOUSLY, THE MARRIAGE, THE commitment. I wanted more than anything to make it work.” Dana's eyes are bright, her smile engaging, as she sips a soda in the shelter and tries to explain what held her in thrall to Ted for so many years. Only the hesitation in her voice betrays her anxiety. “There was a fear of losing him, that he couldn't take care of himself.”

Though Dana believed the beatings were unprovoked and often came without warning, she blamed herself. “I used to think, ‘Maybe I could have done things better. Maybe if I had bought him one more Mont Blanc pen.’ ” In the wake of Nicole Simpson's slaying, Dana now says that she was Ted's “prisoner.” “I still loved him,” she says, trying to explain her servitude. “It didn't go away. I didn't want to face the fact that I was battered.”

IT IS IMPOSSIBLE TO CLASSIFY THE WOMEN WHO are at risk of being slain by a partner. Although the men who kill often abuse alcohol or drugs, suffer from personality disorders, have histories of head injuries or witnessed abuse in their childhood homes, such signs are often masterfully cloaked. “For the most part, these are people who are functioning normally in the real world,” says Bathrick of Men Stopping Violence. “They're not punching out their bosses or jumping in cops' faces. They're just committing crimes in the home.”

The popular tendency is to dismiss or even forgive the act as a “crime of passion.” But that rush of so-called passion is months, even years, in the making. “There are few cases where murder comes out of the blue,” says Sally Goldfarb, senior staff attorney for the NOW Legal Defense and Education Fund. “What we are talking about is domestic violence left unchecked and carried to its ultimate outcome.” Abuse experts also decry the argument that a man's obsessive love can drive him beyond all control. “Men who are violent are rarely completely out of control,” psychologist Browne argues. “If they were, many more women would be dead.”

Some researchers believe there is a physiological factor in domestic abuse. A study conducted by the University of Massachusetts Medical Center's domestic-violence research and treatment center found, for instance, that 61% of men involved in marital violence have signs of severe head trauma. “The typical injuries involve the frontal lobe,” says Al Rosenbaum, the center's director. “The areas we suspect are injured are those involved in impulse control, and reduce an individual's ability to control aggressive impulses.”

Researchers say they can also distinguish two types among the men most likely to kill their wives: the “loose cannon” with impulse-control problems, and those who are calculated and focused, whose heart rate drops even as they prepare to do violence to their partners. The latter group may be the more dangerous. Says Neil Jacobson, a psychology professor at the University of Washington: “Our research

“I didn't want to face the fact I was battered.”

right to do whatever it takes to regain control.”

shows that those men who calm down physiologically when they start arguing with their wives are the most aggressive during arguments."

There may be other psycho-physiological links to violence. It is known, for instance, that alcohol and drug abuse often go hand in hand with spousal abuse. So does mental illness. A 1988 study by Maiuro of Seattle's domestic-violence program documented some level of depression in two-thirds of the men who manifested violent and aggressive behavior. Maiuro is pioneering work with Paxil, an antidepressant that, like Prozac, regulates the brain chemical serotonin. He reports that "it appears to be having some benefits" on his subjects.

Most studies, however, deal not with battering as an aftereffect of biology but of violence as learned behavior. Fully 80% of the male participants in a Minneapolis, Minnesota, violence-control program grew up in homes where they saw or were victims of physical, sexual or other abuse. Women who have witnessed abuse in their childhood homes are also at greater risk of reliving such dramas later in their lives, unless counseling is sought to break the generational cycle. "As a child, if you learn that violence is how you get what you want, you get a dysfunctional view of relationships," says Barbara Schroeder, a domestic-violence counselor in Oak Park, Illinois. "You come to see violence as an O.K. part of a loving relationship."

The cruelest paradox is that when a woman is murdered by a loved one, people are far more inclined to ask, "Why didn't she leave?" than "Why did he do that?" The question of leaving not only reflects an ingrained societal assumption that women bear primary responsibility for halting abuse in a relationship; it also suggests that a battered woman has the power to douse a raging man's anger—and to do it at a moment when her own strength is at an ebb. "It's quite common with women who have been abused that they don't hold themselves in high esteem," says Dr. Allwyn Levine, a Ridgewood, New Jersey, forensic psychiatrist who evaluates abusers for the court system. "Most of these women really feel they deserve it." Furthermore, says Susan Forward, the psychoanalyst who counseled Nicole Simpson on two occasions, "too many therapists will say, 'How did it feel when he was hitting you?' instead of addressing the issue of getting the woman away from the abuser."

Most tragically, a woman may have a self-image that does not allow her to see herself—or those nearby to see her—as a victim. Speaking of her sister Nicole Simpson, Denise Brown told the New York *Times* last week, "She was not a battered woman. My definition of a battered woman is somebody who gets beat up all the time. I don't want people to think it was like that. I know Nicole. She was a very strong-willed person."

Such perceptions are slowly beginning to change, again as a direct result of Simpson's slaying. "Before, women were ashamed," says Peggy Kerns, a Colorado state legislator. "Simpson has almost legitimized the concerns and fears around domestic violence. This case is telling them, 'It's not your fault.'" The women who phoned hot lines last week seemed emboldened to speak openly about the abuse in their lives. "A woman told me right off this week about how she was hit with a bat," says Carole Saylor, a Denver nurse who treats battered women. "Before, there might have been excuses. She would have said that she ran into a wall."

Abusive men are also taking a lesson from the controversy. The hot lines are ringing with calls from men who ask if their own conduct constitutes abusive behavior, or who say that they want to stop battering a loved one but don't know how. Others have been frightened by the charges against O.J. Simpson and voice fears about their own capacity to do harm. "They're worried they could kill," says Rob Gallup, executive director of AMEND, a Denver-based violence prevention and intervention group. "They figure, 'If [O.J.] had this fame and happiness, and chose to kill, then what's to prevent me?'"

EVEN IF DANA IS ABLE TO HOLD TED AT BAY, THE DAMAGE he has inflicted on her both physically and psychologically will never go away. Doctors have told her that her hearing will never be restored and that she is likely to become totally deaf within the decade. She is now brushing up the sign-language skills she learned years ago while working with deaf youngsters. At the moment, she is making do with a single set of hearing aids. Ted stole her other pair.

Dana reflects on her narrow escape. But she knows that her refuge in the shelter is only temporary. As the days go by, she grows increasingly resentful of her past, fearful of her present, and uncertain about her future. "I don't know when I'll be leaving, or where I'll be going."

And Ted is still out there.

—*Reported by*

Ann Blackman/Washington, Wendy Cole/Chicago, Scott Norvell/Atlanta, Elizabeth Rudulph and Andrea Sachs/New York and Richard Woodbury/Denver

Legalization madness

JAMES A. INCIARDI & CHRISTINE A. SAUM

JAMES A. INCIARDI is director of the Center for Drug and Alcohol Studies at the University of Delaware. CHRISTINE A. SAUM is a research associate at the Center.

FRUSTRATED by the government's apparent inability to reduce the supply of illegal drugs on the streets of America, and disquieted by media accounts of innocents victimized by drug-related violence, some policy makers are convinced that the "war on drugs" has failed. In an attempt to find a better solution to the "drug crisis" or, at the very least, to try an alternative strategy, they have proposed legalizing drugs.

They argue that, if marijuana, cocaine, heroin, and other drugs were legalized, several positive things would probably occur: (1) drug prices would fall; (2) users would obtain their drugs at low, government-regulated prices, and they would no longer be forced to resort to crime in order to support their habits; (3) levels of drug-related crime, and particularly violent crime, would significantly decline, resulting in less crowded courts, jails, and prisons (this would allow law-enforcement personnel to focus their energies on the "real criminals" in society); and (4) drug production, distribution, and sale would no longer be controlled by organized crime, and thus such criminal syndicates as the Colombian cocaine "cartels," the Jamaican "posses," and the various "mafias" around the country and the world would be decapitalized, and the violence associated with drug distribution rivalries would be eliminated.

By contrast, the anti-legalization camp argues that violent crime would not necessarily decline in a legalized drug market. In fact, there are three reasons why it might actually increase. First, removing the criminal sanctions against the possession and distribution of illegal drugs would make them more available and attractive and, hence, would create large numbers of new users. Second, an increase in use would lead to a greater number of dysfunctional addicts who could not support themselves, their habits, or their lifestyles through legitimate means. Hence crime would be their only alternative. Third, more users would mean more of the violence associated with the ingestion of drugs.

These divergent points of view tend to persist because the relationships between drugs and crime are quite complex and because the possible outcomes of a legalized drug market are based primarily on speculation. However, it is possible, from a careful review of the existing empirical literature on drugs and violence, to make some educated inferences.

Considering "legalization"

Yet much depends upon what we mean by "legalizing drugs." Would all currently illicit drugs be legalized or would the experiment be limited to just certain ones? True legalization would be akin to selling such drugs as heroin and cocaine on the open market, much like alcohol and tobacco, with a few age-related restrictions. In contrast, there are "medicalization" and "decriminalization" alternatives. Medicalization approaches are of many types, but, in essence, they would allow users to obtain prescriptions for some, or all, currently illegal substances. Decriminalization removes the criminal penalties associated with the possession of small amounts of illegal drugs for personal use, while leaving intact the sanctions for trafficking, distribution, and sale.

But what about crack-cocaine? A quick review of the literature reveals that the legalizers, the decriminalizers, and the medicalizers avoid talking about this particular form of cocaine. Perhaps they do not want to legalize crack out of fear of the drug itself, or of public outrage. Arnold S. Trebach, a professor of law at American University and president of the Drug Policy Foundation, is one of the very few who argues for the full legalization of all drugs, including crack. He explains, however, that most are reluctant to discuss the legalization of crack-cocaine because, "it is a very dangerous drug.... I know that for many people the very thought of making crack legal destroys any inclination they might have had for even thinking about drug-law reform."

There is a related concern associated with the legalization of cocaine. Because crack is easily manufactured from powder cocaine (just add water and baking soda and cook on a stove or in a microwave), many drug-policy reformers hold that no form of cocaine should be legalized. But this weakens the argument that legalization will reduce drug-related violence; for much of this violence would appear to be in the cocaine- and crack-distribution markets.

To better understand the complex relationship between drugs and violence, we will discuss the data in the context of three models developed by Paul J. Goldstein of the University of Illinois at Chicago. They are the "psychopharmacological," "economically compulsive," and "systemic" explanations of violence. The first model holds, correctly in our view, that some individuals may become excitable, irrational, and even violent due to the ingestion of specific drugs. In contrast, taking a more economic approach to the behavior of drug users, the second holds that some drug users engage in violent crime mainly for the sake of supporting their drug use. The third model maintains that drug-related violent crime is simply the result of the drug market under a regime of illegality.

Psychopharmacological violence

The case for legalization rests in part upon the faulty assumption that drugs themselves do not cause violence; rather, so goes the argument, violence is the result of depriving drug addicts of drugs or of the "criminal" trafficking in drugs. But, as researcher Barry Spunt points out, "Users of drugs do get violent when they get high."

Research has documented that chronic users of amphetamines, methamphetamine, and cocaine in particular tend to exhibit hostile and aggressive behaviors. Psychopharmacological violence can also be a product of what is known as "cocaine psychosis." As dose and duration of cocaine use increase, the development of cocaine-related psychopathology is not uncommon. Cocaine psychosis is generally preceded by a transitional period characterized by increased suspiciousness, compulsive behavior, fault finding, and eventually paranoia. When the psychotic state is reached, individuals may experience visual, as well as auditory, hallucinations, with persecutory voices commonly heard. Many believe that they are being followed by police or that family, friends, and others are plotting against them.

Moreover, everyday events are sometimes misinterpreted by cocaine users in ways that support delusional beliefs. When coupled with the irritability and hyperactivity that cocaine tends to generate in almost all of its users, the cocaine-induced paranoia may lead to violent behavior as a means of "self-defense" against imagined persecutors. The violence associated with cocaine psychosis is a common feature in many crack houses across the United States. Violence may also result from the irritability associated with drug-withdrawal syndromes. In addition, some users ingest drugs before committing crimes to both loosen inhibitions and bolster their resolve to break the law.

Acts of violence may result from either periodic or chronic use of a drug. For example, in a study of drug use and psychopathy among Baltimore City jail inmates, researchers at the University of Baltimore reported that cocaine use was related to irritability, resentment, hostility, and assault. They concluded that these indicators of aggression may be a function of drug effects rather than of a predisposition to these behaviors. Similarly, Barry Spunt and his colleagues at National Development and Research Institutes (NDRI) in New York City found that of 269 convicted murderers incarcerated in New York State prisons, 45 percent were high at the time of the offense. Three in 10 believed that the homicide was related to their drug use, challenging conventional beliefs that violence only infrequently occurs as a result of drug consumption.

Even marijuana, which pro-legalizers consider harmless, may have a connection with violence and crime. Spunt and his colleagues attempted to determine the role of marijuana in the crimes of the homicide offenders they interviewed in the New York State prisons. One-third of those who had ever used marijuana had smoked the drug in the 24-hour period prior to the homicide. Moreover, 31 percent of those who considered themselves to be "high" at the time of committing murder felt that the homicide and marijuana were related. William Blount of the University of South Florida interviewed abused women in prisons and shelters for battered women located throughout Florida. He and his colleagues found that 24 percent of those who killed their abusers were marijuana users while only 8 percent of those who did not kill their abusers smoked marijuana.

And alcohol abuse

A point that needs emphasizing is that alcohol, because it is legal, accessible, and inexpensive, is linked to violence to a far greater extent than any illegal drug. For example, in the study just cited, it was found that an impressive 64 percent of those women who eventually killed their abusers were alcohol users (44 percent of those who did not kill their abusers were alcohol users). Indeed, the extent to which alcohol is responsible for violent crimes in comparison with other drugs is apparent from the statistics. For example, Carolyn Block and her colleagues at the Criminal Justice Information Authority in Chicago found that, between 1982 and 1989, the use of alcohol by offenders or victims in local homicides ranged from 18 percent to 32 percent.

Alcohol has, in fact, been consistently linked to homicide. Spunt and his colleagues interviewed 268 homicide offenders incarcerated in New York State correctional facilities to determine the role of alcohol in their crimes: Thirty-one percent of the respondents reported being drunk at the time of the crime and 19 percent believed that the homicide was related to their drinking. More generally, Douglass Murdoch of Quebec's McGill University found that in some 9,000 criminal cases drawn from a multinational sample, 62 percent of violent offenders were drinking shortly before, or at the time of, the offense.

It appears that alcohol reduces the inhibitory control of threat, making it more likely that a person will exhibit violent behaviors normally suppressed by fear. In turn, this reduction of inhibition heightens the probability that intoxicated persons will perpetrate, or become victims of, aggressive behavior.

When analyzing the psychopharmacological model of drugs and violence, most of the discussions focus on the offender and the role of drugs in causing or facilitating crime. But what about the victims? Are the victims of drug- and alcohol-related homicides simply casualties of someone else's substance abuse? In addressing these questions, the data demonstrates that victims are likely to be drug users as well. For example, in an analysis of the 4,298 homicides that occurred in New York City during 1990 and 1991, Kenneth Tardiff of Cornell University Medical College found that the victims of these offenses were 10 to 50 times more likely to be cocaine users than were members of the general population. Of the white female victims, 60 percent in the 25- to 34-year age group had cocaine in their systems; for black females, the figure was 72 percent. Tardiff speculated that the classic symptoms of cocaine use—irritability, paranoia, aggressiveness—may have instigated the violence. In another study of cocaine users in New York City, female high-volume users were found to be victims of violence far more frequently than low-volume and nonusers of cocaine. Studies in numerous other cities and countries have yielded the same general findings—that a great many of the victims of homicide and other forms of violence are drinkers and drug users themselves.

Economically compulsive violence

Supporters of the economically compulsive model of violence argue that in a legalized market, the prices of "expensive drugs" would decline to more affordable levels, and, hence, predatory crimes would become unnecessary. This argument is based on several specious assumptions. First, it assumes that there is empirical support for what has been referred to as the "enslavement theory of addiction." Second, it assumes that people addicted to drugs commit crimes only for the purpose of supporting their habits. Third, it assumes that, in a legalized market, users could obtain as much of the drugs as they wanted whenever they wanted. Finally, it assumes that, if drugs are inexpensive, they will be affordable, and thus crime would be unnecessary.

With respect to the first premise, there has been for the better part of this century a concerted belief among many in the drug-policy field that addicts commit crimes because they are "enslaved" to drugs, and further that, because of the high price of heroin, cocaine, and other illicit chemicals on the black market, users are forced to commit crimes in order to support their drug habits. However, there is no solid empirical evidence to support this contention. From the 1920s through the end of the 1960s, hundreds of studies of the relationship between crime and addiction were conducted. Invariably, when one analysis would support the posture of "enslavement theory," the next would affirm the view that addicts were criminals first and that their drug use was but one more manifestation of their deviant lifestyles. In retrospect, the difficulty lay in the ways that many of the studies had been conducted: Biases and deficiencies in research designs and sampling had rendered their findings of little value.

Studies since the mid 1970s of active drug users on the streets of New York, Miami, Baltimore, and elsewhere have demonstrated that the "enslavement theory" has little basis in reality. All of these studies of the criminal careers of drug users have convincingly documented that, while drug use tends to intensify and perpetuate criminal behavior, it usually does not initiate criminal careers. In fact, the evidence suggests that among the majority of street drug users who are involved in crime, their criminal careers are well established prior to the onset of either narcotics or cocaine use. As such, it would appear that the "inference of causality" that the high price of drugs on the black market itself causes crime—is simply false.

Looking at the second premise, a variety of studies show that addicts commit crimes for reasons other than supporting their drug habit. They do so also for daily living expenses. For example, researchers at the Center for Drug and Alcohol Studies at the University of Delaware who studied crack users on the streets of Miami found that, of the active addicts interviewed, 85 percent of the male and 70 percent of the female interviewees paid for portions of their living expenses through street crime. In fact, one-half of the men and one-fourth of the women paid for 90 percent or more of their living expenses through crime. And, not surprisingly, 96 percent of the men and 99 percent of the women had not held a legal job in the 90-day period before being interviewed for the study.

With respect to the third premise, that in a legalized market users could obtain as much of the drugs as they wanted whenever they wanted, only speculation is possible. More than likely, however, there would be some sort of regulation, and hence black markets for drugs would persist for those whose addictions were beyond the medicalized or legalized allotments. In a decriminalized market, levels of drug-related violence would likely either remain unchanged or increase (if drug use increased).

As for the last premise, that cheap drugs preclude the need to commit crimes to obtain them, the evidence emphatically suggests that this is not the case. Consider crack-cocaine: Although crack "rocks" are available on the illegal market for as little as two dollars in some locales, users are still involved in crime-driven endeavors to support their addictions. For example, researchers Norman S. Miller and Mark S. Gold surveyed 200 consecutive callers to the 1-800-COCAINE hotline who considered themselves to have a problem with crack. They found that, despite the low cost of crack, 63 percent of daily users and 40 percent of non-daily users spent more than $200 per week on the drug. Similarly, interviews conducted by NDRI researchers in New York City with almost 400 drug users contacted in the streets, jails, and treatment programs revealed that almost one-half of them spent over $1,000 a month on crack. The study also documented that crack users—despite the low cost of their drug of choice—spent more money on drugs than did users of heroin, powder cocaine, marijuana, and alcohol.

Systemic violence

It is the supposed systemic violence associated with trafficking in cocaine and crack in America's inner cities that has recently received the attention of drug-policy critics interested in legalizing drugs. Certainly it might appear that, if heroin and cocaine were legal substances, systemic drug-related violence would decline. However, there are two very Important questions in this regard: First, is drug-related violence more often psychopharmacological or systemic? Second, is the great bulk of systemic violence related to the distribution of crack? If most of the drug-related violence is psychopharmacological in nature, and if systemic violence is typically related to crack—the drug generally excluded from consideration when legalization is recommended—then legalizing drugs would probably *not* reduce violent crime.

Regarding the first question, several recent studies conducted in New York City tend to contradict, or at least not support, the notion that legalizing drugs would reduce violent, systemic-related crime. For example, Paul J. Goldstein's ethnographic studies of male and female drug users during the late 1980s found that cocaine-related violence was more often psychopharmacological than systemic. Similarly, Kenneth Tardiff's study of 4,298 New York City homicides found that 31 percent of the victims had used cocaine in the 24-hour period prior to their deaths. One of the conclusions of the study was that the homicides were not necessarily related to drug dealing. In all likelihood, as victims of homicide, the cocaine users may have provoked violence through their irritability, paranoid thinking, and verbal or physical aggression—all of which are among the psychopharmacological effects of cocaine.

Regarding the second question, the illegal drug most associated with systemic violence is crack-cocaine. Of all illicit drugs, crack is the one now responsible for the most homicides. In a study done in New York City in 1988 by Goldstein and his colleagues, crack was found to be connected with 32 percent of all homicides and 60 percent of all drug-related homicides. Furthermore, although there is evidence that crack sellers are more violent than other drug sellers, this violence is not confined to the drug-selling context—violence potentials appear to precede involvement in selling.

Thus, though crack has been blamed for increasing violence in the marketplace, this violence actually stems from the psychopharmacological consequences of crack use. Ansley Hamid, a professor of anthropology at the John Jay College of Criminal Justice in New York, reasons that increases in crack-related violence are due to the deterioration of informal and formal social controls throughout communities that have been destabilized by economic processes and political decisions. If this is the case, does anyone really believe that we can improve these complex social problems through the simple act of legalizing drugs?

Don't just say no

The issue of whether or not legalization would create a multitude of new users also needs to be addressed. It has been shown that many people do not use drugs simply because drugs are illegal. As Mark A. R. Kleiman, author of *Against Excess: Drug Policy for Results,* recently put it: "Illegality by itself tends to suppress consumption, independent of its effect on price, both because some consumers are reluctant to disobey the law and because illegal products are harder to find and less reliable as to quality and labeling than legal ones."

Although there is no way of accurately estimating how many new users there would be if drugs were legalized, there would probably be many. To begin with, there is the historical example of Prohibition. During Prohibition, there was a decrease of 20 percent to 50 percent in the number of alcoholics. These estimates were calculated based on a decline in cirrhosis and other alcohol-related deaths; after Prohibition ended, both of these indicators increased.

Currently, relatively few people are steady users of drugs. The University of Michigan's *Monitoring the Future* study reported in 1995 that only two-tenths of 1 percent of high-school seniors are daily users of either hallucinogens, cocaine, heroin, sedatives, or inhalants. It is the addicts who overwhelmingly consume the bulk of the drug supply—80 percent of all alcohol and almost 100 percent of all heroin. In other words, there are significantly large numbers of non-users who have yet to even try drugs, let alone use them regularly. Of those who begin to use drugs "recreationally," researchers estimate that approximately 10 percent go on to serious, heavy, chronic, compulsive use. Herbert Kleber, the former deputy director of the Office of National Drug Control Policy, recently estimated that cocaine legalization might multiply the number of addicts from the current 2 million to between 18 and 50 million (which are the estimated numbers of problem drinkers and nicotine addicts).

This suggests that drug prohibition seems to be having some very positive effects and that legalizing drugs would not necessarily have a depressant effect on violent crime. With legalization, violent crime would likely escalate; or perhaps some types of systemic violence would decline at the expense of greatly increasing the overall rate of violent crime. Moreover, legalizing drugs would likely increase physical illnesses and compound any existing psychiatric problems among users and their family members. And finally, legalizing drugs would not eliminate the effects of unemployment, inadequate housing, deficient job skills, economic worries, and physical abuse that typically contribute to the use of drugs.

Groups and Roles in Transition

- Marriage and Family (Articles 12–14)
- Sex, Gender, and Gender Relationships (Articles 15–17)
- Communities and Community Action (Articles 18–20)

Primary groups are small, intimate, spontaneous, and personal. In contrast, secondary groups are large, formal, and impersonal. Primary groups include the family, couples, gangs, cliques, teams, and small tribes or rural villages. Primary groups are the main sources that the individual draws upon in developing values and an identity. Secondary groups include most of the organizations and bureaucracies in a modern society and carry out most of its instrumental functions. Often primary groups are formed within secondary groups such as a factory, school, or business.

Urbanization, geographic mobility, centralization, bureaucratization, and other aspects of modernization have had an impact on the nature of groups, the quality of the relationships between people, and individuals' feelings of belonging. The family, in particular, is undergoing radical transformation. The greatly increased participation of women in the paid labor force and their increased careerism have caused severe strains between their work and family roles.

The first section of this unit deals with marriage and family. Everyone seems to agree that the family is in trouble, but there is not much agreement over what to do about it. Some believe that the answer is to return to the traditional family form. Stephanie Coontz argues that this view misunderstands the history of the American family form and ignores the negative aspects of the family system of the 1950s. Her review of family history demonstrates that the traditional family was very nontraditional. One positive aspect of the 1950s family form that does need to be reinstated is the strong role of fathers. In the next essay, David Popenoe assesses the impact of the decline of fatherhood on children. Studies show that mothers and fathers parent in different manners, and both are needed. The mother cannot substitute for the father, Popenoe states. Next, Pepper Schwartz describes the growing number of peer marriages, the type of marriage that feminists have supported but that has been slow in coming.

The next section focuses on sexual behavior, gender relationships, and changing conditions for the two sexes. In "Now for the Truth about Americans and Sex," Philip Elmer-Dewitt reviews a recent national survey of American sexual behavior and points out, among other things, that Americans are more sexually faithful to their spouses than is commonly perceived. In the next selection, counselors Aaron Kipnis and Elizabeth Herron tell how to end the battle between the sexes. They have been conducting gender workshops that focus on improving communications between the sexes. Here they give advice that takes into account the cultural differences between men and women. The next article describes the changing conditions for men in the economy in America and Europe. Its prognosis is pessimistic, especially for those with inadequate educations.

In the next section, the focus shifts to the community level. Just as families are in trouble, so are communities. William Raspberry observes that group demands are cut-

UNIT 3

ting communities and the public interest to ribbons. "The whole society seems to be disintegrating into special interests." He exhorts us to put behind us "the politics of difference, the marketing of disadvantage, the search for enemies" and start to heal the crisis of community. Robert Putnam also plays the community crisis theme, citing statistics on declining civic participation. His major contribution is in showing how volunteerism contributes to the functioning of democratic and social institutions. But he also leads the reader on an interesting search for the explanation of the civic decline. Then Karl Zinsmeister shows the negative side of the suburb as the haven for the troubled family. In reality, the suburbs are not supportive of family values, he says, because they cut social roots, impede neighborly social life, and starve community life.

Looking Ahead: Challenge Questions

What is your opinion of the family form of the 1950s? What family form do you think is ideal?

What is happening to fatherhood, and with what consequences?

What makes for good relations between the sexes?

What factors create community? How can they be brought into being under today's conditions? What are the impediments to community?

The Way We Weren't

The Myth and Reality of the "Traditional" Family

Stephanie Coontz

Families face serious problems today, but proposals to solve them by reviving "traditional" family forms and values miss two points. First, no single traditional family existed to which we could return, and none of the many varieties of families in our past has had any magic formula for protecting its members from the vicissitudes of socioeconomic change, the inequities of class, race, and gender, or the consequences of interpersonal conflict. Violence, child abuse, poverty, and the unequal distribution of resources to women and children have occurred in every period and every type of family.

Second, the strengths that we also find in many families of the past were rooted in different social, cultural, and economic circumstances from those that prevail today. Attempts to reproduce any type of family outside of its original socioeconomic context are doomed to fail.

Colonial Families

American families always have been diverse, and the male breadwinner-female homemaker, nuclear ideal that most people associate with "the" traditional family has predominated for only a small portion of our history. In colonial America, several types of families coexisted or competed. Native American kinship systems subordinated the nuclear family to a much larger network of marital alliances and kin obligations, ensuring that no single family was forced to go it alone. Wealthy settler families from Europe, by contrast, formed independent households that pulled in labor from poorer neighbors and relatives, building their extended family solidarities on the backs of truncated families among indentured servants, slaves, and the poor. Even wealthy families, though, often were disrupted by death; a majority of colonial Americans probably spent some time in a stepfamily. Meanwhile, African Americans, denied the legal protection of marriage and parenthood, built extensive kinship networks and obligations through fictive kin ties, ritual co-parenting or godparenting, adoption of orphans, and complex naming patterns designed to preserve family links across space and time.

The dominant family values of colonial days left no room for sentimentalizing childhood. Colonial mothers, for example, spent far less time doing child care than do modern working women, typically delegating this task to servants or older siblings. Among white families, patriarchal authority was so absolute

From *National Forum: The Phi Kappa Phi Journal,* Summer 1995, pp. 11-14. © 1995 by Stephanie Coontz. Reprinted by permission of the publisher.

that disobedience by wife or child was seen as a small form of treason, theoretically punishable by death, and family relations were based on power, not love.

The Nineteenth-Century Family

With the emergence of a wage-labor system and a national market in the first third of the nineteenth century, white middle-class families became less patriarchal and more child-centered. The ideal of the male breadwinner and the nurturing mother now appeared. But the emergence of domesticity for middle-class women and children depended on its absence among the immigrant, working class, and African American women or children who worked as servants, grew the cotton, or toiled in the textile mills to free middle-class wives from the chores that had occupied their time previously.

Even in the minority of nineteenth-century families who could afford domesticity, though, emotional arrangements were quite different from nostalgic images of "traditional" families. Rigid insistence on separate spheres for men and women made male-female relations extremely stilted, so that women commonly turned to other women, not their husbands, for their most intimate relations. The idea that all of one's passionate feelings should go toward a member of the opposite sex was a twentieth-century invention — closely associated with the emergence of a mass consumer society and promulgated by the very film industry that "traditionalists" now blame for undermining such values.

Early Twentieth-Century Families

Throughout the nineteenth century, at least as much divergence and disruption in the experience of family life existed as does today, even though divorce and unwed motherhood were less common. Indeed, couples who marry today have a better chance of celebrating a fortieth wedding anniversary than at any previous time in history. The life cycles of nineteenth-century youth (in job entry, completion of schooling, age at marriage, and establishment of separate residence) were far more diverse than they became in the early twentieth-century. At the turn of the century a higher proportion of people remained single for their entire lives than at any period since. Not until the 1920s did a bare majority of children come to live in a male breadwinner-female homemaker family, and even at the height of this family form in the 1950s, only 60 percent of American children spent their entire childhoods in such a family.

Not until the 1920s did a bare majority of children come to live in a male breadwinner-female homemaker family . . . .

From about 1900 to the 1920s, the growth of mass production and emergence of a public policy aimed at establishing a family wage led to new ideas about family self-sufficiency, especially in the white middle class and a privileged sector of the working class. The resulting families lost their organic connection to intermediary units in society such as local shops, neighborhood work cultures and churches, ethnic associations, and mutual-aid organizations.

As families related more directly to the state, the market, and the mass media, they also developed a new cult of privacy, along with heightened expectations about the family's role in fostering individual fulfillment. New family values stressed the early independence of children and the romantic coupling of husband and wife, repudiating the intense same-sex ties and mother-infant bonding of earlier years as unhealthy. From this family we get the idea that women are sexual, that youth is attractive, and that marriage should be the center of our emotional fulfillment.

Even aside from its lack of relevance to the lives of most immigrants, Mexican Americans, African Americans, rural families, and the urban poor, big contradictions existed between image and reality in the middle-class family ideal of the early twentieth century. This is the period when many Americans first accepted the idea that the family should be sacred from outside intervention; yet the development of the private, self-sufficient family depended on state intervention in the economy, government regulation of parent-child relations, and state-directed destruction of class and community institutions that hindered the development of family privacy.

Acceptance of a youth and leisure culture sanctioned early marriage and raised expectations about the quality of married life, but also introduced new tensions between the generations and new conflicts between husband and wife over what were adequate levels of financial and emotional support.

The nineteenth-century middle-class ideal of the family as a refuge from the world of work was surprisingly modest compared with emerging twentieth-century demands that the family provide a whole alternative world of satisfaction and intimacy to that of work and neighborhood. Where a family succeeded in doing so, people might find pleasures in the home never before imagined. But the new ideals also increased the possibilities for failure: America has had the highest divorce rate in the world since the turn of the century.

In the 1920s, these contradictions created a sense of foreboding about "the future of the family" that was every bit as widespread and intense as today's. Social scientists and popular commentators of the time hearkened back to the "good old days," bemoaning the sexual revolution, the fragility of nuclear family ties, the cult of youthful romance, the decline of respect for grandparents, and the threat of the "New Woman." But such criticism was sidetracked by the stockmarket crash, the Great Depression of the 1930s, and the advent of World War II.

Domestic violence escalated during the Depression, while murder rates were as high in the 1930s as in the 1980s. Divorce rates fell, but desertion increased and fertility plummeted. The war stimulated a marriage boom, but by the late 1940s one in every three marriages was ending in divorce.

The 1950s Family

At the end of the 1940s, after the hardships of the Depression and war, many Americans revived the nuclear family ideals that had so disturbed commentators during the 1920s. The unprecedented postwar prosperity allowed young families to achieve consumer satisfactions and socioeconomic mobility that would have been inconceivable in earlier days. The 1950s family that resulted from these economic and cultural trends, however, was hardly "traditional." Indeed it is best seen as a historical aberration. For the first time in 100 years, divorce rates dropped, fertility soared, the gap between men's and women's job and educational prospects widened (making middle-class women more dependent on marriage), and the age of marriage fell—to the point that teenage birth rates were almost double what they are today.

Admirers of these very *nontraditional* 1950s family forms and values point out that household arrangements and gender roles were less diverse in the 1950s than today, and marriages more stable. But this was partly because diversity was ruthlessly suppressed and partly because economic and political support systems for socially-sanctioned families were far more generous than they are today. Real wages rose more in any single year of the 1950s than they did in the entire decade of the 1980s; the average thirty-year-old man could buy a median-priced home on 15 to 18 percent of his income. The government funded public investment, home ownership, and job creation at a rate more than triple that of the past two decades, while 40 percent of young men were eligible for veteran's benefits. Forming and maintaining families was far easier than it is today.

Yet the stability of these 1950s families did not guarantee good outcomes for their members. Even though most births occurred within wedlock, almost a third of American children lived in poverty during the 1950s, a higher figure than today. More than 50 percent of black married-couple families were poor. Women were often refused the right to serve on juries, sign contracts, take out credit cards in their own names, or establish legal residence. Wife-battering rates were low, but that was because wife-beating was seldom counted as a crime. Most victims of incest, such as Miss America of 1958, kept the secret of their fathers' abuse until the 1970s or 1980s, when the women's movement became powerful enough to offer them the support denied them in the 1950s.

The Post-1950s Family

In the 1960s, the civil rights, antiwar, and women's liberation movements exposed the racial, economic, and sexual injustices that had been papered over by the Ozzie and Harriet images on television. Their activism made older kinds of public and private oppression unacceptable and helped create the incomplete, flawed, but much-needed reforms of the Great Society. Contrary to the big lie of the past decade that such programs caused our current family dilemmas, those antipoverty and social justice reforms helped overcome many of the family problems that prevailed in the 1950s.

In 1964, after fourteen years of unrivaled family stability and economic prosperity, the poverty rate was still 19 percent; in 1969, after five years of civil rights activism, the rebirth of feminism, and the institution of nontraditional if relatively modest government welfare programs, it was down to 12 percent, a low that has not been seen again since the social welfare cutbacks began in the late 1970s. In 1965, 20 percent of American children still lived in poverty; within five years, that had fallen to 15 percent. Infant mortality was cut in half between 1965 and 1980. The gap in nutrition between low-income Americans and other Americans narrowed significantly, as a direct result of food stamp and school lunch programs. In 1963, 20 percent of Americans living below the poverty line had *never* been examined by a physician; by 1970 this was true of only 8 percent of the poor.

Since 1973, however, real wages have been falling for most Americans. Attempts to counter this through tax revolts and spending freezes have led to drastic cutbacks in government investment programs. Corporations also spend far less on research and job creation than they did in the 1950s and 1960s, though the average compensation to executives has soared. The gap between rich and poor, according to the April 17, 1995, *New York Times*, is higher in the United

States than in any other industrial nation.

Family Stress

These inequities are *not* driven by changes in family forms, contrary to ideologues who persist in confusing correlations with causes; but they certainly exacerbate such changes, and they tend to bring out the worst in *all* families. The result has been an accumulation of stresses on families, alongside some important expansions of personal options. Working couples with children try to balance three full time jobs, as employers and schools cling to policies that assume every employee has a "wife" at home to take care of family matters. Divorce and remarriage have allowed many adults and children to escape from toxic family environments, yet our lack of social support networks and failure to forge new values for sustaining intergenerational obligations have let many children fall through the cracks in the process.

Meanwhile, young people find it harder and harder to form or sustain families. According to an Associated Press report of April 25, 1995, the median income of men aged twenty-five to thirty-four fell by 26 percent between 1972 and 1994, while the proportion of such men with earnings below the poverty level for a family of four more than doubled to 32 percent. The figures are even worse for African American and Latino men. Poor individuals are twice as likely to divorce as more affluent ones, three to four times less likely to marry in the first place, and five to seven times more likely to have a child out of wedlock.

. . . romanticizing "traditional" families and gender roles will not produce the changes . . . that would permit families to develop moral and ethical systems relevant to 1990s realities.

As conservatives insist, there is a moral crisis as well as an economic one in modern America: a pervasive sense of social alienation, new levels of violence, and a decreasing willingness to make sacrifices for others. But romanticizing "traditional" families and gender roles will not produce the changes in job structures, work policies, child care, medical practice, educational preparation, political discourse, and gender inequities that would permit families to develop moral and ethical systems relevant to 1990s realities.

America needs more than a revival of the narrow family obligations of the 1950s, whose (greatly exaggerated) protection for white, middle-class children was achieved only at tremendous cost to the women in those families and to all those who could not or would not aspire to the Ozzie and Harriet ideal. We need a concern for children that goes beyond the question of whether a mother is waiting with cookies when her kids come home from school. We need a moral language that allows us to address something besides people's sexual habits. We need to build values and social institutions that can reconcile people's needs for independence with their equally important rights to dependence, and surely we must reject older solutions that involved balancing these needs on the backs of women. We will not find our answers in nostalgia for a mythical "traditional family."

Stephanie Coontz teaches history and family studies at The Evergreen State College in Olympia, Washington. Her publications include *The Way We Never Were: American Families and the Nostalgia Trap* and *The Way We Really Are: Coming to Terms with America's Changing Families* (both published by Basic Books). She is a recipient of the Washington Governor's Writer's Award and the Dale Richmond Award of the American Academy of Pediatrics.

WHERE'S PAPA?

Disappearing dads are destroying our future

David Popenoe

David Popenoe is a professor of sociology at Rutgers University.

The decline of fatherhood is one of the most basic, unexpected, and extraordinary social trends of our time. Its dimensions can be captured in a single statistic: In just three decades, between 1960 and 1990, the percentage of U.S. children living apart from their biological fathers more than doubled, from 17 percent to 36 percent. By the turn of the century, nearly 50 percent of American children may be going to sleep each evening without being able to say good night to their dads.

No one predicted this trend, few researchers or government agencies have monitored it, and it is not widely discussed, even today. But the decline of fatherhood is a major force behind many of the most disturbing problems that plague American society: crime and delinquency; teenage pregnancy; deteriorating educational achievement; depression, substance abuse, and alienation among adolescents; and the growing number of women and children living in poverty. The current generation of children may be the first in our nation's history to be less well off—psychologically, socially, economically, and morally—than their parents were at the same age. The United States, observes Senator Daniel Patrick Moynihan, "may be the first society in history in which children are distinctly worse off than adults."

Even as this calamity unfolds, our cultural view of fatherhood itself is changing. Few people doubt the fundamental importance of mothers. But fathers? More and more, the question of whether fathers are really necessary is being raised. Fatherhood is said by many to be merely a social role that others—mothers, partners, stepfathers, uncles and aunts, grandparents—can play.

There was a time in the past when fatherlessness was far more common than it is today, but death was to blame, not divorce, desertion, and out-of-wedlock births. In early-17th-century Virginia, only an estimated 31 percent of white children reached age 18 with both parents still alive. That figure climbed to 50 percent by the early 18th century, to 72 percent by the start of the 20th century, and close to its current level by 1940. Today, well over 90 percent of America's youngsters turn 18 with two living parents. Almost all of today's "fatherless" children have fathers who are alive, well, and perfectly capable of shouldering the responsibilities of fatherhood. Who would have thought that so many men would relinquish them?

Not so long ago, social scientists and others dismissed the change in the cause of fatherlessness as irrelevant. Children, it was said, are merely losing their parents in a different way than they used to. You don't hear that very much anymore. A surprising finding of recent research is that it is decidedly worse for a child to lose a father in the modern, voluntary way than through death. The children of divorced and never-married mothers are less successful in life by almost every measure than the children of widowed mothers. The replacement of death by divorce as the prime cause of fatherlessness is a monumental setback in the history of childhood.

Until the 1960s, the falling death rate and the rising divorce rate neutralized each other. In 1900 the percentage of American children living in single-parent families was 8.5 percent. By 1960 it had increased to just 9.1 percent. Virtually no one during those years was writing or thinking about family breakdown, disintegration, or decline.

Indeed, what is most significant about the changing family demography of the first six decades of the 20th century is this: Because the death rate was dropping faster than the divorce rate was rising, more children were living with both of their natural parents by 1960 than at any other time in world history. The figure was close to 80 percent for the generation born in the late 1940s and early 1950s. But then the decline in the death rate slowed, and the divorce rate skyrocketed. "The scale of marital breakdowns in the West since 1960 has no historical precedent that I know of," says Lawrence Stone, a noted Princeton University family historian. "There has been nothing like it for the last 2,000 years, and probably longer."

Consider what has happened to children. Most estimates are that only about 50 percent of the children born during the 1970–84 "baby bust" period will still live with their natural parents by age 17—a staggering drop from nearly 80 percent.

In theory, divorce need not mean disconnection. In reality, it often does. A large survey conducted in the late 1980s found that about one in five divorced fathers had not seen his children in the past year and that fewer than half of divorced fathers saw their children more than several times a year. A 1981 survey of adolescents who were living apart from their fathers found that 52 percent hadn't seen them at all in more than a year; only 16 percent saw their fathers as often as once a week—and the fathers' contact with their children dropped off sharply over time.

The picture grows worse. Just as divorce has overtaken death as the leading cause of fatherlessness, out-of-wedlock births are expected to surpass divorce in the 1990s. They accounted for 30 percent of all births by 1991; by the turn of the century they may account for 40 percent (and 80 percent of minority births). And there is substantial evidence that having an unmarried father is even worse for a child than having a divorced father.

From *Utne Reader,* September/October 1996, pp. 68-71, 104-107. Originally from *The Wilson Quarterly,* Spring 1996. Excerpted from *Life Without Father* by David Popenoe.

Across time and cultures, fathers have always been considered essential—and not just for their sperm. Indeed, no known society ever thought of fathers as potentially unnecessary. Marriage and the nuclear family—mother, father, and children—are the most universal social institutions in existence. In no society has the birth of children out of wedlock been the cultural norm. To the contrary, concern for the legitimacy of children is nearly universal.

Eighty-eight percent of children living with one parent are living with their mother.

In my many years as a sociologist, I have found few other bodies of evidence that lean so much in one direction as this one: On the whole, two parents—a father and a mother—are better for a child than one parent. There are, to be sure, many factors that complicate this simple proposition. We all know of a two-parent family that is truly dysfunctional—the proverbial family from hell. A child can certainly be raised to a fulfilling adulthood by one loving parent who is wholly devoted to the child's well-being. But such exceptions do not invalidate the rule any more than the fact that some three-pack-a-day smokers live to a ripe old age casts doubt on the dangers of cigarettes.

The collapse of children's well-being in the United States has reached breathtaking proportions. Juvenile violent crime has increased from 18,000 arrests in 1960 to 118,000 in 1992, a period in which the total number of young people in the population remained relatively stable. Reports of child neglect and abuse have quadrupled since 1976, when data were first collected. Since 1960, eating disorders and depression have soared among adolescent girls. Teen suicide has tripled. Alcohol and drug abuse among teenagers, although it has leveled off in recent years, continues at a very high rate. Scholastic Aptitude Test scores have declined more than 70 points, and most of the decline cannot be accounted for by the increased academic diversity of students taking the test. Poverty has shifted from the elderly to the young. Of all the nation's poor today, 38 percent are children.

One can think of many explanations for these unhappy developments: the growth of commercialism and consumerism, the influence of television and the mass media, the decline of religion, the widespread availability of guns and addictive drugs, and the decay of social order and neighborhood relationships. None of these causes should be dismissed. But the evidence is now strong that the absence of fathers from the lives of children is one of the most important causes.

What do fathers do? Partly, of course, it is simply being a second adult in the home. Bringing up children is demanding, stressful, and often exhausting. Two adults can support and spell each other; they can also offset each other's deficiencies and build on each other's strengths.

Beyond that, fathers—men—bring an array of unique and irreplaceable qualities that women do not ordinarily bring. Some of these are familiar, if sometimes overlooked or taken for granted. The father as protector, for example, has by no means outlived his usefulness. And he is important as a role model. Teenage boys without fathers are notoriously prone to trouble. The pathway to adulthood for daughters is somewhat easier, but they still must learn from their fathers, as they cannot from their mothers, how to relate to men. They learn from their fathers about heterosexual trust, intimacy, and difference. They learn to appreciate their own femininity from the one male who is most special in their lives (assuming that they love and respect their fathers). Most important, through loving and being loved by their fathers, they learn that they are worthy of love.

Recent research has given us much deeper—and more surprising—insights into the father's role in child rearing. It shows that in almost all of their interactions with children, fathers do things a little differently from mothers. What fathers do—their special parenting style—is not only highly complementary to what mothers do but is by all indications important in its own right.

For example, an often-overlooked dimension of fathering is play. From their children's birth through adolescence, fathers tend to emphasize play more than caretaking. This may be troubling to egalitarian feminists, and it would indeed be wise for most fathers to spend more time in caretaking. Yet the fathers' style of play seems to have unusual significance. It is likely to be both physically stimulating and exciting. With older children it involves more physical games and teamwork that require the competitive testing of physical and mental skills. It frequently resembles an apprenticeship or teaching relationship: Come on, let me show you how.

Mothers generally spend more time playing with their children, but mothers' play tends to take place more at the child's level. Mothers provide the child with the opportunity to direct the play, to be in charge, to proceed at the child's own pace. Kids, at least in the early years, seem to prefer to play with daddy. In one study of 2½-year-olds who were given a choice, more than two-thirds chose to play w ith their fathers.

The way fathers play affects everything from the management of emotions to intelligence and academic achievement. It is particularly important in promoting the essential virtue of self-control. According to one expert, "Children who roughhouse with their fathers . . . usually quickly learn that biting, kicking, and other forms of physical violence are not acceptable." They learn when enough is enough.

Children, a committee assembled by the Board on Children and Families of the National Research Council concluded, "learn critical lessons about how to recognize and deal with highly charged emotions in the context of playing with their fathers. Fathers, in effect, give children practice in regulating their own emotions and recognizing others' emotional clues." A study of convicted murderers in Texas found that 90 percent of them either didn't play as children of played abnormally.

At play and in other realms, fathers tend to stress competition, challenge, initiative, risk taking, and independence. Mothers, as caretakers, stress emotional security and personal safety. On the playground, fathers will try to get the child to swing higher than the person

Of all the children living with their mothers—whether as a result of nonmarital birth or divorce—35 percent *never* see their fathers.

(Statistics from Father Facts 2, *by Wade F. Horn and the National Fatherhood Initiative)*

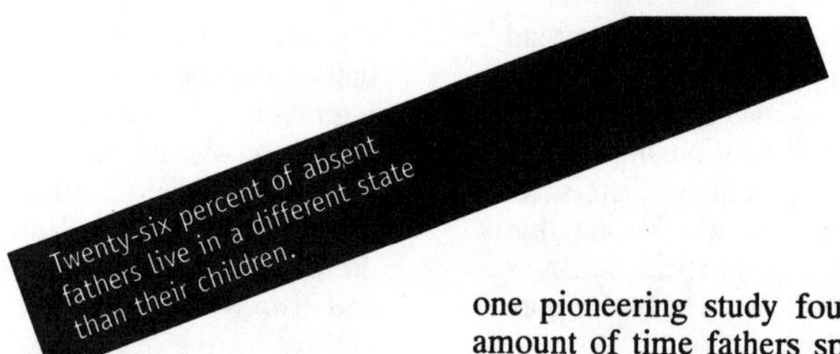

on the next swing, while mothers will worry about an accident. It's sometimes said that fathers express more concern for the child's long-term development, while mothers focus on the child's immediate well-being. It is clear that children have dual needs that must be met. Becoming a mature and competent adult involves the integration of two often-contradictory human desires: for *communion,* or the feeling of being included, connected, and related, and for *agency,* which entails independence, individuality, and self-fulfillment. One without the other is a denuded and impaired humanity, an incomplete realization of human potential.

For many couples, to be sure, these functions are not rigidly divided along standard female-male lines, and there may even be a role reversal. But the exceptions prove the rule. Gender-differentiated parenting is so important that in child rearing by gay and lesbian couples, one partner commonly fills the male role while the other fills the female role.

It is ironic that in our public discussion of fathering, it's seldom acknowledged that fathers have a distinctive role to play. Indeed, it's far more often said that fathers should be more like mothers (and that men generally should be more like women—less aggressive, less competitive). While such things may be said with the best of intentions, the effects are perverse. After all, if fathering is no different from mothering, males can easily be replaced in the home by women. It might even seem better. Already viewed as a burden and obstacle to self-fulfillment, fatherhood thus comes to seem superfluous and unnecessary as well.

We know that fathers have a surprising impact on children. Fathers' involvement seems to be linked to improved quantitative and verbal skills, improved problem-solving ability, and higher academic achievement. Several studies have found that the presence of the father is one of determinants of girls' proficiency in mathematics. And one pioneering study found that the amount of time fathers spent reading was a strong predictor of their daughters' verbal ability.

For sons, who can more directly follow their fathers' example, the results have been even more striking. A number of studies have uncovered a strong relationship between father involvement and the quantitative and mathematical abilities of their sons. Other studies have found a relationship between paternal nurturing and boys' verbal intelligence.

How fathers produce these intellectual benefits is not yet clear. No doubt it is partly a matter of the time and money a man brings to his family. But it is probably also related to the unique mental and behavioral qualities of men; the male sense of play, reasoning, challenge, and problem solving, and the traditional male association with achievement and occupational advancement.

Men also have a vital role to play in promoting cooperation and other "soft" virtues. We don't often think of fathers as teachers of empathy, but involved fathers, it turns out, may be of special importance for the development of this character trait, essential to an ordered society of law-abiding, cooperative, and compassionate adults. Examining the results of a 26-year longitudinal study, a trio of researchers at McGill University reached a "quite astonishing" conclusion: The single most important childhood factor in developing empathy is paternal involvement in child care. Fathers who spent time alone with their children more than twice a week—giving meals, baths, and other basic care—reared the most compassionate adults.

It is not yet clear why fathers are so important in instilling this quality. Perhaps merely by being with their children they provide a model for compassion. Perhaps it has to do with their style of play or mode of reasoning. Perhaps it is somehow related to the fact that fathers typically are the family's main arbiter with the outside world. Or perhaps it is because mothers who receive help from their mates have more time and energy to cultivate the soft virtues. Whatever the reason, it is hard to think of a more important contribution that fathers can make to their children.

Men, too, suffer grievously from the growth of fatherlessness. The world over, young and unattached males have always been a cause for social concern. They can be a danger to themselves and to society. Young unattached men tend to be more aggressive, violent, promiscuous, and prone to substance abuse; they are also more likely to die prematurely through disease, accidents, or self-neglect. They make up the majority of deviants, delinquents, criminals, killers, drug users, vice lords, and miscreants of every kind. Senator Moynihan put it succinctly when he warned that a society full of unattached males "asks for and gets chaos."

Family life—marriage and child rearing—is a civilizing force for men. It encourages them to develop prudence, cooperativeness, honesty, trust, self-sacrifice, and other habits that can lead to success as an economic provider. Marriage also focuses male sexual energy. Having children typically impresses on men the importance of setting a good example. Who hasn't heard at least one man say that he gave up a socially irresponsible way of life when he married and had children?

The civilizing effect of being a father is highlighted by a path-breaking program started in 1982 in one of Cleveland's inner-city neighborhoods by social worker Charles Ballard. Using an intensive social-work approach that includes home visits, parenting programs, and group therapy sessions, he has reunited more than 2,000 absent, unwed fathers with their children through his Institute for Responsible Fatherhood and Family Revitalization.

The standard theory is that if you want inner-city men to be responsible fathers, you first must find them a job. Ballard has stood this theory on its head. His approach is that you first must convince the young men of the importance of being a good father, and then they will be motivated to finish school and find work. An independent evaluation of his approach showed that

More than one-half of children not living with their fathers have never been in their father's home.

it works. Only 12 percent of the young men had full-time work when they entered his program, but 62 percent later found such work, and 12 percent found part-time jobs. Ninety-seven percent of the men he dealt with began to provide financial support for their children, and 71 percent had no additional children out of wedlock.

Marriage by itself, even without children, is also a major civilizing force for men. No other institution save religion (and perhaps the military) places such moral demands on men. To be sure, there is a selection factor in marriage. The men whom women would care to marry already have some of the civilized virtues, and those who are morally beyond the pale have difficulty finding mates. Yet studies have shown that marriage has a civilizing effect independent of the selection factor. Marriage actually promotes health, competence, virtue, and personal well-being. Along with the continued growth of fatherlessness, we can expect to see a nation of men who are at worst morally out of control and at best unhappy, unhealthy, and unfulfilled.

Just as cultural forms can be discarded, dismantled, and declared obsolete, so can they be reinvented. In order to restore marriage and reinstate fathers in the lives of their children, we are somehow going to have to undo the cultural shift of the past few decades toward radical individualism. We are going to have to re-embrace some cultural propositions that throughout history have been universally accepted but that today are unpopular, if not rejected outright.

Marriage must be re-established as a strong social institution. The father's role must also be redefined in a way that neglects neither historical models nor the unique attributes of modern societies, the new roles for women, and the special qualities that men bring to child rearing. Such changes are by no means impossible. Witness the transformations wrought by the civil rights, women's, and environmental movements, and even the campaigns to reduce smoking and drunk driving. What is necessary is for large numbers of adults, and especially our cultural and intellectual leaders, to agree on the importance of change.

There are many practical steps that can be taken. Employers, for example, can reduce the practice of uprooting and relocating married couples with children, provide generous parental leave, and experiment with more flexible forms of work. Religious leaders can reclaim moral ground from the culture of divorce and nonmarriage, resisting the temptation to equate "committed relationships" with marriage. Marriage counselors and family therapists can begin with a bias in favor of marriage, stressing the needs of the marriage at least as much as the needs of the individual. As for the entertainment industry, pressure already is being brought to bear to curtail the glamorization of unwed motherhood, marital infidelity, alternative lifestyles, and sexual promiscuity.

What about divorce? Current laws send the message that marriage is not a socially important relationship that involves a legally binding commitment. We should consider a two-tier system of divorce law: Marriages without minor children would be relatively easy to dissolve, but marriages with young children would be dissolvable only by mutual agreement or on grounds that clearly involve a wrong by one party against the other, such as desertion or physical abuse. Longer waiting periods for divorcing couples with children might also be necessary, combined with some form of mandatory marriage counseling.

Because the causes of the decline of marriage and fatherhood lie mainly in the moral, behavioral, and even spiritual realms, the decline is mostly resistant to public-policy and government cures. All of the Western societies, regardless of governmental system and political persuasion, have been beset by this problem. The decline of marriage is almost as great in Sweden, which has the West's most ambitious welfare state, as it is in the United States, the most laissez-faire of the industrialized nations.

Nevertheless, government policies do have some impact. While the statistical relationship of economic cycles to marriage and divorce is not particularly strong, for example, low wages, unemployment, and poverty have never been friendly to marriage. Government can do something about that. It can also remedy the decline in the value of the income tax exemption for dependent children and erase the tax code's "marriage penalty." As a society, we have decided through a variety of government programs to socialize much of the cost of growing old, but less of the cost of raising children. At the very least, we should strive for generational equity. But more than anything else, parents need time to be with their children, the kind of time that would be afforded by a more generous family-leave policy.

We also should consider providing educational credits or vouchers to parents who leave the paid labor force to raise their young children. These parents are performing an important social service at the risk of damaging their long-term career prospects. Education subsidies, like those in the GI Bill of Rights, would reward parents by helping them resume their careers.

Government policies should be designed to favor married, child-rearing couples. Some critics argue that the federal government should not involve itself in sensitive moral issues or risk stigmatizing alternative lifestyles. But recognizing alternatives doesn't require treating them as equivalent to marriage. The government regularly takes moral positions on a whole range of issues, such as the rights of women, income equality, and race relations. A position on the need for children to have two committed parents, a father and a mother, during their formative years would hardly be a radical departure.

Our social order is fraying badly. We seem, despite notable accomplishments in some areas, to be on a path of decline. The past three decades have seen steeply rising crime rates, growing personal and corporate greed, deteriorating communities, and increasing confusion over moral issues. For most Americans, life has become more anxious, unsettled, and insecure.

In large part, this represents a failure of social values. People can no longer be counted on to conduct themselves according to the virtues of honesty, self-sacrifice, and personal responsibility. In our ever-growing pursuit of the self—self-expression, self-development, self-actualization, and self-fulfillment—we seem to have slipped off many of our larger social obligations.

At the heart of our discontent lies an erosion of personal relationships. People no longer trust others as they once did; they no longer feel the same sense of commitment and obligation to others. In part, this may be an unavoidable product of the modern condition. But it has gone much deeper than that. Some children now to go bed each night wondering whether their father will be there the next morning. Some wonder what happened to their father. And some wonder who he is. What are these children learning at this most basic of all levels about honesty, self-sacrifice, personal responsibility, and trust?

What the decline of fatherhood and marriage in America really means is that slowly, insidiously, and relentlessly our society has been moving in an ominous direction. If we are to move toward a more just and humane society, we must reverse the tide that is pulling fathers apart from their families. Nothing is more important for our children or for our future as a nation.

Modernizing MARRIAGE

The power structure of traditional relationships is a wellspring of resentment that ultimately undermines love. So welcome a new kind of coupling that's more intimate and rewarding to both partners. America's leading sociologist of sex finds that "peer marriage" has arrived—and it works!

Pepper Schwartz, Ph.D.

It is the nature of human relationships that each commitment requires some modifications of totally unfettered individual self-interest. I have spent the past several years studying an emerging type of relationship in which couples have successfully reconstructed gender roles on a genuinely equitable basis. I call these "Peer Marriages." The rule books haven't been written yet; peer couples are making it up as they go along. But this much I have observed: Peer couples trade a frustrated, angry relationship with a spouse for one of deep friendship. They may have somewhat tamer sex lives than couples in traditional marriages. They definitely have fewer external sources of validation. And these couples have a closeness that tends to exclude others. But theirs is a collaboration of love and labor that produces profound intimacy and mutual respect. Traditional couples live in separate spheres and have parallel lives. Above all, peer couples live the same life. In doing so, they have found a new way to make love last.

The dialogue of the decade is the sound of American men and women reshuffling traditional gender relations. The common experience is that women enter the labor force with little or no modification of their traditional duties at home. There, it is said, they work a "second shift." And they break down. While that may be true for the majority, I now know that is not the way it has to be.

If You *ever have to say, "I'll do that if he'll let me," or "I don't know if she'll let me," that should be an alert. It's not a question of what someone lets you do or not do with money, it's what you've arrived at as a couple. Insofar as we let money determine status in the relationship, it corrodes equality and friendship.*

In 1983, my late colleague, sociologist Philip Blumstein, Ph.D., and I published *American Couples*, discussing our decade-long study on the nature of American relationships. During the course of that study I noticed that there were many same-sex couples, and a few heterosexual ones, with an egalitarian relationship that both partners felt was fair and supportive. As a sociologist, I sometimes get trapped by the law of large numbers instead of those in the minority. But over time, I realized there was more I wanted to know about these people, about their success at this aspect of their relationship. How could married couples get past traditions of gender and construct a relationship they both needed? I began reexamining those couples, and sought out more of them to talk to and learn from.

These couples, I discovered, base their marriage on a mix of equity—each person gives in proportion to what he or she receives—and equality—each has equal status and is equally responsible for emotional, economic, and household duties. But these couples have more than their dedication to fairness. They achieve a true companionship and a deeply collaborative marriage. The idea of "peer" is important because it incorporates the notion of friendship. Peer marriages embody a profound psychological connection.

Peering at Peers

In their deep and true partnership based on equality, equity, and intimacy, peer couples, I found, share four important characteristics.

- The partners do not have more than a 60/40 traditional split of household duties and child raising. The couples do a lot of accounting; the division of duties does not happen naturally on account of our training for traditional male and female roles. These couples ask themselves, "What wouldn't get done if I didn't do it?" The important thing is they do not get angry. These partners demonstrate that couples do not have to have a perfect split of responsibilities to lose resentment; what it does take is good will and a great deal of effort and learning skills. We may not be able to jettison our socialization, but we can modify it.
- Both partners believe the other has equal influence over important decisions.

- Both partners feel they have equal control of the family economy and reasonably equal access to discretionary funds. The man does not have automatic veto power. Money is so crucial because in our society, and most societies, we give final authority to the person who is the economic support of the relationship. If you ever have to say, "I'll do that if he'll let me," or "I don't know if she'll let me," that should be an alert. It's not a question of what someone lets you do or not do with money, it's what you've arrived at as a couple. Insofar as we let money determine status in the relationship, it always corrodes equality and friendship.
- Each person's work is given equal weight in the couple's life plans. Whether or not both partners work, they do not systematically sacrifice one person's work for the other's. The person who earns the least is not the person always given the most housework or child care. These couples consciously consider the role of marriage and their relationship in making their life plans. They examine *how* they wanted to be married.

Many couples I talked to believed they were doing this, or believed in the ideology of equality—but in actuality they weren't doing it. They were doing the best they could. Or they knew they weren't doing it now and had deferred it to "some day." Many had plans for it. I call these couples "near peers." Most couples in American culture today are near peers. I compared the peer and near peer couples to traditional couples—those who divide male and female roles into separate spheres of influence and responsibility, with final authority given to the husband.

In my research, I saw each spouse separately and together. I gave them problems to solve and assigned them discussions, all of which were tape-recorded. The peer couples would argue seriously; they had equal standing to do so. One didn't defer to the other. But in the near peers, many of the men would show off to me, to try to show who was really smart.

There are a lot of men who, if approached with the idea of creating a marriage on these terms, would be extremely amenable. Either they do not get reached in time, and then they develop too much of an investment in the way they are living, or, the woman simply never demands it.

About Peer Men

What most undermined near peers was their attachment to the man's income and the man's job. They talked about friendship. But under no circumstances would they endanger his job, even if it meant pulling up stakes every two years, which is very hard on a marriage, or if the man worked days, nights, and weekends for 10 years, or if he traveled to the extent that he essentially put in special guest appearances with his family. Many were fast-track men where both partners agreed this was the way to be.

The second distinguishing factor of the near peers was lack of equal participation in parenthood. Either the woman did not want to give the male full entry into parenthood, or he picked and chose what he would do while extending his work hours.

There are some men who are hell-bent on becoming peer men. They need little guidance in achieving mutual respect, shared responsibility, and joint child raising. There are many more men who want to raise their children, who want an equal partner and a friendship, and who want to enjoy their work but not make it the sole point of their life—but don't know how to do it; many resent not having had a dad in the household while growing up.

I consider both of these groups of men equally available for peer marriages. But without an explicit conversation between the two partners about their values, they will never achieve what they really want. Instead, they get caught up in the provider role because it has been traditionally expected of them.

Most often, it is the woman who has the vision of a peer relationship. Women are more positioned for it because the long-term gains are more apparent to them. It is also women who have the most to lose as parenthood approaches, in terms of what they want in a father for their child. As a result, I believe, it is often the woman's responsibility to get across to her partner the relationship style she wants.

There are a lot of men who, if approached with the idea of creating a marriage on these terms, would be extremely amenable. Either they do not get reached in time, and then they develop too much of an investment in the way they are living, or, the woman simply never demands it. I believe that whoever has the vision has the responsibility to say, "Here's what I need," and "Here is what we've got to solve."

There are those who see peer men as weak, who question anyone's desire to be a peer man, with all that housework and child care—as if those did not have great benefits for the relationship. The truth is, what makes a peer relationship is not "housework"—it's joint purpose. It's creating something together. It's not "child care"—it's knowing and loving your children and being a team on that. It will not kill your work, but it will shape it.

What It Takes

Although the peer couples tended to be dual earners, not all were. There was no hidden hierarchy in these relationships. The couples I talked to were not on an ideological quest, although they embody ideals feminists have talked about. They came to peer relationships from life experience. For many, the peer relationship grew out of a rejection of past experience; it developed only in a second marriage. Many peers had formerly had a traditional marriage that had been unsatisfying and that they did not want to repeat.

Almost half of the peer marriages contained a previously divorced partner. Of the women who were previously divorced, most said they left their first marriage because of inequitable treatment. The previously married men typically said they sought a peer relationship after a devastating period of fighting over property and support in a marriage marked by emotional and financial dependence. The second time around, many fell in love with exactly the sort of independent woman they avoided or felt insecure around when they were younger.

Peer couples are made, not born. Having a peer relationship requires first and foremost having a sense of yourself; you need something to be equal to. It is harder to do when you're very young, a time when many are willing to hand themselves over and change with each partner. When young, you may land in a peer relationship by sheer luck, but you have not been tested in certain ways about yourself. You may still be trying to find out who you are and your first several attempts may be very wrong.

Flexibility is another important trait, the ability to enter a marriage with no ironclad rules about roles, but to see what you are both doing and say, "Here is what I need." It is possible to enter a relationship mistaking your needs, but you must be able to say, "This is not working for us," and reconfigure how it could work. I was shocked by how many women poured out their resentment who never simply said to a partner, "Here's what has to get done; what's a fair amount of responsibility for you to carry?" Many of these women insist they have a close relationship, but they are afraid it would end if they asked for what they believe a relationship should be.

The only way peers get to the position of equal responsibility is by coming to some agreement about values. "Who am I and what do I need?" "Who are you and what do you need?" They have to figure out, make lists of those things they do need, even of things that get taken for granted. And it takes lots of interacting and negotiating. Often the person who brings up a subject is the one who is being taken advantage of. It may seem banal to talk about the executive role of the household—"Who will take the dogs for their shots?" "Are we going to take a trip this summer and who will plan it?"—but that's what makes a household work.

COSTS AND BENEFITS

Many benefits accrue to people in peer marriages:

• Primacy of the relationship. They give priority to their relationship over their work and over all other relationships, even with their children. Their mutual friendship is the most satisfying part of their lives. Each partner feels secure in the other person's regard and support.

• Intimacy. Because they share housework, children, and economic responsibility, partners experience the world in a more similar way, understand the other more accurately, and communicate better. They negotiate more than other couples, share conversational time, and are less often dismissive of the other.

• Commitment. These couples are much more likely to find each other irreplaceable. They describe their relationship as "unique." Their interdependence becomes so deep and so customized that the costs of splitting up become prohibitive.

If peer marriage is so rewarding, why, then, doesn't everyone seek or achieve it?

• There is little or no outside validation for this new type of domestic arrangement, and often outright opposition. Outsiders—from parents and friends to coworkers and bosses—tend to question the couple's philosophy and be unwilling to modify work or other schedules to help a couple share family life. Commonly, a man's parents feel betrayed and see their son as emasculated. His in-laws aren't generally enamored of his peerishness, either. The Good Provider role is still uppermost in their minds.

• There are career costs. Couples need jobs that allow them to coparent. Either they have to wait long enough to have enough clout to manage their work-life this way, or they have to be in jobs that naturally support parenting. Many couples report they have to modify their career ambitions in favor of family aspirations. It's not that one or both partners can't be high-powered lawyers, but it's almost impossible to be pit-bull litigators whose every hour is spent in court or who may be put in another country for two years on a case. Couples can alternate priorities—one partner's job will take priority for a year or so. But there have to be boundaries. There's no formula—it's an art form.

• Peer couples have to define success differently than by the prevailing mode, which is by traditional male and female roles.

• They make others feel excluded and possibly resentful. Most people like married couples to dish their spouse occasionally. Peer couples tend to be dedicated to their relationship and to parenting.

• They face a new sexual dynamic. They benefit from so much everyday intimacy that they may have to go out of their way to put eroticism back into the relationship.

• In the absence of a blueprint for their type of relationship, peer couples face the inexact challenge of figuring out the right thing to do all the time. It's tiring!

GENDERIZED ROMANCE

In the vast majority of couples today, the relationship exists on women's skills. Perhaps the biggest job women carry is to be the expressive member of the couple. Most of the warmth and interaction is transmitted on women's terms. According to sociologist Francesca Cancian, love has been feminized. Our culture overdoses men with information about what women want.

Women see love and self-revelation as synonymous. But men see expressions of their love in mundane little acts like paying the insurance and other caretaking chores. Unfortunately, with the abundant help of gothic novels, we have come to define love almost exclusively on women's terms; flowers and candy have no intrinsic meaning for most men. The result is that neither partner feels understood or appreciated, and men are judged emotionally incompetent. Ultimately, men stop trying to get credit for their style of loving and withdraw from the intimacy sweepstakes. So women do much of the relationship work—and then resent the lack of reciprocity. Eventually, resentment can thwart desire altogether.

The prevailing definition of love, I believe, is too narrow. By not taking pleasure in utilitarian displays of affection, two partners unnecessarily create emotional limitations for their relationship and reduce the amount of love they can receive from one another.

Because both are living the same life, peer couples are more likely to merge male and female styles of communication and affection. They learn love in each other's terms. Men as well as women are responsible for generating discussion and warmth. And women appreciate men's affectional style. The yearnings for affection that continually arose in my interviews with traditional wives rarely surfaces in peer relationships.

I find that there is a "new romance" being born, and its hallmark is the ability of a couple to relate to each other in each other's terms. If, like peers, a couple lives life the same way, then they both understand intimacy and romance on similar terms.

ROMANCE REDEFINED

One of the most significant differences between peer marriages and traditional marriages is in the role of sex. Peer marriages are built on commonality, traditional marriages on differences, especially power differences—hero and heroine. Traditional sexual tension is anchored on difference; male leadership and control of sex is regarded as inherently erotic. The man's power and status over the woman turns them both on.

Of course, all long-term relationships involve a diminution of sexual interest. Familiarity is simply not as erotic as newness and the desire to be loved and accepted—and to reconcile after difficulties. In traditional marriages, passion is typically kept aroused by disappointment, fear of loss, anger, and other types of negative emotions. Couples may have good sex, but they finish still feeling that they don't know their partner. There is an ultimate loneliness even after making love.

Peer marriages get the ability for romance and for comfortable and happy sex unencumbered by anger. And for intimacy unencumbered by distance and lack of personal knowledge about the other partner.

In traditional marriages, couples have to work against massive odds to achieve intimacy. Any security that may be attained is fleeting. There may even be more sexual frequency when that is the only way a partner can even hope to touch the other person emotionally. Whether it is more deeply satisfying is very open to question. Many women I have interviewed, in this study and others, report that the only time that their husband shows himself to be emotionally needy is during sex. It is the only time that he allows himself to be vulnerable.

Very passionate relationships are often tortured relationships. Volatile people manufacture a great deal of adrenaline, and it adds an edge to sex. Anger in one partner fuels the desire of the insecure partner to be ratified by the one who loves less. It can be thrilling and sexually explosive to get that affirmation, but it goes away right afterwards. It may make for a passionate relationship, but it makes for a lousy marriage and an insecure life. One important conclusion is that passion is not the highest and best valuation of a couple's sex life.

Passion can be evoked upon occasion, but what peer couples are really superb at is romance—the good and lasting romance of equals. When more than one person has expressive skills and uses them, more positive exchange takes place, and satisfaction is greater. Peer couples:

- are dedicated to being a couple, over and above being a family;
- display physical and verbal affection;
- spend nonutilitarian time together;
- exchange conversation and gifts to show that the partner is valued;
- celebrate special days that mark the relationship's beginning, history, and progress.

All couples need these elements to enjoy romance; peer couples are more likely to because both partners take responsibility for them.

If there is a downside to peer relationships, it's that in their strong affinity for one another, peer couples have to fight against the tendency for sex to become a residual category. They must specifically cultivate this part of the relationship. The biggest problem, I found, was that peer couples report that they are not having sex as often as they used to.

'Peer couples can merge male and female styles of affection. They learn to love in each other's terms.'

Of course, for all couples, life gets in the way. But if the vulnerability of traditional couples is anger and resentment, the vulnerability of peer couples is keeping the spark alive. These couples are getting so much from their relationship with each other that they do not need sex to get all their emotional needs met.

Peer couples have to work at eroticism and at ways of coming together sexually. The challenge is to take off the buddy mantle and find erotic ways to play with each other. It may mean going off by themselves for a weekend, or they may want to put on costumes and play out individual fantasies; they have to take a break from the negotiated partnership and from the communal self.

The positive side of this is that peer marriage actually frees partners to bring their own private, uncovered self into the relationship and display it in the bedroom. What is more, there can be more innovation. Equalizing the initiation and leadership responsibilities in sex doubles the creativity that can be brought to bear. Many peer couples speak of this.

What typically happens in traditional marriages, I have long observed, is that the woman makes the children her real emotional community—in place of her partner. In a sense, he just seeds the family and visits it. He does not have the same relationship to the child, and he does not have the same relationship to the relationship that his partner does.

By contrast, peer spouses have built up a real friendship and investment in each other's life; they keep in the front of their mind that their relationship is about the marriage. The children are part of the marriage, but the marriage is not part of the children. Keeping that fact straight is important both for the fluidity and validity of the marriage, but also for safeguarding that child. No doubt, children are best protected by a strong and happy marriage of two parents.

Similarity, when prized, can be exciting. Hierarchy and domination are not essential for arousal.

The Future of Love

Peer marriage, like any other type of marriage, is not a panacea for all things emotional and intimate. There are lots of ways people can be disappointed in each other. They can grow differently. They can come up with strongly held values that do not mesh. Peer marriage is not a guarantee. But it increases the chances that both partners will find emotional rewards, that they will create a stable partnership for parenting, and that love will last without resentment.

It is the direction marriages are going to move in. I am happy to report that some couples have achieved it. There will always be some people who find solace, security, and love in a junior role in the relationship. And some who truly want to further only one income. There will be some who have no desire to know one another in the intimate way I have described.

I do not see all of us in the same kind of relationship. We're all too different from one another for that. I wouldn't sentence everyone to the same kind of roles in marriages. But peer marriage will become a predominant cultural theme and perhaps the predominant type of marriage in the very near future.

Now for the Truth About Americans and SEX

The first comprehensive survey since Kinsey smashes some of our most intimate myths

PHILIP ELMER-DEWITT

Is there a living, breathing adult who hasn't at times felt the nagging suspicion that in bedrooms across the country, on kitchen tables, in limos and other venues too scintillating to mention, other folks are having more sex, livelier sex, better sex? Maybe even that quiet couple right next door is having more fun in bed, and more often. Such thoughts spring, no doubt, from a primal anxiety deep within the human psyche. It has probably haunted men and women since the serpent pointed Eve toward the forbidden fruit and urged her to get with the program.

Still, it's hard to imagine a culture more conducive to feelings of sexual inadequacy than America in the 1990s. Tune in to the soaps. Flip through the magazines. Listen to Oprah. Lurk in the seamier corners of cyberspace. What do you see and hear? An endless succession of young, hard bodies preparing for, recovering from or engaging in constant, relentless copulation. Sex is everywhere in America—and in the ads, films, TV shows and music videos it exports abroad. Although we know that not every ZIP code is a Beverly Hills, 90210, and not every small town a Peyton Place, the impression that is branded on our collective subconscious is that life in the twilight of the 20th century is a sexual banquet to which everyone else has been invited.

Just how good is America's sex life? Nobody knows for sure. Don't believe the magazine polls that have Americans mating energetically two or three times a week. Those surveys are inflated from the start by the people who fill them out: *Playboy* subscribers, for example, who brag about their sex lives in reader-survey cards. Even the famous Kinsey studies—which caused such a scandal in the late 1940s and early '50s by reporting that half of American men had extramarital affairs—were deeply flawed. Although Alfred Kinsey was a biologist by training (his expertise was the gall wasp), he compromised science and took his human subjects where he could find them: in boardinghouses, college fraternities, prisons and mental wards. For 14 years he collared hitchhikers who passed through town and quizzed them mercilessly. It was hardly a random cross section.

Now, more than 40 years after Kinsey, we finally have some answers. A team of researchers based at the University of Chicago has released the long-awaited results of what is probably the first truly scientific survey of who does what with whom in America and just how often they do it.

The findings—based on face-to-face interviews with a random sample of nearly 3,500 Americans, ages 18 to 59, selected using techniques honed through decades of political and consumer polling—will smash a lot of myths. "Whether the numbers are reassuring or alarming depends on where you sit," warns Edward Laumann, the University of Chicago sociologist who led the research team. While the scientists found that the spirit of the sexual revolution is alive

54% of men think about sex daily. 19% of women do

and well in some quarters—they found that about 17% of American men and 3% of women have had sex with at least 21 partners—the overall impression is that the sex lives of most Americans are about as exciting as a peanut-butter-and-jelly sandwich.

Among the key findings:

- Americans fall into three groups. One-third have sex twice a week or more, one-third a few times a month, and one-third a few times a year or not at all.
- Americans are largely monogamous. The vast majority (83%) have one or zero sexual partners a year. Over a lifetime, a typical man has six partners; a woman, two.
- Married couples have the most sex and are the most likely to have orgasms when they do. Nearly 40% of married people say they have sex twice a week, compared with 25% for singles.
- Most Americans don't go in for the kinky stuff. Asked to rank their favorite sex acts, almost everybody (96%) found vaginal sex "very or somewhat appealing." Oral sex ranked a distant third, after an activity that many may not have realized was a sex act: "Watching partner undress."
- Adultery is the exception in America, not the rule. Nearly 75% of married men and 85% of married women say they have never been unfaithful.
- There are a lot fewer active homosexuals in America than the oft-repeated 1 in 10. Only 2.7% of men and 1.3% of women report that they had homosexual sex in the past year.

THE FULL RESULTS OF THE NEW SURVEY ARE scheduled to be published next week as *The Social Organization of Sexuality* (University of Chicago; $49.95), a thick, scientific tome co-authored by Laumann, two Chicago colleagues—Robert Michael and Stuart Michaels—and John Gagnon, a sociologist from the State University of New York at Stony Brook. A thinner companion volume, *Sex in America: A Definitive Survey* (Little, Brown; $22.95), written with New York *Times* science reporter Gina Kolata, will be in bookstores this week.

But when the subject is sex, who wants to wait for the full results? Even before the news broke last week, critics and pundits were happy to put their spin on the study.

"It doesn't ring true," insisted Jackie Collins, author of *The Bitch, The Stud* and other potboilers. "Where are the deviants? Where are the flashers? Where are the sex maniacs I see on TV every day?"

"I'm delighted to hear that all this talk about rampant infidelity was wildly inflated," declared postfeminist writer Camille Paglia. "But if they're saying the sexual revolution never happened, that's ridiculous."

"Positively, outrageously stupid and unbelievable," growled *Penthouse* publisher Bob Guccione. "I would say five partners a year is the average for men."

"Totally predictable," deadpanned Erica Jong, author of the 1973 sex fantasy *Fear of Flying*. "Americans are more interested in money than sex."

"Our Puritan roots are deep," said *Playboy* founder Hugh Hefner, striking a philosophical note. "We're fascinated by sex and afraid of it."

"Two partners? I mean, come on!" sneered *Cosmopolitan* editor Helen Gurley Brown. "We advise our Cosmo girls that when people ask how many partners you've had, the correct answer is always three, though there may have been more."

Europeans seemed less surprised—one way or the other—by the results of the survey. The low numbers tend to confirm the Continental caricature of Americans as flashy and bold onscreen but prone to paralysis in bed. Besides, the findings were pretty much in line with recent studies conducted in England and France that also found low rates of homosexuality and high

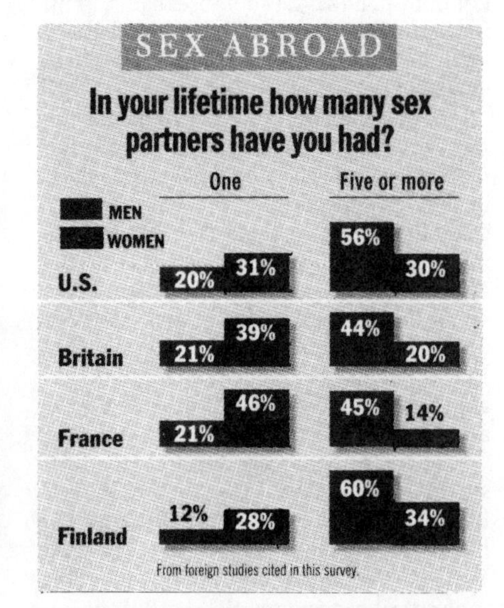

rates of marital fidelity. (The French will be gratified by what a comparison of these surveys shows: that the average Frenchman and -woman has sex about twice as often as Americans do.)

If the study is as accurate as it purports to be, the results will be in line with the experience of most Americans. For many, in fact, they will come as a relief. "A lot of people think something is wrong with them when they don't have sexual feelings," says Toby, a 32-year-old graduate student from Syracuse, New York, who, like 3% of adult Americans (according to the survey), has never had sex. "These findings may be liberating for a lot of people. They may say, 'Thank God, I'm not as weird as I thought.' "

Scientists, on the whole, praise the study. "Any new research is welcome if it is well done," says Dr. William Masters, co-author of the landmark 1966 study Human Sexual Response. By all accounts, this one was very well done. But, like every statistical survey, it has its weaknesses. Researchers caution that the sample was too limited to reveal much about small subgroups of the population—gay Hispanics, for example. The omission of people over 59 is regrettable, says Shirley Zussman, past president of the American Association of Sex Educators, Counselors and Therapists: "The older population is more sexually active than a 19-year-old thinks, and it's good for both 19-year-olds and those over 59 to know that."

The Chicago scientists admit to another possible defect: "There is no way to get around the fact some people might conceal information," says Stuart Michaels of the Chicago team, whose expertise is designing questions to get at those subjects people are most reluctant to discuss. The biggest

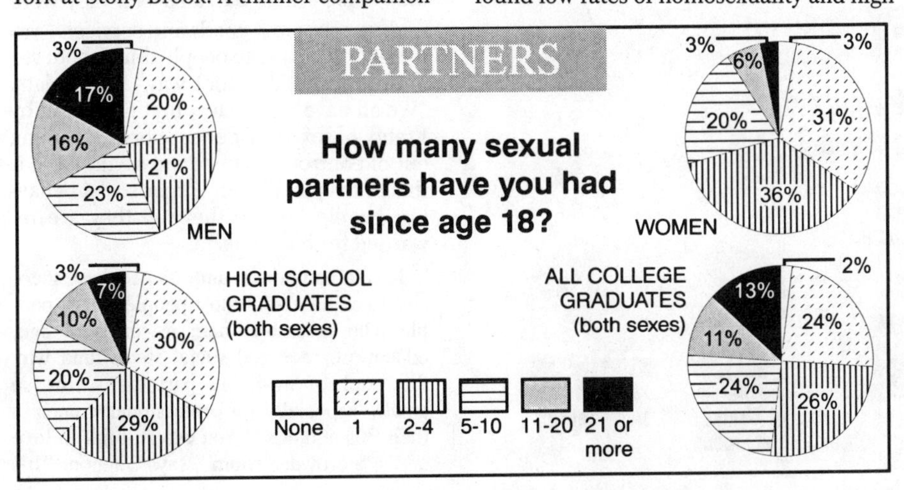

The 2nd most appealing sex act: seeing partner undress

hot button, he says, is homosexuality. "This is a stigmatized group. There is probably a lot more homosexual activity going on than we could get people to talk about."

It was, in large part, to talk about homosexual activity that the study was originally proposed. The project was conceived in 1987 as a response to the AIDS crisis. To track the spread of the AIDS virus—and to mount an effective campaign against it—government researchers needed good data about how much risky sexual behavior (anal sex, for example) was really going on. But when they looked for scientific data about sex, they found little besides Kinsey and Masters and Johnson.

So the National Institutes of Heath issued a formal request for a proposal, tactfully giving it the bland title "Social and Behavioral Aspects of Fertility Related Behavior" in an attempt to slip under the radar of right-wing politicians. But the euphemism fooled no one—least of all Jesse Helms. In the Reagan and Bush era, any government funding for sex research was suspect, and the Senator from North Carolina was soon lobbying to have the project killed. The Chicago team redesigned the study several times to assuage conservative critics, dropping the questions about masturbation and agreeing to curtail the interview once it was clear that a subject was not at high risk of contracting AIDS. But to no avail. In September 1991 the Senate voted 66 to 34 to cut off funding.

The vote turned out to be the best thing that could have happened—at least from the point of view of the insatiably curious. The Chicago team quickly rounded up support from private sources, including the Robert Wood Johnson, Rockefeller and Ford foundations. And freed of political constraints, they were able to take the survey beyond behavior related to AIDS transmission to tackle the things inquiring minds really want to know: Who is having sex with whom? How often do they do it? And when they are behind closed doors, what exactly do they do?

The report confirms much of what is generally accepted as conventional wisdom. Kids *do* have sex earlier now: by 15, half of all black males have done it; by 17, the white kids have caught up to them. There *was* a lot of free sex in the '60s: the percentage of adults who have racked up 21 or more sex partners is significantly higher among the fortysomething boomers than among other Americans. And AIDS *has* put a crimp in some people's sex lives: 76% of those who have had five or more partners in the past year say they have changed their sexual behavior, by either slowing down, getting tested or using condoms faithfully.

But the report is also packed with delicious surprises. Take masturbation, for example. The myth is that folks are more likely to masturbate if they don't have a sex partner. According to the study, however, the people who masturbate the most are the ones who have the most sex. "If you're having sex a lot, you're thinking about sex a lot," says Gagnon. "It's more like Keynes (wealth begets wealth) and less like Adam Smith (if you spend it on this, you can't spend it on that)."

Or take oral sex. Not surprisingly, both men and women preferred receiving it to giving it. But who would have guessed that so many white, college-educated men would have done it (about 80%) and so few blacks (51%)? Skip Long, a 33-year-old African American from Raleigh, North Carolina, thinks his race's discomfort with oral sex may owe much to religious teaching and the legacy of slavery: according to local legend, it was something slaves were required to do for their masters. Camille Paglia is convinced that oral sex is a culturally acquired preference that a generation of college students picked up in the '70s from seeing Linda Lovelace do it in *Deep Throat,* one of the first—and last—X-rated movies that men and women went to see together. "They saw it demonstrated on the screen, and all of a sudden it was on the map," says Paglia. "Next thing you knew, it was in *Cosmo* with rules about how to do it."

More intriguing twists emerge when sexual behavior is charted by religious affiliation. Roman Catholics are the most likely to be virgins (4%) and Jews to have the most sex partners (34% have had 10 or more). The women most likely to achieve orgasm each and every time (32%) are, believe it or not, conservative Protestants. But Catholics edge out mainline Protestants in frequency of intercourse. Says Father Andrew Greeley, the sociologist-priest and writer of racy romances: "I think the church will be surprised at how often Catholics have sex and how much they enjoy it."

Among women, 29% always had an orgasm during sex

But to concentrate on the raw numbers is to miss the study's most important contribution. Wherever possible, the authors put those figures in a social context, drawing on what they know about how people act out social scripts, how they are influenced by their social networks and how they make sexual bargains as if they were trading economic goods and services. "We were trying to make people think about sex in an entirely different way," says Kolata. "We all have this image, first presented by Freud, of sex as a riderless horse, galloping out of control. What we are saying here is that sex is just like any other social behavior: people behave the way they are rewarded for behaving."

Kolata and her co-authors use these theories to explain why most people marry people who resemble them in terms of age, education, race and social status, and why the pool of available partners seems so small—especially for professional women in their 30s and 40s. "You can still fall in love across a crowded room," says Gagnon. "It's

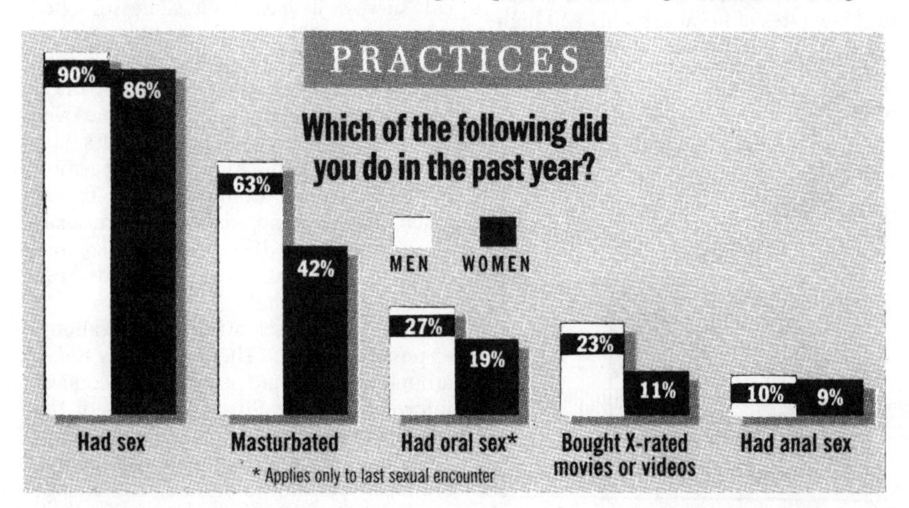

Of married people, 94% were faithful in the past year

just that society determines whom you're in the room with."

That insight, applied to AIDS, leads the Chicago team to a conclusion that is sure to get them into trouble. America's AIDS policy, they say, has been largely misdirected. Although AIDS spread quickly among intravenous drug users and homosexuals, the social circles these groups travel in are so rigidly circumscribed that it is unlikely to spread widely in the heterosexual population. Rather than pretend that AIDS affects everyone, they say, the government would be better advised to concentrate its efforts on those most at risk.

That's a conclusion that will not sit well with AIDS activists or with many health-policy makers. "Their message is shocking and flies against the whole history of this epidemic," says Dr. June Osborn, former chair of the National Commission on AIDS. "They're saying we don't have to worry if we're white, heterosexual adults. That gets the public off the hook and may keep parents from talking to their kids about sex. The fact is, teens are at enormous risk for experimentation."

Other groups will find plenty here to make a fuss about. Interracial couples are likely to take offense at the authors' characterization of mixed-race marriages as unlikely to succeed. And right-to-life activists who believe abortion is widely used as a cruel form of birth control are likely to be unconvinced by the finding that 72% of the women who have an abortion have only one.

Elsewhere in the study, the perceptual gulf between the sexes is reminiscent of the scene in *Annie Hall* where Woody Allen tells his psychiatrist that he and Annie have sex "hardly ever, maybe three times a week," and she tells hers that they do it "constantly; I'd say three times a week." In the Chicago study, 54% of the men say they think about sex every day or several times a day. By contrast, 67% of the women say they think about it only a few times a week or a few times a month. The disconnect is even greater when the subject turns to forced sex. According to the report, 22% of women say they have been forced to do sexual things they didn't want to, usually by someone they loved. But only 3% of men admit to ever forcing themselves on women. Apparently men and women have very different ideas about what constitutes voluntary sex.

But the basic message of *Sex in America* is that men and women have found a way to come to terms with each other's sexuality—and it is called marriage. "Our study," write the authors, "clearly shows that no matter how sexually active people are before and between marriages ... marriage is such a powerful social institution that, essentially, married people are all alike—they are faithful to their partners as long as the marriage is intact."

Americans, it seems, have come full circle. It's easy to forget that as recently as 1948, Norman Mailer was still using the word fug in his novels. There may have been a sexual revolution—at least for those college-educated whites who came of age with John Updike's swinging *Couples*, Philip Roth's priapic *Portnoy* and Jong's *Fear of Flying*—but the revolution turned out to have a beginning, a middle and an end. "From the time of the Pill to Rock Hudson's death, people had a sense of freedom," says Judith Krantz, author of *Scruples*. "That's gone."

It was the first survey—Kinsey's—that got prudish America to talk about sex, read about sex and eventually watch sex at the movies and even try a few things (at least once). Kinsey's methods may have been less than perfect, but he had an eye for the quirky, the fringe, the bizarre. The new report, by contrast, is a remarkably conservative document. It puts the fringe on the fringe and concentrates on the heartland: where life, apparently, is ruled by marriage, monogamy and the missionary position. The irony is that the report Jesse Helms worked so hard to stop has arrived at a conclusion that should make him proud. And it may even make the rest of us a bit less anxious about what's going on in that bedroom next door. ***—Reported by Wendy Cole/Chicago, John F. Dickerson/New York and Martha Smilgis/Los Angeles***

Ending the battle between the sexes

First separate, then communicate

Aaron R. Kipnis & Elizabeth Herron

SPECIAL TO *UTNE READER*

Aaron R. Kipnis, Ph.D., *is a consultant to numerous organizations concerned with gender issues and lectures for many clinical training institutes. He is author of the critically acclaimed book* Knights Without Armor: A Practical Guide for Men in Quest of Masculine Soul, *(Jeremy Tarcher, 1991, Putnam, 1992).*

Elizabeth Herron, M.A., *specializes in women's empowerment and gender reconciliation. She is co-director of the Santa Barbara Institute for Gender Studies and is co-author of* Gender War/Gender Peace; The Quest for Love and Justice Between Women and Men *(Feb. 1994, Morrow).*

Have you noticed that American men and women seem angrier at one another than ever? Belligerent superpowers have buried the hatchet, but the war between the sexes continues unabated. On every television talk show, women and men trade increasingly bitter accusations. We feel the tension in our homes, in our workplaces, and in our universities.

The Clarence Thomas-Anita Hill controversy and the incidents at the Navy's Tailhook convention brought the question of sexual harassment into the foreground of national awareness, but it now appears that these flaps have merely fueled male-female resentment instead of sparking a productive dialogue that might enhance understanding between the sexes.

Relations between women and men are rapidly changing. Often, however, these changes are seen to benefit one sex at the expense of the other, and the mistrust that results creates resentment. Most men and women seem unable to entertain the idea that the two sexes' differing perspectives on many issues can be equally valid. So polarization grows instead of reconciliation, as many women and men fire ever bigger and better-aimed missiles across the gender gap. On both sides there's a dearth of compassion about the predicaments of the other sex.

For example:

• Women feel sexually harassed; men feel their courting behavior is often misunderstood.

• Women fear men's power to wound them physically; men fear women's power to wound them emotionally.

• Women say men aren't sensitive enough; men say women are too emotional.

• Women feel men don't do their fair share of housework and child care; men feel that women don't feel as much pressure to provide the family's income and do home maintenance.

• Many women feel morally superior to men; many men feel that they are more logical and just than women.

• Women say men have destroyed the environment; men say the women's movement has destroyed the traditional family.

• Men are often afraid to speak about the times that they feel victimized and powerless; women frequently deny their real power.

• Women feel that men don't listen; men feel that women talk too much.

• Women resent being paid less than men; men are concerned about the occupational hazards and stress that lead to their significantly shorter life spans.

• Men are concerned about unfairness in custody and visitation rights; women are concerned about fathers who shirk their child support payments.

It is very difficult to accept the idea that so many conflicting perspectives could all have intrinsic value. Many of us fear that listening to the story of another will

From *Utne Reader*, January/February 1993, pp. 69-76. © 1993 by Aaron R. Kipnis, Ph.D., and Elizabeth Herron, M.A. Reprinted by permission.

somehow weaken our own voice, our own initiative, even our own identity. The fear keeps us locked in adversarial thinking and patterns of blame and alienation. In this frightened absence of empathy, devaluation of the other sex grows.

In an attempt to address some of the discord between the sexes, we have been conducting gender workshops around the country. We invite men and women to spend some time in all-male and all-female groups, talking about the opposite sex. Then we bring the two groups into an encounter with one another. In one of our mixed groups this spring, Susan, a 35-year-old advertising executive, told the men, "Most men these days are insensitive jerks. When are men going to get it that we are coming to work to make a living, not to get laid? Anita Hill was obviously telling the truth. Most of the women I work with have been harassed as well."

Michael, her co-worker, replied, "Then why didn't she tell him ten years ago that what he was doing was offensive? How are we supposed to know where your boundaries are if you laugh at our jokes, smile when you're angry, and never confront us in the direct way a man would? How am I supposed to learn what's not OK with you, if the first time I hear about it is at a grievance hearing?"

We've heard many permutations of this same conversation:

Gina, a 32-year-old school teacher in Washington, D.C., asks, "Why don't men ever take *no* for an answer?"

Arthur, a 40-year-old construction foreman, replies that in his experience, "some women *do* in fact say no when they mean yes. Women seem to believe that men should do all the pursuing in the mating dance. But then if we don't read her silent signals right, we're the bad guys. If we get it right, though, then we're heroes."

Many men agree that they are in a double bind. They are labeled aggressive jerks if they come on strong, but are rejected as wimps if they don't. Women feel a similar double bind. They are accused of being teases if they make themselves attractive but reject the advances of men. Paradoxically, however, as Donna, a fortyish divorcée, reports, "When I am up front about my desires, men often head for the hills."

As Deborah Tannen, author of the best-seller about male-female language styles *You Just Don't Understand*, has observed, men and women often have entirely different styles of communication. How many of us have jokingly speculated that men and women actually come from different planets? But miscommunication alone is not the source of all our sorrow.

Men have an ancient history of enmity toward women. For centuries, many believed women to be the cause of our legendary fall from God's grace. "How can he be clean that is born of woman?" asks the Bible. Martin Luther wrote that "God created Adam Lord of all living things, but Eve spoiled it all." The "enlightened" '60s brought us Abbie Hoffman, who said: "The only alliance I would make with the women's liberation movement is in bed." And from the religious right, Jerry Falwell still characterizes feminism as a "satanic attack" on the American family.

In turn, many feel the women's movement devalues the role of men. Marilyn French, author of *The Women's Room*, said, "All men are rapists and that's all they are." In response to the emerging men's movement, Betty Friedan commented, "Oh God, sick . . . I'd hoped by now men were strong enough to accept their vulnerability and to be authentic without aping Neanderthal cavemen."

This hostility to the men's movement is somewhat paradoxical. Those who are intimately involved with the movement say that it is primarily dedicated to ending war and racism, increasing environmental awareness, healing men's lives and reducing violence, promoting responsible fatherhood, and creating equal partnerships with women—all things with which feminism is ideologically aligned. Yet leaders of the men's movement often evoke indignant responses from women. A prominent woman attorney tells us, "I've been waiting 20 years for men to hear our message. Now instead of joining us at last, they're starting their *own* movement. And now they want us to hear that they're wounded too. It makes me sick."

On the other hand, a leader of the men's movement says, "I was a feminist for 15 years. Recently, I realized that all the men I know are struggling just as much as women. Also, I'm tired of all the male-bashing. I just can't listen to women's issues anymore while passively watching so many men go down the tubes."

Some of our gender conflict is an inevitable by-product of the positive growth that has occurred in our society over the last generation. The traditional gender roles of previous generations imprisoned many women and men in soul-killing routines. Women felt dependent and disenfranchised; men felt distanced from feelings, family, and their capacity for self-care.

With almost 70 percent of women now in the work force, calls from Barbara Bush and Marilyn Quayle for women to return to the home full time seem ludicrous, not to mention financially impossible. In addition, these calls for the traditional nuclear family ignore the fact that increasing numbers of men now want to downshift from full-time work in order to spend more time at home. So if we can't go back to the old heroic model of masculinity and the old domestic ideal of femininity, how then do we weave a new social fabric out of the broken strands of worn-out sexual stereotypes?

Numerous participants in the well-established women's movement, as well as numbers of men in the smaller but growing men's movement, have been discovering the strength, healing, power, and sense of security that come from being involved with a same-sex group. Women and men have different social, psychological, and biological realities and receive different behavioral training from infancy through adulthood.

In most pre-technological societies, women and men both participate in same-sex social and ceremonial groups. The process of becoming a woman or a man usually begins with some form of ritual initiation. At the onset of puberty, young men and women are brought into the men's and women's lodges, where they gain a deep sense of gender identity.

Even in our own culture, women and men have traditionally had places to meet apart from members of the other sex. For generations, women have gathered over coffee or quilts; men have bonded at work and in taverns. But in our modern society, most heterosexuals believe that a member of the opposite sex is supposed to fulfill all their emotional and social needs. Most young people today are not taught to respect and honor the differences of the other gender, and they arrive at adulthood both mystified and distrustful, worried about the other sex's power to affect them. In fact, most cross-gender conflict is essentially *conflict between different cultures.* Looking at the gender war from this perspective may help us develop solutions to our dilemmas.

In recent decades, cultural anthropologists have come to believe that people are more productive members of society when they can retain their own cultural identity within the framework of the larger culture. As a consequence, the old American "melting pot" theory of cultural assimilation has evolved into a new theory of diversity, whose model might be the "tossed salad." In this ideal, each subculture retains its essential identity, while coexisting within the same social container.

Applying this idea to men and women, we can see the problems with the trend of the past several decades toward a sex-role melting pot. In our quest for gender equality through sameness, we are losing both the beauty of our diversity and our tolerance for differences. Just as a monoculture is not as environmentally stable or rich as a diverse natural ecosystem, androgyny denies the fact that sexual differences are healthy.

In the past, perceived differences between men and women have been used to promote discrimination, devaluation, and subjugation. As a result, many "we're all the same" proponents—New Agers and humanistic social theorists, for example—are justifiably suspicious of discussions that seek to restore awareness of our differences. But pretending that differences do not exist is not the way to end discrimination toward either sex.

Our present challenge is to acknowledge the value of our differing experiences as men and women, and to find ways to reap this harvest in the spirit of true equality. Carol Tavris, in her book *The Mismeasure of Women*, suggests that instead of "regarding cultural and reproductive differences as problems to be eliminated, we should aim to eliminate *the unequal consequences that follow from them.*"

Some habits are hard to change, even with an egalitarian awareness. Who can draw the line between what is socially conditioned and what is natural? It may not be possible, or even desirable, to do so. What seems more important is that women and men start understanding each other's different cultures and granting one another greater freedom to experiment with whatever roles or lifestyles attract them.

Lisa, a 29-year-old social worker from New York participating in one of our gender workshops, told us, "Both Joel [her husband] and I work full time. But it always seems to be me who ends up having to change my schedule when Gabe, our son, has a doctor's appointment or a teacher conference, is sick at home or has to be picked up after school. It's simply taken for granted that in most cases my time is less important than his. I know Joel tries really hard to be an engaged father. But the truth is that I feel I'm always on the front line when it comes to the responsibilities of parenting and keeping the home together. It's just not fair."

Joel responds by acknowledging that Lisa's complaint is justified; but he says, "I handle all the home maintenance, fix the cars, do all the banking and bookkeeping and all the yard work as well. These things aren't hobbies. I also work more overtime than Lisa. Where am I supposed to find the time to equally co-parent too? Is Lisa going to start mowing the lawn or help me build the new bathroom? Not likely."

In many cases of male-female conflict, as with Lisa and Joel, there are two differing but *equally valid* points of view. Yet in books, the media, and in women's and men's groups, we only hear about most issues from a woman's point of view or from a man's. This is at the root of the escalating war between the sexes.

For us, the starting point in the quest for gender peace is for men and women to spend more time with members of the same sex. We have found that many men form intimate friendships in same-sex groups. In addition to supporting their well-being, these connections can take some of the pressure off their relationships with women. Men in close friendships no longer expect women to satisfy *all* their emotional needs. And when women meet in groups they support one another's need for connection and also for empowerment in the world. Women then no longer expect men to provide their sense of self-worth. So these same-sex groups can enhance not only the participants' individual lives, but their relationships with members of the other sex as well.

If men and women *remain* separated, however, we risk losing perspective and continuing the domination or scapegoating of the other sex. In women's groups, male-bashing has been running rampant for years. At a recent lecture we gave at a major university, a young male psychology student said, "This is the first time in three years on campus that I have heard anyone say a single positive thing about men or masculinity."

Many women voice the same complaint about their experiences in male-dominated workplaces. Gail, a middle management executive, says, "When I make proposals to the all-male board of directors, I catch the little condescending smirks and glances the men give one another. They don't pull that shit when my male colleagues speak. If they're that rude in front of me, I can only imagine how degrading their comments are when they meet in private."

There are few arenas today in which women and men can safely come together on common ground to frankly discuss our rapidly changing ideas about gender justice. Instead of more sniping from the sidelines, what is needed is for groups of women and men to communicate directly with one another. When we take this *next step* and make a commitment to spend time apart and then meet with each other, then we can begin to build a true social, political, and spiritual equality. This process also instills a greater appreciation for the unique gifts each sex has to contribute.

Husband-and-wife team James Sniechowski and Judith Sherven conduct gender reconciliation meetings—similar to the meetings we've been holding around the country—each month in Southern California. In a recent group of 25 people (11 women, 14 men), participants were invited to explore questions like: What did you learn about being a man/woman from your mother? From your father? Sniechowski reports that, "even though, for the most part, the men and women revealed their confusions, mistrust, heartbreaks, and bewilderments, the room quickly filled with a poignant beauty." As one woman said of the meeting, "When I listen to the burdens we suffer, it helps me soften my heart toward them." On another occasion a man said, "My image of women shifts as I realize they've been through some of the same stuff I have."

Discussions such as these give us an opportunity to really hear one another and, perhaps, discover that many of our disagreements come from equally valid, if different, points of view. What many women regard as intimacy feels suffocating and invasive to men. What many men regard as masculine strength feels isolating and distant to women. Through blame and condemnation, women and men shame one another. Through compassionate communication, however, we can help one another. This mutual empowerment is in the best interests of both sexes, because when one sex suffers, the other does too.

Toward the end of our meetings, men and women inevitably become more accountable for the ways in which they contribute to the problem. Gina said, "I've never really heard the men's point of view on all this before. I must admit that I rarely give men clear signals when they say or do something that offends me."

Arthur then said, "All my life I've been trained that my job as a man is to keep pursuing until 'no' is changed to 'yes, yes, yes.' But I hear it that when a woman says no, they want me to respect it. I get it now that what I thought was just a normal part of the dance is experienced as harassment by some women. But you know, it seems that if we're ever going to get together now, more women are going to have to start making the first moves."

After getting support from their same-sex groups and then listening to feedback from the whole group, Joel and Lisa realize that if they are both going to work full time they need to get outside help with family tasks, rather than continuing to blame and shame one another for not doing more.

Gender partnership based on strong, interactive, separate but equal gender identities can support the needs of both sexes. Becoming more affirming or supportive of our same sex doesn't have to lead to hostility toward the other sex. In fact, the acknowledgment that gender diversity is healthy may help all of us to become more tolerant toward other kinds of differences in our society.

Through gender reconciliation—both formal workshops and informal discussions—the sexes can support each other, instead of blaming one sex for not meeting the other's expectations. Men and women clearly have the capacity to move away from the sex-war rhetoric that is dividing us as well as the courage necessary to create forums for communication that can unite and heal us.

Boys and girls need regular opportunities in school to openly discuss their differing views on dating, sex, and gender roles. In universities, established women's studies courses could be complemented with men's studies, and classes in the two fields could be brought together from time to time to deepen students' understanding of both sexes. The informal discussion group is another useful format in which men and women everywhere can directly communicate with each other (see *Utne Reader* issue no. 44 [March/April 1991]). In the workplace the struggle for gender understanding needs to go beyond the simple setting up of guidelines about harassment; it is essential that women and men regularly discuss their differing views on gender issues. Outside help is often needed in structuring such discussions and getting them under way. Our organization, the Santa Barbara Institute for Gender Studies, trains and provides "reconciliation facilitators" for that purpose.

These forums must be fair. Discussions of women's wage equity must also include men's job safety. Discussions about reproductive rights, custody rights, or parental leave must consider the rights of both mothers and fathers—and the needs of the children. Affirmative action to balance the male-dominated political and economic leadership must also bring balance to the female-dominated primary-education and social-welfare systems.

We call for both sexes to come to the negotiating table from a new position of increased strength and self-esteem. Men and women do not need to become more like one another, merely more deeply themselves. But gender understanding is only a step on the long road that must ultimately lead to fundamental institutional change. We would hope, for example, that in the near future men and women will stop arguing about whether women should go into combat and concentrate instead on how to end war. The skills and basic attitudes that will lead to gender peace are the very ones we need in order to meet the other needs of our time—social, political, and environmental—with committed action.

MEN

Tomorrow's second sex

The signs are everywhere in America and Europe: more women at work; girls doing better in school; debate about "feminisation" in America's politics; its "million-man march" last year. This article summarises the evidence of a growing social problem: uneducated, unmarried, unemployed men

THESE four pages nail the following arguments to the door of debate:

- that boys are doing worse than girls at every age in school, except university where girls are narrowing the gap;
- that women dominate the jobs that are growing, while men (especially those with the least education) are trapped in jobs that are declining;
- that, for some reason, men are not even trying to do "women's work";
- that there is a loose connection between work and marriage: joblessness reduces the attractiveness of men as marriage partners;
- that men do not necessarily adopt "social behaviour" (obeying the law; looking after women and children) if left to themselves; rather, they seem to learn it through some combination of work and marriage (this is a matter of anthropological observation rather than statistical proof);
- and hence, putting these claims together, that men pose a growing problem. They are failing at school, at work and in families. Their failure shows up in crime and unemployment figures. The problem seems to be related in some way to male behaviour and instincts. It is more than merely a matter of economic adjustment. And (considering the growth in "knowledge-based" employment) it is likely to get worse.

The problem is already far worse in some areas than others. Over the past 30 years, professional men have been less badly affected by economic change than their unskilled brethren. And (to the limited extent they want to) some have added so-called "New Man" attitudes to their traditional breadwinning role. They have adjusted reasonably well to social and economic change. But unskilled men have lost on both counts. Traditional family values—husband winning the bread, wife watching the bairns—tend to be strongest (at least in theory) at the poorer end of the labour market. But American working-class men are increasingly unable (or unwilling) to support families; in Europe, high unemployment has fallen on such men disproportionately. And because providing for a family has been central to men's social role, finding a substitute for steady work will be an immensely hard—conceivably an impossible—task.

Trouble in class

The trouble with men appears early: at school. Though men take up half or more university places in most countries (America is an exception), at primary and secondary school girls are increasingly outperforming boys. In England and Wales, for example, girls score higher than boys in tests conducted at seven, nine, eleven and—which is less often realised—at five. In America, boys are much more likely than girls to be held back a grade and twice as likely to drop out of high school.

Both the reasons for this discrepancy and its true extent are hotly debated. In some subjects at some ages boys still do better than girls (for example, mathematics at 16). Traditionally, boys have done less well than girls before puberty but used to catch up afterwards. What is new now is that boys are no longer catching up. English and Welsh 16-year-olds take a series of tests known as GCSEs. A standard measurement is the percentage of children who achieve grades A, B, or C in five or more subjects; 48.1% of girls achieve this, compared with 39% of boys. In some of Britain's poorer areas, the disparity is greater. In Hackney, a poor part of east London, for example, a mere 14.9% of boys reached this standard, compared with 30.2% of girls.

The pattern is repeated all over Europe. In 1995, in the European Union, 124 girls got general leaving certificates to every 100 boys. The boys' narrow lead in vocational certificates—they took 5% more—does not close the gap. Girls also tend to stay in school longer: Austria and Switzerland apart, in every West-European country, more girls than boys stay on in education beyond school-leaving age (though the boys who do stay are slightly more likely to go to university, taking 51% of places).

Trouble at work

Because jobs are increasingly "knowledge-based", this disparity in educational attainment is bound to be reflected in employment once today's schoolchildren become adults. This does not necessarily mean that girls have better job prospects than boys; other factors, including sex discrimination at work, may intervene. But it does mean that girls are improving their job prospects relative to boys. Moreover, the job market is already moving the girls' way.

Between 1980 and 1992, women accounted for three-fifths of the increase in the American workforce and two-thirds of the increase in the European one. Between 1990 and 1993, in ten of the then 12 EU members, women's share of unemployment fell.

But the problem for men is not just that women are taking more jobs; it is that a significant proportion of men are dropping out of the job market altogether as women enter it. In the 1960s, almost all men worked and less than half of women. Not so now. The percentage of working-age men in the EU outside the labour force rose from just 8% in 1968 to 22% in 1993. For women, the trend was reversed, falling from 58% to 44% over the same period.

In America, the pattern is slightly different: while women's labour-force participation has risen from 43% in 1970 to about 60% now, men's has dropped relatively little from 80% to 75% (though there is an important exception: male high-school dropouts—those completing fewer than 12 years of school; in 1970, 86% were either working

Boys and girls at school

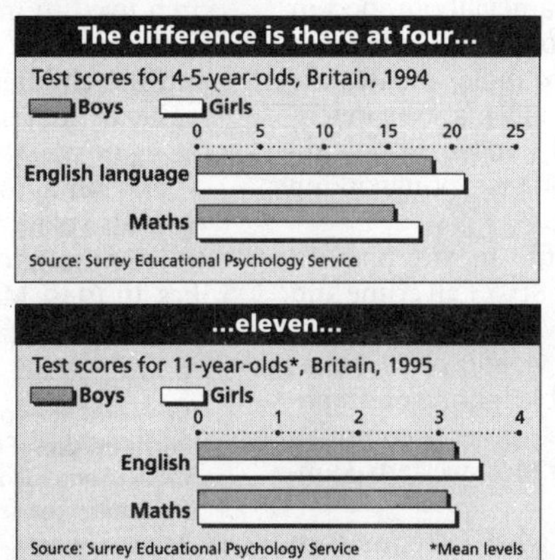

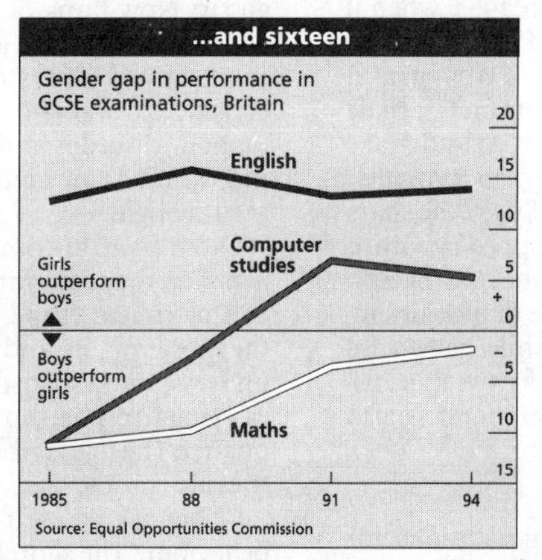

Men and women at work

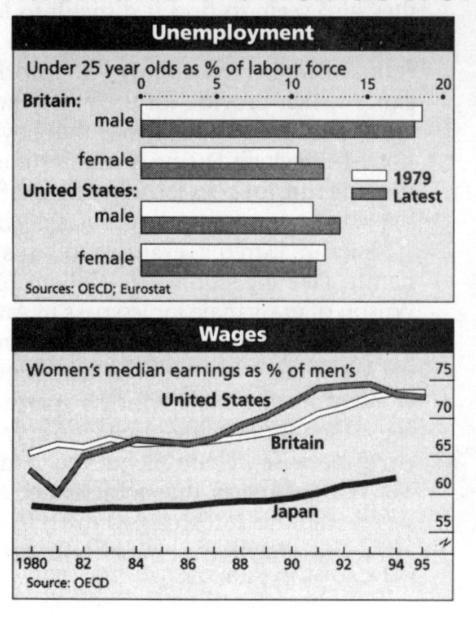

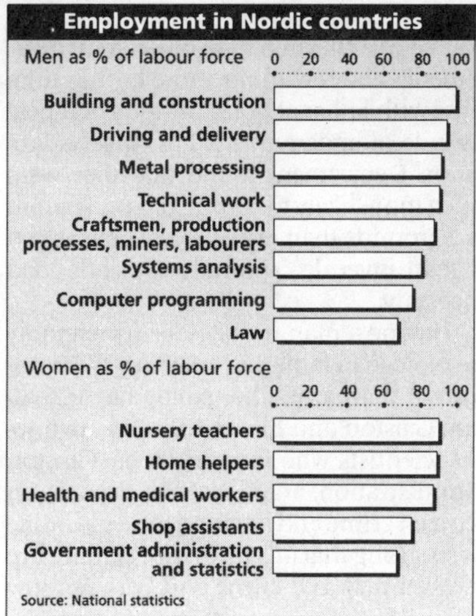

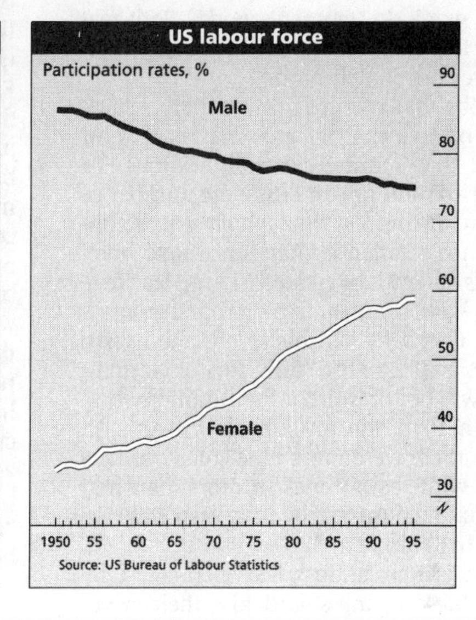

or looking for work; by 1993, only 72% were). If its employment trends continue, America will be employing nearly as many women as men by 2005.

Overall, then, the picture in the West is as follows: the labour market is increasingly friendly to women (though men still make more money and are more likely to be in work); but there are growing numbers of men outside the labour market in a way that women have been accustomed to but men are not.

The future for men looks bleaker, even if you disregard what is going on in schools. Western occupational surveys show that for the foreseeable future new job-growth will be in work typically done by women. America's Bureau of Labour, for example, forecasts that the five fastest-growing kinds of work between now and 2002 will be residential care, computer and data processing, health services, child care, and business services. Women dominate all these activities: their share of employment in them is respectively: 79%, 68%, 70%, 70% and 51%. In contrast, the five sectors declining fastest will be footwear, ammunition-making, shipbuilding, leather-working and photographic supplies. They are man's work. Men account for at least two-thirds of the workforce in all the categories.

There are numerous explanations for female success in growing businesses: women tend to be better educated; they stay in jobs longer (especially women with children); low-paid jobs are growing quickly and women are readier to accept them than men, who still see themselves as a family's breadwinner; women tend to have the social skills needed for jobs in services (though whether because of nature or nurture is disputed).

For men, the obvious response to such economic shifts would be to move into the bits of the economy which are expanding. But they are not doing this. Social theorists may like to claim that concepts like "men's work" or "women's work" are outmoded. The choices people make in the labour force tell a different story. Even in Nordic countries, which have made sexual neutrality a principle of national policy, sex segregation is the norm. Local government, state-run day-care centres, schools and social services are run by women. Men weld cars and take out the rubbish. The pattern is unchanging. In America's "administrative services" (ie, office work) men accounted for only 19.8% of the workforce in 1985 and 20.5% in 1995.

Why should this be? Part of the explanation no doubt lies in the advantages that women have in the workforce—especially their willingness to accept lower-paid jobs. But there seems to be more to it than that. Men continue to spurn even well-paid work that is dominated by women. Less than 5% of America's registered nurses, for example, are men, though the average starting salary of a registered nurse is a comfortable $30,000-35,000. Women account for 96% of America's licensed practical nurses, a responsible but not especially highly-skilled job that pays a full-time worker about $23,000.

The picture is similar in Europe. In Britain, the proportion of men in nursing, 11.6%, has budged little since 1984, when it was 10.2%. As the EU noted in a report on "Occupational Segregation of Women and Men in the European Community", male manual workers are "willing to undertake low-paid and low-skilled jobs provided they are not feminised." In Spain, the share of male office clerks has dropped by a third since 1980—even though this has been a fast-growing field at a time of high unemployment; women, meanwhile, have withdrawn from the textile and footwear industries. In these areas, job separation seems to have increased.

Blue-collar blues

So women are catching up with men for economic reasons ("women's jobs" are growing faster than men's) and social ones (men won't do "women's work"). Both reasons hit unskilled and ill-educated men disproportionately hard.

Jobs that require some tertiary education or training are growing faster than those that require no qualifications. In America's ten largest cities, the number of jobs requiring less than a high-school education has fallen by half since 1970; two-thirds of new jobs created in America since 1989 have been professional and managerial. Germany's Ministry of Labour estimates that by 2010, only 10% of German jobs will be appropriate for unskilled workers. In 1976, the proportion was 35%. In 1970, there were more blue-collar workers than white-collar ones in more than half the OECD countries; by 1990, that was true only in Spain.

In principle, unskilled men could accept these changes and kiss their wives goodbye on the doorstep as the little woman goes off to work at the nursing home. In reality, that is not happening. Despite huge social change during the past 30 years, traditional sexual attitudes retain a stubborn hold. A survey for the EU found that more than two-thirds of Europeans (ranging from 85% in Germany to 60% in Denmark) thought it better for the mother of a young child to stay at home than the father. Mothers, said this survey, should take care of nappies, clothes and food; fathers are for money, sport and punishment.

Trouble at home

Among the poor, this combination of traditional sexual attitudes and male unemployment has been deadly to two groups: men in general in high unemployment areas and, especially, young unemployed men there. The reason is that the combination has set off a spiral of harmful and sometimes uncontrollable consequences which is tearing the web that ties together work, family and law-abiding behaviour.

Consider for a moment a neighbourhood in which most working-age women are not in paid jobs. This may conjure up a picture of tidy homes, children at play and gossip. Now think of a neighbourhood in which most men are jobless. The picture is more sinister. Areas of male idleness are considered, and often are, places of deterioration, disorder and danger. Non-working women are mothers; non-working men, a blight.

Men tend to commit most crimes. In America, they commit 81% of all crime and 87% of violent crime. Adolescent boys are the most volatile and violent of all. Those under 24 are responsible for half of America's violent crime; those under 18 commit a quarter. The figures for most western countries are comparable.

Now ask yourself what restrains such behaviour? The short answer is: a two-parent home. Without belabouring the complexity of family policies, two-parent families are demonstrably better at raising trouble-free children than one-parent ones. Fatherless boys commit more crimes than those with father at home; a study of repeat juvenile offenders by the Los Angeles Probation Department found that they were much more likely to come from one-parent backgrounds than either the average child or than juvenile criminals who offended once only.

Having a man in the house (preferably the biological father) is, it seems, more important than any other single factor. William Galston and Elaine Kamarck, two social scientists who worked in the Clinton administration, argue that the connection between crime and having a father at home "is so strong that [it] erases the relationship between race and crime and between low income and crime." That is why it is a worry that, in America in 1991, just 50.8% of children lived in traditional nuclear families (families where both parents were present and the children were the biological offspring of both parents). Among Hispanics, the figure was 38%; among blacks, 27%.

But family is not the end of the matter. Work also plays a part, both in its own right and as a means of keeping men tied to families. In 1949, Margaret Mead, an eminent anthropologist, argued that

> In every known human society, everywhere in the world, the young male learns that when he grows up, one of the things which he must do in order to be a full member of society is to provide food for some female and her young... Every known human society rests firmly on the learned nurturing behaviour of men.

When men find it impossible to provide, they also seem to find it difficult to learn the nurturing bits. They may retreat into fundamentalist masculinity—the world of gangs which provide for their members a kind of rule-based behaviour that boys do not get elsewhere. For everyone else (and, in the long run, for boys too), the effects of failing to learn nurturing are universally bad.

For an extreme example of this dynamic, take the studies by William Julius Wilson of mass male joblessness in American inner cities*. Here, for the first time in the West, most men are not working and women are the breadwinners (partly because they are working more than ever and partly because welfare cheques go to them). Mr. Wilson argues that joblessness, espe-

* "When Work Disappears", by William Julius Wilson. Alfred A. Knopf. 352 pages. $26.

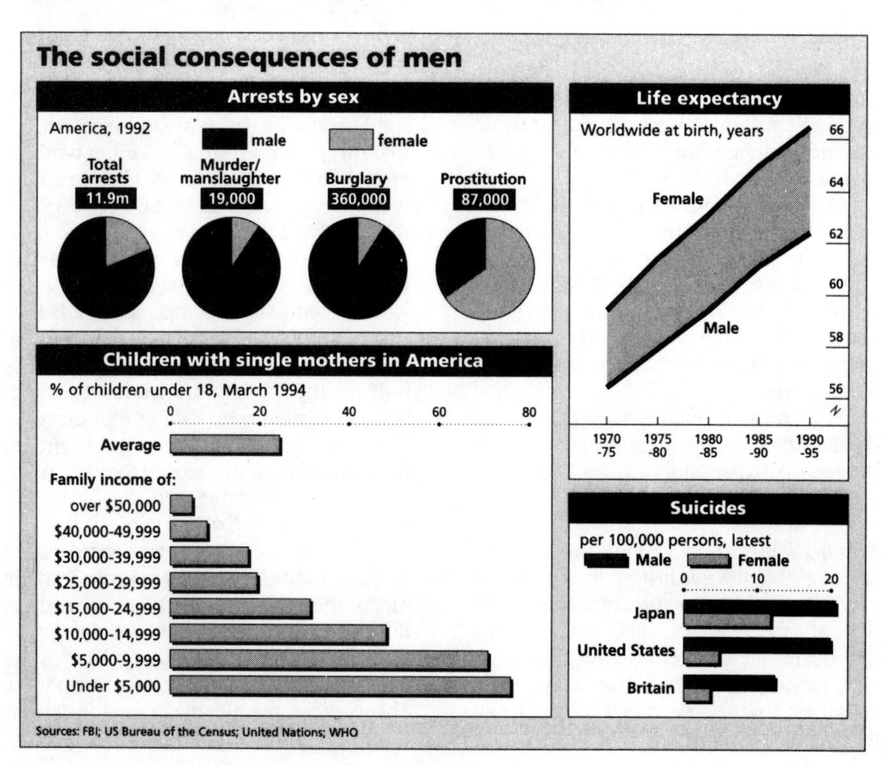

Sources: FBI; US Bureau of the Census; United Nations; WHO

cially among young men, not poverty is a prime cause of the disintegration of inner cities: "high rates of neighbourhood poverty are less likely to trigger problems of social organisation if the residents are working." Mass unemployment, he claims, destroys the institutions that enforce social behaviour—small firms, clubs, informal networks and, above all, the family.

Mr. Wilson demonstrates that, for men, employment is strongly linked to marriage and fatherhood; for a woman, children and work are separate (often competing) worlds. Men who cannot support a family are much less likely to form one; their attractiveness as a marriage partner sags. Among the inner-city blacks in Chicago whom he studied, almost 60% of those aged 18-44 have never been married and marriage rates are even lower among jobless black fathers than among employed black fathers. And this is true in America as a whole. Black men born in the early 1940s, who came of age during an era of full employment, were more than twice as likely to marry as those born since the late 1950s, who joined the workforce at a time when blue-collar jobs were falling.

So, in the language of social science, "The uncertainties in the labour market are carrying over into uncertainties in the marriage market," as John Ermisch of Britain's Economic and Social Research Centre puts it. Work, it seems, helps determine other aspects of men's lives.

Obviously, the impact upon a neighbourhood of large numbers of jobless single men is influenced by other factors, especially the prevalence of sophisticated deadly weapons. Europe has nothing like Chicago's South Side, the so-called "black belt" that is Mr. Wilson's particular area of study. Compared with the South Side, drug use is lower in European cities; gun ownership is much lower; crime is lower; fear of crime is lower. The living standards of the poorest are generally higher. Schools do not have metal detectors.

All the same, if male joblessness is the crucial factor in neighbourhood decline (as Mr. Wilson claims), then Europe faces similar problems, though they may be bottled up. Not only is average unemployment more than twice as high as in America, but a jobless European stays that way longer. Of Europe's 19m unemployed, more than half have been out of work for a year or more, compared with about 10% of Americans. And young people make up a disproportionate share of Europe's unemployed (except in Germany) and take even longer to get a job. In France, unemployment among under 25-year-olds is 27%, compared with a national average of 12.5%. More than 60% of European youths who are long-term unemployed have never had work; 40% will take two years or more to find their first job. You do not need a sociology degree to worry about the effects of so many young men with nothing to do.

The Japanese solution? No, thanks

There is one rich country that does not have these problems; where just 1% of children are born to single mothers; where crime is low; where marriage rates are relatively high—and where the labour market is rigged in favour of men. This is Japan. In any recession, the "office flowers" are made redundant first. Women are expected to give up jobs on marriage. In the professions, there is not so much a glass ceiling as a concrete one: hard to miss, painful to hit.

The trouble is that the West is unlikely to copy such a system, which is showing cracks in Japan, too. Just because men—or some of them—are struggling in work and at home, women are not about to stride back to the past, accepting the kitchen and nursery as their allotted spheres.

That is all to the good. But there has been a loser in women's march to something closer to equality and that is the man in the blue-collar uniform. Many of the gains that the West has made through enhancing the economic position of women will be tarnished if the male labourer is pushed to the margins. Once known as the salt of the earth, at the moment his troubles are making countries lose their savour.

Crisis of Community

MAKE AMERICA WORK FOR AMERICANS

WILLIAM RASPBERRY, *Columnist for* Washington Post

Delivered at the Landon Lecture Series at Kansas State University, Manhattan, Kansas, April 13, 1995

I'VE been writing a good deal of late about the violence in our streets, the apathy in our schools, and the hopelessness among our young people — the crisis in our community. But this morning, I want to talk about a deeper, more pervasive and ultimately more serious crisis. Let me call it our crisis of community.

America has a crisis of community that is as deep and wide as it is unnoticed. And it threatens to destroy our solidarity as a nation, in much the same fashion as a similar crisis in community has ripped apart the former Soviet Union and what used to be Yugoslavia.

I refer, of course, to the gender wars newly resurrected by the latest battles in the Clarence Thomas/Anita Hill holy wars; to the ethnic battles over university canons and multi-culturalism, to the political warfare that makes party advantage more important than the success of the nation, and to the racial animosities and suspicions fueled by everything from the rantings of Khalid Abdul Muhammed to the O.J. Simpson trial to Charles Murray's pseudo-intellectual call for racial abandonment.

But when I express my fear that we are coming unglued, I'm thinking about far more than these things.

I'm talking about more even than the normal give and take among the various sectors and ideologies of the society. I am talking about our growing inability to act — even to *think* — in the interest of the nation.

It's almost as though there IS no national interest, apart from the aggregate interests of the various components. The whole society seems to be disintegrating into special interests.

And not just in politics. College campuses are being ripped apart by the insistence of one group after another on proving their victimization at the hands of white males, and therefore their right to special exemptions and privileges.

One example of what I'm talking about: A few years ago, the Federal Aviation Administration adopted a rule that would bar emergency exit row seating to passengers who are blind, deaf, obese, frail or otherwise likely to inhibit movement during an emergency evacuation. Common sense? Only if you think of the common interests of all the passengers.

Surely it is reasonable to have those emergency seats occupied by people who can hear the instructions of the crew, read the directions for operating the emergency doors and assist other passengers in their escape.

But some organizations representing the deaf, blind and otherwise disabled reacted to the regulation only as a form of discrimination against their clients who, they insist, have a "right" to the emergency seats.

It is true that the majority must never be allowed to run roughshod over the rights of minorities. That is one of the tenets of the American system. But the notion of fairness to particular groups as an element of fairness to the whole has been perverted into a wholesale jockeying for group advantage.

Mutual fairness, with regard to both rights and responsibilities, can be the glue that bonds this polyglot society into a nation. Single-minded pursuit of group advantage threatens to rip us apart at the seams. The struggle for group advantage has us so preoccupied with one another's ethnicity that we are losing our ability to deal with each other as fellow humans.

What are we to make of this dismaying evidence that the relationships among us are getting worse — even among our college students? I believe two things are happening, and that they reinforce one another. The first is the racism and bigotry that never went away, even though it was relatively quiet for a time.

The second is what has been called the politics of difference. There is a pattern I have seen repeated on campuses across America. A black group, perhaps motivated by some combination of discomfort and rejection, goes looking (always successfully) for demonstrable evidence of racism.

I used to marvel at this search. Of course there was racism on campus, but what was the point of PROSPECTING for it, as though panning for gold?

I mean, where was the assay office to which one took these nuggets of racism and traded them in for something of value?

Well, it turns out that there IS such an assay office. It's called the Administration Building. Turn in enough nuggets and you get your reward: a Black Student Union, a special course offering, an African American wing in a preferred dormitory — whatever. All it takes is proof that you are a victim.

But despite the reports one hears these days, college students aren't exactly stupid. They are bright enough to see that there are rewards in the politics of difference, in demonstrated victimism. So the victories won by black students become models for similar prizes for gay students or Hispanic students or female students, all of whom gather up their nuggets of victimism and take them to the administration building for redemption.

Cornell University, one of the finest institutions in America, has a dormitory called Ujamaa College, a residence for black students; Akwe:kon, a dorm for Native Americans, and also the Latin Living Center.

That's the trend when the accent is on difference. And finally, it turns out that everybody gets something out of the politics of difference except white males, who start to feel sorry for themselves.

And if they can't find anyone to reward them for their sense of being slighted, they may turn to behavior that was once unthinkable — the "acting out" that manifests itself in incivility, reactionary politics, open bigotry and, on occasion, violence.

Every gain by minority groups justifies the sense of victimism on the part of white males, and every repugnant act of white males becomes a new nugget for a minority to take to the assay office.

Two things get lost in this sad ritual. The first is that the administration seldom gives up any of its own power: the gains of one group of students are extracted from other groups of students, who then must play up their own disadvantage to wrest some small advantage from another group. The administration's power remains intact.

The second overlooked aspect is that the process turns the campus into warring factions — each, no doubt, imagining itself as the moral successor to the heroes of the Civil Rights Movement. There's a difference, though. Dr. King's constantly repeated goal was not special advantage but unity. His dream was not of a time when blacks would finally overcome whites; his dream was that we should overcome, black and white together.

His hope was not that we should celebrate our differences but that we should recognize the relative unimportance of these differences. The differences do not *seem* unimportant, of course. Sometimes we seem to notice ONLY our differences.

That's why I find it helpful to look at what used to be the Soviet Union and what used to be Yugoslavia. From this distance, it seems clear that the similarities between the Serbs and Croats and other ethnic neighbors in Bosnia-Herzegovinia should outweigh their differences.

They share the history of a place and indeed many were intermarried. But now that Yugoslavia has broken up, even the marriages have been ripped apart.

I find myself wishing these erstwhile Yugoslavs could see for themselves what distance makes clear to us. And I wish we could learn to see ourselves as from a distance. Maybe we'd learn to appreciate how great are our similarities and how trivial our differences, and get OUR act together.

A "Star Trek" episode of some years ago makes my point. Capt. Kirk and his crew rescue a humanoid who, on his left side, is completely black. His right side, it turns out, is altogether white.

They are in the process of trying to learn the origins of this stranger — Lokai, he is called — when they are confronted by a similar humanoid named Bele — this one black on his right side and white on the left. The Enterprise crew, of course, can hardly tell them apart. But the humanoids can see themselves only as complete opposites — which, of course in one sense they are. And not just opposites. Though they are from the same planet, they are also sworn enemies.

I won't try to tell you the whole episode, but let me recall this much. Lokai is thought to be a political traitor, and Bele, an official of their home planet's Commission on Political Traitors, has been chasing him throughout the galaxy for a thousand years.

Lokai tries to convince the Enterprise crew that Bele and his kind are murderous oppressors. Bele counters that Lokai and his kind are ungrateful savages. The Enterprise crew decides to travel near the strangers' planet.

When they come within sensor range they are surprised to learn there is no sapient life there. The cities are intact, vegetation and lower animals abound, but the people are dead. They have annihilated each other. These two have survived only because they happened to be in the business of chasing each other down.

And what do they do when they learn what has happened to their planet? They lunge at each other in furious battle. Though the Enterprise crew is appalled, Kirk is unable to convince the two enemies of the futility of their war.

"To expect sense from two mentalities of such extreme viewpoints is not logical," says Spock. "They are playing out the drama of which they have become the captives, just as their compatriots did."

"But their people are dead," Sulu says slowly. "How can it matter to them now which one is right?"

"It does to them," says Spock. "And at the same time, in a sense, it doesn't. A thousand years of hating and running have become all of life."

We don't learn from this "Star Trek" episode the nature of the original problem between these warring humanoids, though we can be certain each felt fully justified in continuing the war. They had made a mistake that too many of us make in real life: They had forgotten the difference between problems and enemies.

And so have we. Virtually every issue that strikes us as urgent or important is made more intractable by our insistence on seeing it as a matter of us against them.

Give us a problem, and we'll find an enemy. Is the U.S. economy in trouble? Make the Japanese the enemy. Are we concerned about the discouraged and dangerous underclass? Blame white racists.

Members of my own profession seem unable to tell a story, no matter how significant, unless they can transform it into a case of one person, or one group, against another — unless they can make it a matter of enemies.

It is not so much that the enemies we identify are innocent as that identifying and pursuing them takes time and attention away from the search for solutions.

It was no trouble at all to come up with evidence that the Japanese were hurting the American economy through predatory pricing, product dumping and nonreciprocity, and certainly all these things merited attention.

But the U.S. auto industry improved its position relative to Japan's auto industry not when we all became expert at bashing our Japanese enemy but when Detroit started making better cars.

And that's the point. The failure to distinguish between the enemy and the problem has us looking balefully at one another instead of jointly attacking the problem which, in most cases, is as much a problem for us as for those we attack.

Take the current fight over affirmative action, for instance. Politicians who lack the imagination to address the *problem* settle for giving us each other to attack. White men — particularly those with a high school education or less — are not imagining things when they feel less secure economically than their fathers were. But they make a mistake when they suppose that their jobs have somehow been handed over to black people in the name of affirmative action. More likely those jobs are in Taiwan or Singapore or have gone up in the smoke of corporate mergers and downsizing. We've got a problem, and we waste our time assaulting enemies.

Honest communication about the problem might lead us to look for ways to restore our industrial base, expand our economy, improve the quality of our products and put our people to work. Focusing on enemies produces stirring speeches and little else.

You've heard the speeches. You've watched as communities have been ripped apart by those who deliver these speeches. There's how Teresa Heinz, widow of the late Pennsylvania senator, described them in a recent speech:

> ". . . critical of everything, impossible to please, indifferent to nuance, incapable of compromise. They laud perfection but oddly never see it in anybody but themselves. They are right all the time, eager to say I told you so, and relentlessly unforgiving. They occasionally may mean well, but the effect of even their good intentions is to destroy. They corrode self-confidence and good will; they cultivate guilt; they rule by fear and ridicule.
>
> "They are creatures of opportunity as much as of principles, extremists of the left and the right who feed on our fear and promote it, who dress up their opponents in ugly costumes, who drive a bitter wedge between us and the Other, the one not like us, the one who sees the world just a shade differently. . . . They demonize us by our parts and tear our country into pieces."

My own formulation is less eloquent; they focus on enemies rather than on problems. They forget that, at the end of the day, when we've all taken our unfair shots at one another, this simple truth remains: The *problem* is the problem.

Our politicians and our factional leaders never miss an opportunity to list the atrocities the *enemy* has committed against us. But nothing changes.

Sometimes we're not even sure what we want to change, or what we want the people we call enemies to do. We say we want things to get better, when sometimes I think we only want to score points.

We say we want a society in which all of us can live together as brothers and sisters, and the whole time we are saying it we are busy creating another group of barriers to place between us.

It's a strange sort of progress we have made since the death of Dr. King. We have "progressed" to the point where we are embarrassed to speak of brotherhood, of black and white together, of our shared status as Americans.

That's not an accusation; it's a confession. All of us are capable of getting so caught up in the distance that remains to be run that we forget to give ourselves full credit for the distance we've come.

Yet, every now and then, we manage to overcome our embarrassment and see things as from a distance. In that spirit, I'd like to share something I wrote a while back — something I still believe but something I may have trouble saying again.

Here it is: The immigration applications, the legal and illegal dodges for getting into this country, the longings you hear in virtually every other part of the world all attest to two astounding facts.

The first, widely accepted though not always with good grace, is that "everybody" wants to be an American. The second, of which we take almost no notice, is that virtually anybody can *become* an American.

To see just how extraordinary a fact that is, imagine hearing anyone — black, white or Asian — saying he wants to "become Japanese." It sounds like a joke. One can *live* in Japan (or Ghana or Sweden or Mexico) — can live there permanently, and prosper. But it's essentially impossible to imagine anyone born anywhere else becoming anything else — except American.

It's a thought that crosses my mind whenever I hear demands that the government protect the ethnic or language heritage of particular groups: when African Americans demand that the *public* schools adopt an Afrocentric curriculum, for instance, or when immigrants from Latin America are sworn in as American citizens — in Spanish.

It crossed my mind again when I came across Jim Sleeper's essay, "In Defense of Civic Culture."

I won't try to characterize Sleeper's piece or to summarize its recommendations. [the Washington-based Progressive Foundation] I won't even tell you I agree with everything Sleeper has to say on the subject of race and ethnicity.

But he says some things that echo my own feelings, especially when I ponder the extraordinary possibility of becoming American.

He acknowledges the obvious: that the America that counted my great-great-grandfather as only three-fifths of a human being has never been free of ethnic and racial bigotry, and that that bigotry has sometimes achieved the status of law, of philosophy — even of religion.

But he notes something else: that America is one of the few places on the globe where accusation of such bigotry is a serious indictment. Even when America has been at its ugliest in fact — slavery, the slaughter of Native Americans, the internment of the Japanese and the full range of private and public atrocities, "yet always America held out the promise that, as Ralph Waldo Emerson put it, 'in this asylum of all nations, the energy of . . . all the European tribes [and] of the Africans, and of the Polynesians will construct a new race.'"

The civic culture Sleeper writes about includes this notion of Americans as a new and different race, but it also entails what he describes as characteristic American virtues: tolerance, optimism, self-restraint, self-reliance, reason, public-mindedness — virtues that are "taught and caught in the daily life of local institutions and in the examples set by neighbors, co-workers and public leaders."

It is, he suggests, the internalizing of these virtues that defines "becoming American."

But the transformation works both ways. If people from an awesome range of colors, cultures and ethnicities have become Americans, so has America become what it is (and continues to become) by absorbing and embracing these myriad influences.

Some of us are angry, and ought to be, that our academic texts and teachings still disregard or underestimate our part of these influences.

Some of us are disappointed that what we bring to the smorgasbord is often undervalued, even brutally rejected.

But surely the cure is in working for greater inclusion, not cultural isolation. That's what observers as different as Sleeper, Arthur Schlesinger and John Gardner have been saying. That's what Gary Trudeau was saying in that hilarious (and sobering) series of "Doonesbury" strips that ended with black students — already having attained their separate courses and dormitories — demanding, at last, separate drinking fountains. Sleeper's insight is that there is nothing "natural" or automatic about those values and attitudes that used to be called "the American way." Educators must teach them, he says, and also "teach that self-esteem is enhanced not simply through pride in one's own cultural origins but, more importantly, by taking pride in one's mastery of civic virtues and graces that all Americans share and admire in building our society."

Critics of this view will argue that Sleeper's virtuous and graceful American is a figment, that America is a deeply — perhaps irredeemably — racist society.

I prefer to think that Americans are still becoming Americans, just as America is still becoming America.

How can we accelerate that becoming? By recognizing its importance, by understanding that hating and running must not become all of life, and by working to grasp the difference between problems and enemies.

Confront a difficulty as a problem, and you have taken the first steps toward creating the climate for change.

Confront it as the work of enemies and you create the necessity for DEFEATING someone, of intimidating someone, of browbeating someone into doing something against his will.

Enemies have to be sought out, branded and punished. Which, naturally, gives them one more reason to find an opportunity to strike back at us. And the beat goes on.

Problems, on the other hand, admit of cooperative solutions that can help build community.

Searching for enemies is most often a pessimist's game, calculated less to resolve difficulties than to establish that the difficulties are someone else's fault. Identifying *problems* is by its very nature optimistic and healing. The whole point of delineating problems is to fashion solutions.

Maybe that's what President Clinton had in mind when he called on America to bring back "the old spirit of partnership, of optimism, of renewed dedication to common efforts."

"We need," he said, "an array of devoted, visionary, healing leaders throughout this nation, willing to work in their communities to end the long years of denial and neglect and divisiveness and blame, to give the American people their country back."

And that is precisely what we need. America has had enough of the politics of difference, the marketing of disadvantage, the search for enemies. It's about time we started to work on what may be the most important problem we face:

How to heal our crisis of community and make America work — not for blacks or whites or women or gays; not for ethnics; not for Christians, Moslems or Jews — but for Americans.

THE STRANGE DISAPPEARANCE OF CIVIC AMERICA

ROBERT D. PUTNAM

For the last year or so, I have been wrestling with a difficult mystery. It is a classic brainteaser, with a corpus delicti, a crime scene strewn with clues, and many potential suspects. As in all good detective stories, however, some plausible miscreants turn out to have impeccable alibis, and some important clues hint at portentous developments that occurred before the curtain rose.

The mystery concerns the strange disappearance of social capital and civic engagement in America. By "social capital," I mean features of social life—networks, norms, and trust—that enable participants to act together more effectively to pursue shared objectives. (Whether or not their shared goals are praiseworthy is, of course, entirely another matter.) I use the term "civic engagement" to refer to people's connections with the life of their communities, not only with politics.

Although I am not yet sure that I have solved the mystery, I have assembled evidence that clarifies what happened. An important clue, as we shall see, involves differences among generations. Americans who came of age during the Depression and World War II have been far more deeply engaged in the life of their communities than the generations that have followed them. The passing of this "long civic generation" appears to be an important proximate cause of the decline of our civic life. This discovery does not in itself crack the case, but when combined with other data it points strongly to one suspect against whom I shall presently bring an indictment.

A more extended version of this article, complete with references, appears in the Winter 1995 issue of PS, a publication of the American Political Science Association. This work, originally delivered as the inaugural Ithiel de Sola Pool Lecture, builds on Putnam's earlier articles, "Bowling Alone: America's Declining Social Capital," Journal of Democracy *(January 1995) and "The Prosperous Community,"* TAP *(Spring 1993).*

Evidence for the decline of social capital and civic engagement comes from a number of independent sources. Surveys of average Americans in 1965, 1975, and 1985, in which they recorded every single activity during a day—so-called "time-budget" studies—indicate that since 1965 time spent on informal socializing and visiting is down (perhaps by one-quarter) and time devoted to clubs and organizations is down even more sharply (by roughly half). Membership records of such diverse organizations as the PTA, the Elks club, the League of Women Voters, the Red Cross, labor unions, and even bowling leagues show that participation in many conventional voluntary associations has declined by roughly 25 percent to 50 percent over the last two to three decades. Surveys show sharp declines in many measures of collective political participation, including attending a rally or speech (off 36 percent between 1973 and 1993), attending a meeting on town or school affairs (off 39 percent), or working for a political party (off 56 percent).

Some of the most reliable evidence about trends comes from the General Social Survey (GSS), conducted nearly every year for more than two decades. The GSS demonstrates, at all levels of education and among both men and women, a drop of roughly one-quarter in group membership since 1974 and a drop of roughly one-third in social trust since 1972. (Trust in political authorities, indeed in many social institutions, has also declined sharply over the last three decades, but that is conceptually a distinct trend.) Slumping membership has afflicted all sorts of groups, from sports clubs and professional associations to literary discussion groups and labor unions. Only nationality groups, hobby and garden clubs, and the catch-all category of "other" seem to have resisted the ebbing tide. Gallup polls report that church attendance fell by roughly 15 percent during the 1960s and has remained at that lower level ever since, while data from the National Opinion

Research Center suggest that the decline continued during the 1970s and 1980s and by now amounts to roughly 30 percent. A more complete audit of American social capital would need to account for apparent countertrends. Some observers believe, for example, that support groups and neighborhood watch groups are proliferating, and few deny that the last several decades have witnessed explosive growth in interest groups represented in Washington. The growth of such "mailing list" organizations as the American Association of Retired People and the Sierra Club, although highly significant in political (and commercial) terms, is not really a counterexample to the supposed decline in social connectedness, however, since these are not really associations in which members meet one another. Their members' ties are to common symbols and ideologies, but not to each other. Similarly, although most secondary associations are not-for-profit, most prominent nonprofits (from Harvard University to the Ford Foundation to the Metropolitan Opera) are bureaucracies, not secondary associations, so the growth of the "third sector" is not tantamount to a growth in social connectedness. With due regard to various kinds of counterevidence, I believe that the weight of available evidence confirms that Americans today are significantly less engaged with their communities than was true a generation ago.

Of course, American civil society is not moribund. Many good people across the land work hard every day to keep their communities vital. Indeed, evidence suggests that America still outranks many other countries in the degree of our community involvement and social trust. But if we examine our lives, not our aspirations, and if we compare ourselves not with other countries but with our parents, the best available evidence suggests that we are less connected with one another.

Reversing this trend depends, at least in part, on understanding the causes of the strange malady afflicting American civic life. This is the mystery I seek to unravel here: Why, beginning in the 1960s and accelerating in the 1970s and 1980s, did the fabric of American community life begin to fray? Why are more Americans bowling alone?

The Usual Suspects

Many possible answers have been suggested for this puzzle:

- busy-ness and time pressure;
- economic hard times (or, according to alternative theories, material affluence);
- residential mobility;
- suburbanization;
- the movement of women into the paid labor force and the stresses of two-career families;
- disruption of marriage and family ties;
- changes in the structure of the American economy, such as the rise of chain stores, branch firms, and the service sector;
- the sixties (most of which actually happened in the seventies); including
    - Vietnam, Watergate, and disillusion with public life; and
    - the cultural revolt against authority (sex, drugs, and so on);
- growth of the welfare state;
- the civil rights revolution;
- television, the electronic revolution, and other technological changes.

The classic questions posed by a detective are means, motive, and opportunity. A solution, even a partial one, to our mystery must pass analogous tests.

Is the proposed explanatory factor correlated with trust and civic engagement? If not, that factor probably does not belong in the lineup. For example, if working women turn out to be more engaged in community life than housewives, it would be harder to attribute the downturn in community organizations to the rise of two-career families.

Is the correlation spurious? If parents, for example, were more likely than childless people to be joiners, that might be an important clue. However, if the correlation between parental status and civic engagement turned out to be entirely spurious, due to the effects of (say) age, we would have to remove the declining birth rate from our list of suspects.

Is the proposed explanatory factor changing in the relevant way? Suppose, for instance, that people who often move have shallower community roots. That could be an important part of the answer to our mystery *only if* residential mobility itself had risen during this period.

Is the proposed explanatory factor vulnerable to the claim that it might be the result *of civic disengagement, not the* cause? For example, even if newspaper readership were closely correlated with civic engagement across individuals and across time, we would need to weigh the degree to which reduced newspaper circulation is the result (not the cause) of disengagement.

Against those benchmarks, let us weigh the evidence. But first we must acknowledge a trend that only complicates our task.

Education Deepens the Mystery

Education is by far the strongest correlate that I have discovered of civic engagement in all its forms, including social trust and membership in many different types of groups. In fact, the effects of education become greater and greater as we move up the educational ladder. The four years of education between 14 and 18 total years have *ten times more impact* on trust and membership than the first four years of formal education. This curvilinear pattern applies to both men and women, and to all races and generations.

Sorting out just why education has such a massive effect on social connectedness would require a book in itself. Education is in part a proxy for social class and economic differences, but when income, social status, and education are used together to predict trust and group membership, education continues to be the primary influence. So, well-educated people are much more likely to be joiners and trusters, partly because they are better off economically, but mostly because of the skills, resources, and inclinations that were imparted to them at home and in school.

The expansion of high schools and colleges earlier this century has had an enormous impact on the educational composition of the adult population during just the last two decades. Since 1972 the proportion of adults with fewer than 12 years of education has been cut in half, falling from 40 percent to 18 percent, while the proportion with more than 12 years has nearly doubled, rising from 28 percent to 50 percent, as the generation of Americans educated around the turn of this century (most of whom did not finish high school) died off and were replaced by the baby boomers and their successors (most of whom attended college).

So here we have two facts—education boosts civic engagement sharply, and educational levels have risen massively—that only deepen our central mystery. By itself, the rise in educational levels should have increased social capital during the last 20 years by 15–20 percent, even assuming that the effects of education were merely linear.... By contrast, however, the actual GSS figures show a net decline since the early 1970s of roughly the same magnitude (trust by about 20–25 percent, memberships by about 15–20 percent). The relative declines in social capital are similar within each educational category—roughly 25 percent in group memberships and roughly 30 percent in social trust since the early 1970s, and probably even more since the early 1960s.

While this first investigative foray leaves us more mystified than before, we may nevertheless draw two useful conclusions. First, we need to take account of educational differences in our exploration of other factors to be sure that we do not confuse their effects with the consequences of education. And, second, the mysterious disengagement of the last quarter century seems to have afflicted all educational strata in our society, whether they have had graduate education or did not finish high school.

Mobility and Suburbanization

Many studies have found that residential stability and such related phenomena as homeownership are associated with greater civic engagement. At an earlier stage in this investigation I observed that "mobility, like frequent re-potting of plants, tends to disrupt root systems, and it takes time for an uprooted individual to put down new roots." I must now report, however, that further inquiry fully exonerates residential mobility from any responsibility for our fading civic engagement.

Data from the U.S. Bureau of the Census 1995 (and earlier years) show that rates of residential mobility have been remarkably constant over the last half century. In fact, to the extent that there has been any change at all, both long-distance and short-distance mobility have declined over the last five decades. During the 1950s, 20 percent of Americans changed residence each year and 6.9 percent annually moved across county borders; during the 1990s, the comparable figures are 17 percent and 6.6 percent. Americans, in short, are today slightly more rooted residentially than a generation ago. The verdict on mobility is unequivocal: This theory is simply wrong....

Pressures of Time and Money

Americans certainly *feel* busier now than a generation ago: The proportion of us who report feeling "always rushed" jumped by half between the mid-1960s and the mid-1990s. Probably the most obvious suspect behind our tendency to drop out of community affairs is pervasive busy-ness. And lurking nearby in the shadows are economic pressures so much discussed nowadays, from job insecurity to declining real wages.

Yet, however culpable busy-ness and economic insecurity may appear at first glance, it is hard to find incriminating evidence. In the first place, time-budget studies do not confirm the thesis that Americans are, on average, working longer than a generation ago. On the contrary, a new study by

John Robinson and Geoffrey Godbey of the University of Maryland reports a five hour per week *gain* in free time for the average American between 1965 and 1985, due partly to reduced time spent on housework and partly to earlier retirement. Their claim that Americans have more leisure time now than several decades ago is, to be sure, contested by other observers, notably Juliet Schor, who in her 1991 book *The Overworked American* reports evidence that work hours are lengthening, especially for women.

But whatever the resolution of that controversy, other data call into question whether longer hours at work lead to lessened involvement in civic life or reduced social trust. Results from the GSS show that employed people belong to somewhat more groups than those outside the paid labor force. Even more striking is the fact that among workers, longer hours are linked to more civic engagement. The patterns among men and women on this score are not identical: Women who work part-time appear to be somewhat more civicly engaged and socially trusting than either those who work full-time or those who do not work outside the home at all—an intriguing anomaly, though not relevant to our basic puzzle, since female part-time workers constitute a relatively small fraction of the American population, and the fraction is growing, up from about 8 percent to about 10 percent between the early 1970s and early 1990s.

But what do workaholics do less? Robinson reports that, unsurprisingly, people who spend more time at work do feel more rushed, and these harried souls do spend less time eating, sleeping, reading books, engaging in hobbies, and just doing nothing. Compared to the rest of the population, they also spend a lot less time watching television, almost 30 percent less. However, they do not spend less time on organizational activity. In short, those who work longer forego *Nightline*, but not the Kiwanis club; *ER*, but not the Red Cross.

So hard work does not *prevent* civic engagement. Moreover, the nationwide falloff in joining and trusting is perfectly mirrored among full-time workers, among part-time workers, and among those outside the paid labor force. So if people are dropping out of community life, long hours do not seem to be the reason. . . .

The Changing Role of Women

Most of our mothers were housewives, and most of them invested heavily in social capital formation—a jargony way of referring to untold unpaid hours in church suppers, PTA meetings, neighborhood coffee klatches, and visits to friends and relatives. The movement of women out of the home and into the paid labor force is probably the most portentous social change of the last half century. However welcome and overdue the feminist revolution may be, it is hard to believe that it has had no impact on social connectedness. Could this be the primary reason for the decline of social capital over the last generation?

Some patterns in the survey evidence seem to support this claim. All things considered, women belong to somewhat fewer voluntary associations than men do. On the other hand, time-budget studies suggest that women spend more time on those groups and more time in informal social connecting than men. Although the absolute declines in joining and trusting are approximately equivalent among men and women, the relative declines are somewhat greater among women. Controlling for education, memberships among men have declined at a rate of about 10-15 percent a decade, compared to about 20-25 percent a decade for women. The time-budget data, too, strongly suggest that the decline in organizational involvement in recent years is concentrated among women. These sorts of facts, coupled with the obvious transformation in the professional role of women over this same period, led me in previous work to suppose that the emergence of two-career families might be the most important single factor in the erosion of social capital.

As we saw earlier, however, work status itself seems to have little net impact on group membership or on trust. Housewives belong to different types of groups than do working women (more PTAs, for example, and fewer professional associations), but in the aggregate working women are actually members of slightly more voluntary associations (though housewives, according to Robinson and Godbey, spend more time on them). Moreover, the overall declines in civic engagement are somewhat greater among housewives than among employed women. Comparison of time-budget data between 1965 and 1985 seems to show that employed women as a group are actually spending more time on organizations than before, while housewives are spending less. This same study suggests that the major decline in informal socializing since 1965 has also been concentrated among housewives. The central fact, of course, is that the overall trends are down for all categories of women (and for men, too, even bachelors), but the figures suggest that women who work full-time actually may have been more resistant to the slump than those who do not.

Thus, although women appear to have borne a

disproportionate share of the decline in civic engagement over the last two decades, it is not easy to find any micro-level data that tie that fact directly to their entry into the labor force....

Marriage and Family

Another widely discussed social trend that more or less coincides with the downturn in civic engagement is the breakdown of the traditional family unit—mom, dad, and the kids. Since the family itself is, by some accounts, a key form of social capital, perhaps its eclipse is part of the explanation for the reduction in joining and trusting in the wider community. What does the evidence show?

First of all, evidence of the loosening of family bonds is unequivocal. In addition to the century-long increase in divorce rates (which accelerated from the mid-1960s to the mid-1970s and then leveled off), and the more recent increase in single-parent families, the incidence of one-person households has more than doubled since 1950, in part because of the rising number of widows living alone. The net effect of all these changes, as reflected in the General Social Survey, is that the proportion of all American adults currently unmarried climbed from 28 percent in 1974 to 48 percent in 1994.

Second, married men and women do rank somewhat higher on both our measures of social capital. That is, controlling for education, age, race, and so on, single people—both men and women, divorced, separated, and never married—are significantly less trusting and less engaged civicly than married people. (Multivariate analysis hints that one major reason why divorce lowers connectedness is that it lowers family income, which in turn reduces civic engagement.) Roughly speaking, married men and women are about a third more trusting and belong to about 15–25 percent more groups than comparable single men and women. (Widows and widowers are more like married people than single people in this comparison.)

> The decline of marriage is probably an accessory to the crime, not the main villain.

In short, successful marriage, especially if the family includes children, is statistically associated with greater social trust and civic engagement. Thus, some part of the decline in both trust and membership is tied to the decline in marriage. To be sure, the direction of causality behind this correlation may be complicated, since it is conceivable that loners and paranoids are harder to live with. If so, divorce may in some degree be the consequence, not the cause, of lower social capital. Probably the most reasonable summary of these arrays of data, however, is that the decline in successful marriage is a significant, though modest part of the reason for declining trust and lower group membership. On the other hand, changes in family structure cannot be a major part of our story, since the overall declines in joining and trusting are substantial even among the happily married. My own verdict (based in part on additional evidence to be introduced later) is that the disintegration of marriage is probably an accessory to the crime, but not the major villain of the piece....

Generational Effects

Our efforts thus far to identify the major sources of civic disengagement have been singularly unfruitful. In all our statistical analyses, however, one factor, second only to education, stands out as a predictor of all forms of civic engagement and trust. That factor is age. Older people belong to more organizations than young people, and they are less misanthropic. Older Americans also vote more often and read newspapers more frequently, two other forms of civic engagement closely correlated with joining and trusting.

"Civic Engagement by Age" shows the basic pattern. Civic involvement appears to rise more or less steadily from early adulthood toward a plateau in middle age, from which it declines only late in life. This humpback pattern seems naturally to represent the arc of life's engagements. That, at least, was how I first interpreted the data. But that would be a fundamental misreading of the most important clue in our whole whodunit.

Evidence from the General Social Survey enables us to follow individual cohorts as they age. If the rising lines in the figure indeed represent deepening civic engagement with age, we should be able to track this same deepening engagement as we follow, for example, the first of the baby boomers, born in 1947, as they aged from 25 in 1972 (the first year of the GSS) to 47 in 1994 (the latest year available). Startlingly, however, such an analysis, repeated for successive birth cohorts, produces virtually no evidence of such life cycle changes in civic engagement. In fact, as various generations moved through the period between 1972 and 1994, their levels of trust and membership more often fell than rose, reflecting a more or less simultaneous decline in civic engagement among young and old alike, particularly during the second half of the 1980s. . . .

The central paradox posed by these patterns is this: Older people are consistently more engaged and trusting than younger people, yet we do not become more engaged and trusting as we age. What's going on here?

Time and age are notoriously ambiguous in their effects on social behavior. Social scientists have learned to distinguish three contrasting phenomena:

Life cycle effects represent differences attributable to stage of life. In this case individuals change as they age, but since the effects of aging are, in the aggregate, neatly balanced by the "demographic metabolism" of births and deaths, life cycle effects produce no aggregate change. Everyone's close-focus eyesight worsens as we age, but the aggregate demand for reading glasses changes little.

Period effects affect all people who live through a given era, regardless of their age. Period effects can produce both individual and aggregate change, often quickly and enduringly, without any age-related differences. The sharp drop in trust in government between 1965 and 1975, for example, was almost entirely this sort of period effect, as Americans of all ages changed their minds about their leaders' trustworthiness. Similarly, as just noted, a modest portion of the decline in social capital during the 1980s appears to be a period effect.

Generational effects affect all people born at the same time. Like life cycle effects (and unlike typical period effects), generational effects show up as disparities among age groups at a single point in time, but like period effects (and unlike life cycle effects) generational effects produce real social change, as successive generations, enduringly "imprinted" with divergent outlooks, enter and leave the population. In pure generational effects, no individual ever changes, but society does.

Returning to our conundrum, how could older people today be more engaged and trusting, if they did not become more engaged and trusting as they aged? The key to this paradox, as David Butler and Donald Stokes observed in another context, is to ask, not *how old people are*, but *when they were young*. . . .

The Long Civic Generation

As we begin moving along this queue from left to right—from those raised around the turn of the century to those raised during the Roaring Twenties, and so on—we find relatively high and unevenly rising levels of civic engagement and social trust. Then rather abruptly, however, we encounter signs of reduced community involvement, starting with men and women born in the early 1930s. Remarkably, this downward trend in joining, trusting, voting, and newspaper reading continues almost uninterruptedly for nearly 40 years. The trajectories for the various different indicators of civic engagement are strikingly parallel: Each shows a high, sometimes rising plateau for people born and raised during the first third of the century; each shows a turning point in the cohorts around 1930; and each then shows a more or less constant decline down to the cohorts born during the 1960s.

By any standard, these intergenerational differences are extraordinary. Compare, for example, the generation born in the early 1920s with the generation of their grandchildren born in the late 1960s. Controlling for educational disparities, members of the generation born in the 1920s belong to almost twice as many civic associations as those born in the late 1960s (roughly 1.9 memberships per capita, compared to roughly 1.1 memberships per capita). The grandparents are more than twice as likely to trust other people (50-60 percent compared with 25 percent for the grandchildren). They vote at nearly double the rate of the most recent cohorts (roughly 75 percent compared with 40-45 percent), and they read newspapers almost three times as often (70-80 percent read a paper daily compared with 25-30 percent). And bear in mind that we have found no evidence that the youngest generation will come to match their grandparents' higher levels of civic engagement as they grow older.

Thus, read not as life cycle effects, but rather as generational effects, the age-related patterns in our data suggest a radically different interpretation of our basic puzzle. Deciphered with this key, the [evidence] depicts a long "civic" generation, born roughly between 1910 and 1940, a broad group of people substantially more engaged in community affairs and substantially more trusting than those younger than they. (Members of the 1910-1940 generation also seem more civic than their elders, at least to judge by the outlooks of relatively few men and women born in the late nineteenth century who appeared in our samples.) The culminating point of this civic generation is the cohort born in 1925-1930, who attended grade school during the Great Depression, spent World War II in high school (or on the battlefield), first voted in 1948 or 1952, set up housekeeping in the 1950s, and watched their first television when they were in their late twenties. Since national surveying began, this cohort has been exceptionally civic: voting more, joining more, reading newspapers more, trusting more. . . .

In short, the most parsimonious interpretation of the age-related differences in civic engagement is that they represent a powerful reduction in civic engagement among Americans who came of age in the decades after World War II, as well as some modest additional disengagement that affected all cohorts during the 1980s. These patterns hint that being raised after World War II was a quite different experience from being raised before that watershed. It is as though the postwar generations were exposed to some mysterious X-ray that permanently and increasingly rendered them less likely to connect with the community. Whatever that force might have been, it—rather than anything that happened during the 1970s and 1980s—accounts for most of the civic disengagement that lies at the core of our mystery.

But if this reinterpretation of our puzzle is correct, why did it take so long for the effects of that mysterious X-ray to become manifest? If the underlying causes of civic disengagement can be traced to the 1940s and 1950s, why did the effects become conspicuous in PTA meetings and Masonic lodges, in the volunteer lists of the Red Cross and the Boy Scouts, and in polling stations and church pews and bowling alleys across the land only during the 1960s, 1970s, and 1980s?

The visible effects of this generational disengagement were delayed by two important factors. First, the postwar boom in college enrollments raised levels of civic engagement, offsetting the generational trends. As Warren E. Miller and J. Merrill Shanks observe in their as yet unpublished book, *The American Voter Reconsidered*, the postwar expansion of educational opportunities "forestalled a cataclysmic drop" in voting turnout, and it had a similar delaying effect on civic disengagement more generally.

Second, the full effects of generational developments generally appear several decades after their onset, because it takes that long for a given generation to become numerically dominant in the adult population. Only after the mid-1960s did significant numbers of the "post-civic generation" reach adulthood supplanting older, more civic cohorts. . . .

In short, the very decades that have seen a national deterioration in social capital are the same decades during which the numerical dominance of a trusting and civic generation has been replaced by the dominion of "post-civic" cohorts. Moreover, although the long civic generation has enjoyed unprecedented life expectancy, allowing its members to contribute more than their share to American social capital in recent decades, they are now passing from the scene. Even the youngest members of that generation will reach retirement age within the next few years. Thus, a generational analysis leads almost inevitably to the conclusion that the national slump in trust and engagement is likely to continue, regardless of whether the more modest "period effect" depression of the 1980s continues.

Our Prime Suspect

To say that civic disengagement in contemporary America is in large measure generational merely reformulates our central puzzle. We now know that much of the cause of our lonely bowling probably dates to the 1940s and 1950s, rather than to the 1960s and 1970s. What could have been the mysterious anticivic "X-ray" that affected Americans who came of age after World War II and whose effects progressively deepened at least into the 1970s?

Our new formulation of the puzzle opens the possibility that the zeitgeist of national unity, patriotism, and shared sacrifice that culminated in 1945 might have reinforced civic-mindedness. On the other hand, it is hard to assign any consistent role to the Cold War and the Bomb, since the anticivic trend appears to have deepened steadily from the 1940s to the 1970s, in no obvious harmony with the rhythms of world affairs. Nor is it easy to construct an interpretation of the data on generational differences in which the cultural vicissitudes of the sixties could play a significant role. Neither can economic adversity or affluence easily be tied to the generational decline in civic engagement, since the slump seems to have affected in equal measure those who came of age in the placid fifties, the booming sixties, and the busted seventies.

I have discovered only one prominent suspect against whom circumstantial evidence can be mounted, and in this case, it turns out, some directly incriminating evidence has also turned up. This is not the occasion to lay out the full case for the prosecution, nor to review rebuttal evidence for the defense, but I want to present evidence that justifies indictment.

The culprit is television.

First, the timing fits. The long civic generation was the last cohort of Americans to grow up without television, for television flashed into American society like lightning in the 1950s. In 1950 barely 10 percent of American homes had television sets, but by 1959, 90 percent did, probably the fastest diffusion of a major technological innovation ever recorded. The reverberations from this lightning bolt continued for decades, as viewing hours grew by 17-20 percent during the 1960s and by an additional 7-8 percent during the 1970s. In the early years, TV watching was concentrated among the less educated sectors of the population, but during the 1970s the viewing time of the more educated sectors of the population began to converge upward. Television viewing increases with age, particularly upon retirement, but each generation since the introduction of television has begun its life cycle at a higher starting point. By 1995 viewing per TV household was more than 50 percent higher than it had been in the 1950s.

Most studies estimate that the average American now watches roughly four hours per day (excluding periods in which television is merely playing in the background). Even a more conservative estimate of three hours means that television absorbs 40 percent of the average American's free time, an increase of about one-third since 1965. Moreover, multiple sets have proliferated: By the late 1980s three-quarters of all U.S. homes had more than one set, and these numbers too are rising steadily, allowing ever more private viewing. Robinson and Godbey are surely right to conclude that "television is the 800-pound gorilla of leisure time." This massive change in the way Americans spend their days and nights occurred precisely during the years of generational civic disengagement.

Evidence of a link between the arrival of television and the erosion of social connections is, however, not merely circumstantial. The links between civic engagement and television viewing can be instructively compared with the links between civic engagement and newspaper reading. The basic contrast is straightforward: Newspaper reading is associated with high social capital, TV viewing with low social capital.

Controlling for education, income, age, race, place of residence, work status, and gender, TV viewing is strongly and negatively related to social trust and group membership, whereas the same correlations with newspaper reading are positive. Within every educational category, heavy readers are avid joiners, whereas heavy viewers are more likely to be loners. In fact, more detailed analysis suggests that heavy TV watching is one important reason *why* less educated people are less engaged in the life of their communities. Controlling for differential TV exposure significantly reduces the correlation between education and engagement.

Viewing and reading are themselves uncorrelated—some people do lots of both, some do little of either—but "pure readers" (that is, people who watch less TV than average and read more newspapers than average) belong to 76 percent more civic organizations than "pure viewers" (controlling for education, as always). Precisely the same pattern applies to other indicators of civic engagement, including social trust and voting turnout. "Pure readers," for example, are 55 percent more trusting than "pure viewers."

In other words, each hour spent viewing television is associated with less social trust and less group membership, while each hour reading a newspaper is associated with more. An increase in television viewing of the magnitude that the U.S. has experienced in the last four decades might directly account for as much as one-quarter to one-half of the total drop in social capital, even without taking into account, for example, the indirect effects of television viewing on newspaper readership or the cumulative effects of lifetime viewing hours. Newspaper circulation (per household) has dropped by more than half since its peak in 1947. To be sure, it is not clear which way the tie between newspaper reading and civic involvement works, since disengagement might itself dampen one's interest in community news. But the two trends are clearly linked.

How Might TV Destroy Social Capital?

Time displacement. Even though there are only 24 hours in everyone's day, most forms of social and media participation are positively correlated. People who listen to lots of classical music are more likely, not less likely, than others to attend Cubs games. Television is the principal exception to this generalization—the only leisure activity that seems to inhibit participation outside the home. TV watching comes at the expense of nearly every social activity outside the home, especially social gatherings and informal conversations. TV viewers are homebodies.

Most studies that report a negative correlation between television watching and community

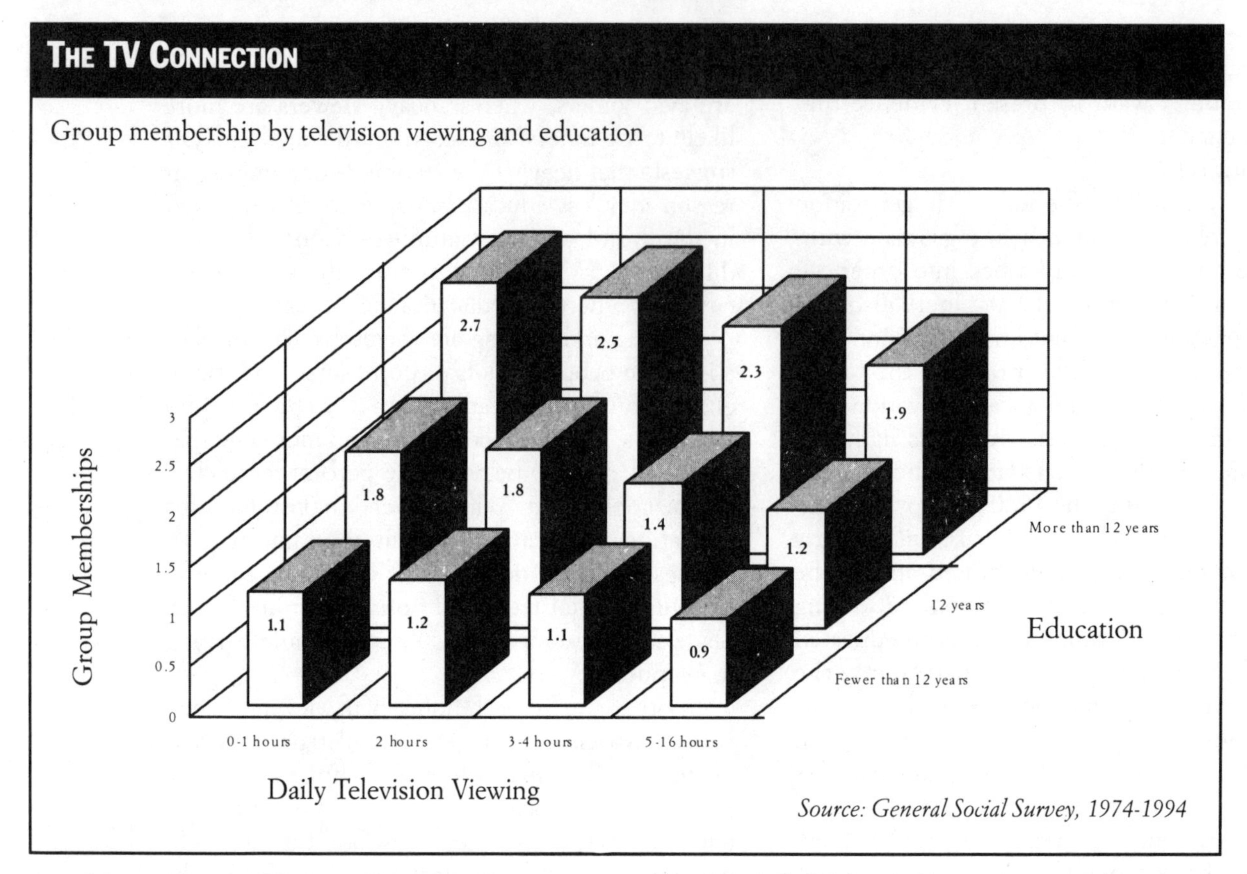

involvement (see "The TV Connection") are ambiguous with respect to causality, because they merely compare different individuals at a single time. However, one important quasi-experimental study of the introduction of television in three Canadian towns found the same pattern at the aggregate level across time. A major effect of television's arrival was the reduction in participation in social, recreational, and community activities among people of all ages. In short, television privatizes our leisure time.

Effects on the outlooks of viewers. An impressive body of literature suggests that heavy watchers of TV are unusually skeptical about the benevolence of other people—overestimating crime rates, for example. This body of literature has generated much debate about the underlying causal patterns, with skeptics suggesting that misanthropy may foster couch-potato behavior rather than the reverse. While awaiting better experimental evidence, however, a reasonable interim judgment is that heavy television watching may well increase pessimism about human nature. Perhaps too, as social critics have long argued, both the medium and the message have more basic effects on our ways of interacting with the world and with one another. Television may induce passivity, as Neil Postman has claimed.

Effects on children. TV consumes an extraordinary part of children's lives, about 40 hours per week on average. Viewing is especially high among pre-adolescents, but it remains high among younger adolescents: Time-budget studies suggest that among youngsters aged 9 to 14 television consumes as much time as all other discretionary activities combined, including playing, hobbies, clubs, outdoor activities, informal visiting, and just hanging out. The effects of television on childhood socialization have, of course, been hotly debated for more than three decades. The most reasonable conclusion from a welter of sometimes conflicting results appears to be that heavy television watching probably increases aggressiveness (although perhaps not actual violence), that it probably reduces school achievement, and that it is statistically associated with "psychosocial malfunctioning," although how much of this effect is self-selection and how much causal remains much debated. The evidence is, as I have said, not yet enough to convict, but the defense has a lot of explaining to do.

More than two decades ago, just as the first signs of disengagement were beginning to appear in American politics, the political scientist Ithiel de Sola Pool observed that the central issue would be—it was then too soon to judge, as he rightly noted—

whether the development represented a temporary change in the weather or a more enduring change in the climate. It now appears that much of the change whose initial signs he spotted did in fact reflect a climatic shift.

Moreover, just as the erosion of the ozone layer was detected only many years after the proliferation of the chlorofluorocarbons that caused it, so too the erosion of America's social capital became visible only several decades after the underlying process had begun. Like Minerva's owl that flies at dusk, we come to appreciate how important the long civic generation has been to American community life just as its members are retiring. Unless America experiences a dramatic upward boost in civic engagement (a favorable "period effect") in the next few years, Americans in 2010 will join, trust, and vote even less than we do today.

In an astonishingly prescient book, *Technologies without Borders*, published in 1991 after his death, Pool concluded that the electronic revolution in communications technology was the first major technological advance in centuries that would have a profoundly decentralizing and fragmenting effect on society and culture. He hoped that the result might be "community without contiguity." As a classic liberal, he welcomed the benefits of technological change for individual freedom, and in part, I share that enthusiasm. Those of us who bemoan the decline of community in contemporary America need to be sensitive to the liberating gains achieved during the same decades. We need to avoid an uncritical nostalgia for the fifties. On the other hand, some of the same freedom-friendly technologies whose rise Pool predicted may indeed be undermining our connections with one another and with our communities. Pool defended what he called "soft technological determinism" because he recognized that social values can condition the effects of technology. This perspective invites us not merely to consider how technology is privatizing our lives—if, as it seems to me, it is—but to ask whether we like the result, and if not, what we might do about it. Those are questions we should, of course, be asking together, not alone.

Are Today's Suburbs Really Family-Friendly?

Karl Zinsmeister

Karl Zinsmeister is DeWitt Wallace Fellow of the American Enterprise Institute and editor in chief of The American Enterprise.

Suburbs are where the majority of Americans live, and Americans are not idiots. The virtues of suburbs are many: They are generally safe and peaceful. They are economically affordable (on their outer edges at least). They provide their residents with lots of physical comfort, ample space, and wide consumer choices. Suburban politics tends to be reasonably responsive.

Suburbia is particularly favored today by families raising children. "Naturally family-friendly terrain" is how most of us think of such places. Suburbs are also considered the homeland of conservatism, of social traditionalism, of America's Norman Rockwell virtues.

There are many good and understandable reasons for these impressions. Yet the truth is that suburbs as they have been built in this country over the last 50 years are much less conservative, traditional, and family-friendly than we sometimes imagine. In "the innumerable suburbs that have sprung up since the Second World War," the great conservative thinker Robert Nisbet once wrote, "there is little more sense of community than there is in a housing project."

If we step back and compare suburbs to the ways people have lived for most of human history, we see that suburbia is actually a fairly radical social experiment, and one directly linked to many modern woes. The hurried life, the disappearance of family time, the weakening of generational links, our ignorance of history, our lack of local ties, an exaggerated focus on money, the anonymity of community life, the rise of radical feminism, the decline of civic action, the tyrannical dominance of TV and pop culture over leisure time—all of these problems have been fed, and in some cases instigated, by suburbanization, in ways that few people anticipated a generation ago when mass suburbs were first being created.

The Suburban Appeal

Modern tract suburbs are completely understandable as a reaction against the barbarism that has overtaken most American cities. My family and I are ourselves refugees from a major East Coast city, and I completely sympathize with people who say that our big urban centers have become unlivable for families. They have. That's why we left. So as I describe what I see as the problems of suburbs today, don't imagine that I think most parents should be raising their children in downtown Chicago or Los Angeles or Manhattan. That would represent jumping from the frying pan into the fire for most families.

But before we discuss possible alternatives to suburbia, let's begin by simply noting that today's typical suburban development is desirable to families not so much for what it *is* as for what it *isn't:* It is not dangerous, not dirty, not racially tense, not uncivil, and not plagued with broken-down public services and disastrous schools as most of today's cities are. Nor is life in the average suburb as economically tenuous as life can be in small town or countryside. The suburb is thus a kind of *anti*-location which, while hardly ideal, is well hedged against the opposing rural and urban risks of modernity. The single most important thing suburbs offer their residents is security—more physical security than cities, and

From *The American Enterprise*, November/December 1996, pp. 36-40, 104.

greater economic security than the average small town.

Deep in the heart of most parents raising children, though, is a hope that their community will be not only safe and inexpensive but also *neighborly.* Friendship, informal cooperation, and mutual aid are tremendously important both to children and to the adults taking care of them. Alas, most suburbs have turned out to be much less neighborly places than one would hope.

Perhaps the best aspect of today's suburban diffusion is the opportunities it provides for individual choice. Professor Robert Fishman describes today's large metro areas as "*à la carte* cities." While no single locality offers a full set of community services, an enormous range of possibilities is available within an acceptable car trip's distance. You pick your mall, your office park, your residential street, your child's daycare and school. There are the secondary choices of exercise club, video store, medical clinic, car repair station, and favorite ethnic restaurants. You assemble all these into a daily travel package, and that is your "community."

The advantage of this is that everyone's "community" is highly personalized. One person likes racquetball, sushi, the Gap, and colonial ramblers. Another wants a weightlifting club, steakhouses, Sears, and a split-level ranch house. They can live across the street from each other.

An obvious disadvantage of this life is that we become slaves to our cars, with traffic, smog, and congestion now worse in many suburbs than they were in dense downtowns. This threatens the very ease, moderate pace, and pleasantness that drew us to the suburbs in the first place.

A deeper, less obvious price of modern suburbanization is that none of our fellow residents shares more than just a small piece of common turf with us. Since the "community" assembled by your next door neighbor is frequently very different from your own, the occupants of suburbia are separated and partitioned off from each other. Denied a common physical community, we must seek out human connections in other places instead: in our shopping, our workplaces, or in the artificial "neighborhood" of television shows.

The mass-produced tracts of suburbia never pretended to be great incubators of individualism. But the original suburbanites, fleeing what they called the "chilliness" of modern urban life, did hope that suburbs would satisfy the natural hunger for human connection—as small towns and old-style city neighborhoods had in the pre-modern era. In practice, though, suburbs turned out to be relatively inhospitable places for individualism and community life both.

A Neighborhood's Physical Structure Influences its Social Life

Urban historian Lewis Mumford once described the spatial structure of a town as the "container" for its social system. A neighborhood's physical layout not only *reflects* the values and preferences of residents, he says, it helps *form* them as well. Winston Churchill made the same point with his observation that "we shape our buildings, then they shape us."

There is no doubt that suburbanization has changed social interaction in America. "The classic suburb is less a community than an agglomeration of houses, shops, and offices connected to one another by cars," writes architect Andres Duany.

The best foundation for strong community life is regular personal contact among residents. The traditional neighborhood, with its easy daily interactions, provides many such opportunities. Take the humble sidewalk (something few modern suburbs have). The exchanges that take place when neighbors randomly cross paths on foot are categorically different from those that occur around office coffee pots or at evening dinner parties. For one thing, they are less guarded. When you bump into someone on your way back from the mail box they see you unmasked—with your hair rumpled, or a cross word for your misbehaving child on your lips, or an easy arm around your spouse. Meeting on territory where you both know every cracked brick, loud dog, and weedy flowerbox, there are ample opportunities for forming communal alliances. Family matters and talk about how to improve the local community will be the foremost topics of discussion. You learn something about a neighbor each time you encounter him or her in this way. If you live in such a way as to miss these casual meetings (or so that they take place through two panes of auto glass instead of face to face), you are deprived of a golden chance for familiarity and closeness.

Suburbs Cut Roots

Suburbanization also disrupts community life by encouraging neighborhood turnover. The suburban boom that began after World War II brought a sharp increase in the moving rate of American families (particularly the rate of long distance moves). Each year now, nearly *one out of five* American families relocates to a new home. That's about twice the rate that prevails in most industrial countries, and the majority of our movers are suburbanites.

The original suburbanites, fleeing what they called the "chilliness" of urban life, had hoped that suburbs would satisfy the natural hunger for human connection.

The 1990 census showed that nearly half of all Americans age 5 or older were living in a different house than they'd occupied in 1985. In the course of a lifetime, the average American will occupy 13 different residences. This is particularly hard on children. Any childhood that includes several moves, studies show, carries real risks.

Adults suffer too. People living in transient areas show many more signs of psychological disturbance. A strong correlation exists, for instance, between the rate of family migration in a locality and the number of divorces that occur. Suicide levels are also linked to the frequency of residential turnover.

The very nature of the suburban day encourages rootlessness. It's been calculated that typical suburbanites will travel the equivalent of more than 20 times around the globe in a lifetime of commuting. Not only breadwinners but also all other family members are affected. As a Hungarian visitor to this country once remarked, "It is very interesting, in America the children are being brought up in moving vehicles." A whole generation now exists that has never known a more settled life.

In a short story, G. K. Chesterton described a protagonist who "found himself in some strange way weary of every moment," yet "hungry for the next." That encapsulates a certain modern personality, one that easy-in/easy-out suburban residences may feed. For transitoriness, as author Max Lerner notes, is built right into the suburb:

> As a man moves from production line to foreman to shop superintendent, or from salesman to division manager to sales manager, he also moves from one type of suburb to another.... By stages the family moves from court apartment to ranch house, and adopts new ways of behavior, new standards of tastes, and new circles of friends.

There is a wonderful aspect to this process, namely its social and class openness. But in its course, families end up putting on and taking off neighborhoods and neighbors like so much old clothing—and there is a price attached to that.

In any case, economic migration accounts for only a small part of our suburban house trading. Though "a lot of people think Americans move for reasons like jobs," says Jeanne Woodward of the U.S. Census Bureau, official figures actually show that only one in five relocations is employment related. In the largest number of cases—about half—families pull up stakes simply to find "a house that fits them better." This consumer's hunt for a more perfect physical environment drives the suburban whirl. One early suburbanite captured its essence when he remarked to *Harper's* in 1953 that "After all, this is only the first wife, first car, first house, first kids—wait 'til we get going."

Towns Without History, Towns Without Values

In communities where there is little shared history, residents often lack unity and direction. After studying the disastrous record of the "urban renewal" and New Town projects in Europe and America that knocked down existing older neighborhoods and replaced them with new "improved" ones, David Riesman concludes that "there were values concealed in the most seemingly depressed urban conglomerations which were lost in the move to the more hygienic and aseptic planned communities—much as farmers for a long time failed to realize that worms and other 'varmint' were essential to a well-nourished soil."

Like a lot of modern experiments, the suburban experiment has had a certain machine-like artificiality to it. At its core lies the social engineer's confidence that if you don't like society the way it is, you only need manufacture a new society. Never mind all the social evolution, and individual trials and errors, and intangible little inheritances from history that go into making existing communities work. Just get a plan and some money and build a *new* structure.

Let's not forget that the government played a huge role in funneling people into suburbia. For one thing, the feds poured vast sums into low-interest FHA and VA mortgages that could only be used for new houses, not for renovating existing homes in small towns or urban neighborhoods. And in the 1960s and '70s the government bulldozed thousands of acres of downtowns in the name of urban renewal, with only the most half-baked ideas of how to regenerate the flattened homes and businesses. Governments also spent billions building new suburban roads and freeways to feed commuters to the new locations. This was a direct subsidy for greenfield suburban construction, and, worse, the new thoroughfares also wrecked many existing communities. City neighborhoods were slashed by elevated highways, and outlying towns had the life snuffed out of them by beltways and controlled-access interstates.

Making the World Over

America is a progressive, future-oriented society always ready to try doing things a new way. This gives our nation a dynamic quality we should be profoundly grateful for. But this impulse, like all impulses, needs to be governed. The liberal, make-the-world-over mentality that serves us so well in, for instance, the realm of technology can cause great damage if applied without caution in the world of human relationships. When it comes to things like building workable communities, raising children, and defining our obligations to fellow citizens, a lot more humility is called for. People have irreducible natures and needs that cannot be "improved" or manipulated in the ways that products and institutions can—as failed social makeovers ranging from the Great Society to today's attempted redefinitions of the family make clear.

Americans accept suburbs, but they aren't the people's choice. When Gallup asks where we would most like to live, only 25 percent of us select a suburb.

The idea of relocating most of the American population within one generation to brave new communities only just carved out of hay fields was a bit heady—somewhat reminiscent of modern efforts to end drug abuse with pills, to stop racism with school buses, to eliminate poverty by writing checks. Tract suburbs, like much else that emerged in the decades after World War II, were a technocratic response to human problems that were mostly moral and economic.

And with no past, no inherited standards, no evolved wisdom wormed down into their cores, suburbs lacked a base on which to build cooperative feeling. So when the late-twentieth-century winds of materialism, selfishness, anonymity, and rat-race workaholism whistled through American society, families in many suburbanized communities flapped in the breeze.

Feeding Giantism

Big new suburban developments often extinguish existing community life when they move into a locality. With the arrival of commuter towns and their highways, "villages and open country settlements that have lived more or less aloof from the large center nearby are in a short space of time incorporated into an urban community," writes historian Amos Hawley. "Village institutions are replaced."

Consider Gwinnett County, Georgia. Situated some 30 miles northeast of Atlanta, Gwinnett County contained fewer than 45,000 people in 1960. The startling suburban boom that has since taken place around Atlanta increased county population to more than 300,000 in 1990. By the year 2000 the county may have 450,000 residents.

That is more people than many cities contain. Yet the suburbanites of Gwinnett County have little sense of community identity. When the *New York Times* purchased the *Gwinnett County News* and tried to compete with the *Atlanta Constitution*, it found that the suburban residents had no attachment to Gwinnett County and could not care less about having their own newspaper. County residents have no common attachments beyond the impersonal, homogenized ones of shopping in the same malls, rooting for the same mercenary athletes, consuming the same metropolitan media, and living standardized managerial/professional lives hardly different from those lived in Fairfax, Virginia; Orange County, California; northern New Jersey; or scores of other similar places.

One reason suburbanization tends to depress community life is because of the sheer giantism it encourages. "Where the leading metropolises of the early twentieth century—New York, London, or Berlin—covered perhaps 100 square miles, the new suburban city routinely encompasses two to three *thousand* square miles," notes Robert Fishman. "Developments of cluster-housing are as large as townships; office parks are set amid hundreds of acres of landscaped grounds; and malls dwarf some of the downtowns they have replaced."

It isn't only the physical territories that have become massive in the era of suburbanization, but the institutions as well. At the end of World War II, the average local school system had fewer than 250 students. Today it has more than 2,500. There are lots of big suburbs with single schools containing 3,000–4,000 students. This fosters neither individual character nor civic unity.

Suburbanization has created megalopolises that sprawl indistinguishably across vast territories of southern California, New Jersey, and Florida. It has turned separate cities like Washington, D.C., and Baltimore, or Dallas and Fort Worth, into huge octopuses. This is alienating, and not even efficient. Just as our bloated federal government now needs downsizing and decentralizing, so too would our metropolitan areas benefit from being broken up into smaller local communities. But to increase local self-determination we must stop thinking about our home communities in the way suburbanization has taught us to—as cogs in a much bigger metro-area machine, as waystations to be moved through as we trade up the income ladder, as commute endpoints.

Suburbanization and Children

Young singles and the elderly may do fine in anonymous suburbs—they have time to motor about the *à la carte* city, grazing on its sprawled choices, and they may actually appreciate suburbia's exaggerated privacy and absence of neighborly involvement. But for child-rearers, the typical tract suburb can make life miserable.

More than any other human beings, parents and children need human attachments. In his study of American ethnic groups, demographer William Petersen analyzes the unusually low child delinquency rates of certain nationalities, particularly Asian Americans, and says their secret is that "parents' responsibility for rearing their offspring is to some degree borne by the whole ethnic community." In traditional communities, neighbors watch out for trouble and offer aid and encouragement to families. Children are expected to take direction respectfully from all adults. Relations between parents and offspring, and between husbands and wives, are subject to informal social regulation. If mistreatment or neglect occurs, ostracism and sanctions will come from the whole community.

This kind of community-wide interest in the young still exists in certain settings. Studies show, for instance, that kids growing up in small towns get to know far more adults well, and in more varied ways, than do metropolitan children. Urban children can receive the same benefits, says historian Jane Jacobs, if they have "the opportunity (in modern life it has become a privilege) of playing and growing up in a daily world composed of both men and women...on lively, diversified city sidewalks." These days, though, only relatively few children experience either small-town streets or secure city sidewalks.

While home life is the foundation of a healthy upbringing, a child's world needs to extend further than one quarter-acre. "It is an uncommon family which can provide all of the things a complete and well-rounded community could offer—facilities, a sense of participation in ongoing community activities, places where teenagers and adults can spontaneously meet, and a chance to observe life as it is lived outside of living room and yard," writes Rutgers University professor David Popenoe. Good mothers and fathers know there are times when children need to be shooed out of the house and toward the sandlot, park, or pond.

If community life is sparse, even the best mothers and fathers will have trouble rounding out their child's personality. "In this respect," asserts author Mihaly Csikszentmihalyi, "families living in today's richest suburbs are barely better off than families living in the slums. What can a strong, vital, intelligent 15-year-old do in your typical suburb? ...What is available is either too artificial or too simple." There are no nearby stores, or bustling workshops, or true natural areas, nor even any ice rinks, libraries, or barber shops that can be reached without a car and chauffeur.

Lots of suburban homes have turned into little more than evening leisure centers and weekend crash pads.

In suburbia, writer James Howard Kunstler observes, school-age youngsters have little public realm to explore and places themselves in. [See] "The City Life We've Lost" [*The American Enterprise,* November/December 1996, page 51], and read the childhood recollections of cruising sidewalks, window-shopping downtown, walking to gyms, and lazing in parks. Contrast those activities to a typical childhood in suburbia today—where often the only "public realm" easily available to older kids is *television.* Which is why TV has become such a childhood addiction.

Suburbanization and Parents

Suburbs also lack support systems for parents—the reason so many suburban childraisers show signs of exhaustion today. Civilizing the barbarians we call children is tough work. Parents of preschoolers in particular need sympathetic ears and slaps on the back. They need people next door who can watch Junior in a pinch. Yet suburbanization has cut families off from the relatives and neighbors who used to be nearby to help with these things.

Suburbanization also finalized the separation of homes from workplaces and businesses. This is now partially reversible, thanks to technology that opens up fresh possibilities for work at home (as I know from personal experience). But old ways of thinking, zoning and labor-law prohibitions, and other barriers need to be broken down.

In separating homes from worksites, suburbanization accelerated the disconnection of men from home life. Right from the beginning, suburbs became daytime ghettoes for women and children. Moreover, those mothers have had to carry out their duties in isolation. Often the only way for them to escape their suburban quarantine is to get in a car—a discouraging prospect for people who already feel as if they are "half woman, half station wagon" from chauffeuring children to distant schools, doctors, and lessons. When the *Wall Street Journal* recently asked one harried "soccer mom" what single social reform would do the most to improve her quality of life, she replied, only half facetiously, "If they lowered the driving age to ten." This particular woman has put 40,000 miles on her minivan in the past 18 months squiring her three girls (ages 11 to 16) to events.

That isn't a great prescription for enjoying parenthood. No wonder lots of new mothers feel like climbing the walls. Too many find their suburban residences strangely unnatural places to nurture children and make a home.

For an early picture of this, consider David Riesman's description of an interview with a mother living in an upper-middle-class suburb of Chicago in the 1950s:

> Her husband had been transferred to Chicago from a southern city and had been encouraged by his company to buy a large house for entertaining customers. Customers, however, seldom came, as the husband was on the road much of the time. The wife and three children hardly ever went downtown (they had no Chicago contacts anyway), and after making sporadic efforts to make the rounds of theater and musical activities in the suburbs, and to make friends there, they found themselves more and more often staying home, eating outdoors in good weather and looking at TV in bed. Observing that "there is not much formal entertaining back and forth," the wife feared she was almost losing her conversational skills.

Is it surprising that women eventually revolted, and that the family boom which began with so much optimism after World War II petered out rather quickly? The kaffeeklatsches, block organizations, and neighborhood social groups that so impressed early suburban chroniclers had mostly disappeared by the time the first generation of suburban kids left home.

The loneliness of average suburbs poisoned many Americans on the possibilities of life at home. In 1963 Betty Friedan published her anguished, angry cry from the heart of suburban Westchester County—complaining in *The Feminine Mystique* that suburban mothers felt painfully abandoned and out of the mainstream. Her book struck a chord among influential upper-middle-class women, sold well nationwide, and launched a zealous feminist reaction against the very idea of home-making and child-rearing as occupations. Friedan hit upon a real problem, and one still with us: the trivialization of domestic life as it was shunted into sterile suburban camps. Americans who would prefer that their wives and daughters not follow Friedan down the path to NOW-style feminism would do well to think hard about how the current structure of our suburban communities feeds this problem.

The emergence of an alienated feminism wasn't the only warning of popular dissatisfaction with the suburban formula. A more concrete bit of evidence was the way our childbearing rate fell, like a marble off a table, from 3.8 lifetime births per mother in 1957 to 1.8 in 1975. Next, parents of infants and toddlers began to leave home during the day in droves. Instead of raising their preschoolers themselves, mothers and fathers have been handing them over in record numbers to hired caregivers. No longer even nurseries, lots of suburban homes have turned into little more than evening leisure centers and weekend crash pads.

Suburbs are Modernism in Bricks and Mortar

Though they began with the idea of bolstering family life, suburbs offered little protection when "self-fulfillment" became the cultural fashion in the 1960s. Judged by its practical effects, suburbia's most enduring products were, in swift succession: the absent working father, then a feminist backlash, followed by the absent working mother, all leading to today's unnerving legacy of daycare centers, split families, and childhood pathology. A heck of a lot more than suburbanization went into causing our family troubles. But suburbanization's erosion of neighborly supports and sanctions dumped a whole lot of straw onto the camel's back.

When the first suburban runaways set up their little arcadias out at the end of the paper routes, they had fantasized that they were migrating to close-knit towns where precious antediluvian values remained intact for them to recover. Actually, they were decamping into the perfect vacuum of commandeered corn fields, and importing all the transitory, impersonal, busy modern habits they imagined they were fleeing. Their new neighbors were all fellow émigrés, and together they permanently enshrined the most rootless aspects of twentieth-century life in a powerful new physical arrangement.

Suburbs didn't counter the defects that industrialism had introduced into our communities, they institutionalized them on virgin ground. The solitary commute, the ever-glowing boob tube, the TV dinner, the mall-based vacation, the Nintendo/Walkman childhood, the three-car family schedule, the no-fault divorce—these are all hallmarks of suburban life today. Suburban high schools are plagued by sexual misbehavior, cheating, and drug use, just as inner-city high schools are, and with less excuse. The suburban insulation against social decay has proved to be thin indeed.

Are There Alternatives?

So: Contrary to popular belief that suburbia is a bastion of traditionalism, it actually grew directly out of the social engineering mania that hit high tide in this country between the 1930s and 1970s. And the processes by which suburbs were formed pushed much of their competition out of the running. By vacuuming the most productive citizens out of established small towns and urban neighborhoods, suburbanization made alternate residential forms less viable. Our rural areas grew that much ricketier, and our cities got even colder and crueler. As a result, many families now feel they have no alternative to life in suburbia. So they make do, living their lives and raising their children with much less social support than families have traditionally enjoyed.

Even against their damaged competition, though, suburbs aren't the people's choice. When the latest Gallup poll asked Americans what kind of place they would most like to live in, only 25 percent chose a suburb. An equal number chose a farm. Just 13 percent said a city. The largest number by far—37 percent—wanted to live in a small town. People accept suburbs, but they aren't particularly enthusiastic about them.

In the essay before this one, Allan Carlson asks if there is any reasonable *alternative* to suburbs for average families living in the 1990s. That is the crucial question. We sometimes assume today that the only alternative to tract living is to be stacked up in apartment towers in grim inner cities. But that, thankfully, is not the only other choice. The *natural* community where humans have lived for most of their 10,000 years of history is the village. Sometimes the village has been a small town surrounded by wheat fields and forests. Other times it has been an urban village, a fairly self-contained 15-minute-walk community located within a larger city.

These kinds of villages still exist in America today—in towns and small cities, in a few lucky places within large cities, even within the more sensibly arranged parts of suburbia. These well-knitted, family-friendly communities are threatened, but they have hardly disappeared.

Surely conservatives and liberals alike can agree on the benefits of preserving and reviving these village-like communities. With the benefit of new technology, fresh services, expanded wealth, and other wholesome aspects of modernism, we ought to be able to encourage the rise of towns that have more of the traditional virtues of cohesive old-style neighborhoods. In the fascinating material you will find on pages 41 to 50 (and elsewhere in this magazine), there are concrete suggestions as to how communities might be built and rebuilt in the future so as to make this happen.

But first we must get beyond the idea that our current patterns of suburban living are paradise for children, families, and local patriotism. For the truth is, when we shifted from the gritty but distinctly communitarian life of rural small towns and tight-knit city neighborhoods to life in scattered houses, malls, and "bedroom communities" (an oxymoron if there ever was one), there was a sharp falloff in the connections among Americans. Mostly by accident, our flight from dangerous cities and stagnant rural areas became mixed up with the modern retreat from family and civic obligations. And the tragedy is that, while no one intended it, suburbs made both kinds of flight remarkably easy.

We've recently begun to correct errors of modernism in many other parts of American society—by recovering our respect for traditional families, for instance, by re-emphasizing personal responsibility, by retreating from the hubris of the Great Society. Our isolating suburbs need fixing too. And good alternatives do exist.

Stratification and Social Inequalities

People are ranked in many different ways: by physical strength, education, wealth, or other characteristics. Those who are rated highly usually enjoy special privileges and opportunities. They often have power over others, special status, and prestige. The differences among people constitute their life chances, or the probability that an individual or group will be able to obtain the valued and desired goods in a society. These differences are referred to as stratification, or the system of structured inequalities in social relationships.

In most industrialized societies, income is one of the most important divisions among people, whereas in agricultural societies, kinship has a major influence on life chances. Karl Marx described stratification in different terms. He used the term social class to refer to two distinct groups: those who control the means of production and those who do not. These groups overlap extensively with the rich and poor. This unit examines the life chances of the rich and the poor and of various disadvantaged groups, because they best demonstrate the crucial features of the stratification system in the United States.

The first section of this unit deals with the differences between classes and between income levels. It is well known that income inequality is increasing, but few know,

UNIT 4

as Robert Frank and Philip Cook point out, that one aspect of increasing inequality is the spread of winner-take-all markets, where rewards depend on relative performance rather than absolute performance. In these markets, such as sports, acting, the professions, and corporate management, small increments of talent have great value and receive great rewards, while slightly less talented performers are paid very little. Next, Nancy Gibbs tells the stories of hardworking Americans who are barely making ends meet. Many are close to the edge and wonder what happened to the American dream. The major way out of economic hardship is through college education, but even college graduates are having a tough time getting good jobs. These hardship stories pale to insignificance next to the stories of ghetto life told by Jonathan Kozol in "Poverty's Children: Growing Up in the South Bronx." In the ghetto, survival is the issue.

The second section discusses the American welfare system. Daniel Huff in "Upside-Down Welfare" describes the welfare system for the rich, and D. Stanley Eitzen analyzes the likely impacts of recent welfare reform. Huff looks at the rich not as winners in the business competition game but as winners in the welfare game. When all government gifts, tax breaks, and benefits are added up, most of them go to the upper and middle classes, not to the poor, he writes. Eitzen defends welfare against its attackers. He argues that welfare does not deserve much of the blame it receives for illegitimacy and poverty, that many programs that are being cut back or dropped have produced good results, and that dismantling welfare will have terrible social costs.

The most poignant inequality in America is the gap between blacks and whites. According to "Whites' Myths about Blacks," whites' attitudes toward blacks have greatly improved, but whites still believe in several unflattering myths about blacks that the essay aims to correct. Another racial problem is the perception of whites that reverse discrimination makes them the new victims. In the first of a two-part article, "Affirmative Action: It Benefits Everyone" and "Let's Get Rid of It," Jesse Jackson challenges this perception, which often is based on the false identification of affirmative action with quotas. In fact, quotas are only mandated by the courts as a last resort in special cases of noncompliance. Jackson shows why affirmative action is a good policy under current conditions. His thesis is that the whole society will benefit from it. Then, Armstrong Williams argues in the same article that affirmative action should be dropped because "as a permanent institution, [it] is patronizing, degrading, and self-defeating." Williams contends that "Americans need policies that promise more progress than affirmative action has delivered."

The last section deals with sex inequalities. Women have made considerable progress occupationally, including their climb into the ranks of middle management. However, women are still badly underrepresented in top management levels. Lisa Mainiero interviewed 55 women who have made it to the top and reports on how they did it. In fact, the paths of these women reveal a leadership style that is superior in a number of ways (including communication patterns) to male leadership styles. In the final unit selection, Toni Nelson reviews the worldwide prevalence of violence and abuse against women. Female genital mutilation (FGM) is one aspect of unequal rights for women that has come to the attention of the American public because some women who have sought political asylum in the United States plead that they would face FGM if they returned to their countries.

This unit portrays tremendous differences in wealth and life chances among people. Systems of inequality affect what people do and when and how they do it. An important purpose of this unit is to help you become more aware of how stratification operates in social life.

Looking Ahead: Challenge Questions

Explain why you believe technology will reduce or increase social inequalities.

Why is stratification such an important theme in sociology?

What social groups are likely to rise in the stratification system in the next decade? Which groups will fall? Why?

How does stratification along income lines differ from stratification along racial or sex lines?

Winner Take All . . .

has become the rule in corporate management, sports, the professions. Where does that leave the rest of society?

Robert H. Frank and Philip J. Cook

ROBERT H. FRANK is Goldwin Smith Professor of Economics at Cornell University. PHILIP J. COOK is ITT/Sanford Professor of Public Policy at Duke University.

Growing income inequality is the major economic story of our time. Scarcely a week passes in which the national media fail to report new evidence of the growing gap between rich and poor in America. The top 1 percent of U.S. earners have captured more than 40 percent of all economic growth since 1973, a period during which the median wage actually fell 15 percent in real terms and the bottom fifth of all earners fared even worse.

The number of workers earning more than $120,000 in 1990 dollars rose from 500,000 to 1 million between the last two censuses, and, although a portion of that growth resulted from increases in the size and average earnings of the professions, more than two-thirds were the result of increased inequality in earnings. CEOs of top corporations now earn more than 120 times as much as the average worker, up from only 35 times in the mid-1970s.

There remains little agreement on the causes of these changes. Some commentators mention changes in public policy, citing tax cuts for the wealthy and program cuts for the poor. Others emphasize the decline of labor unions, the downsizing of corporations, and the growing impact of foreign trade. Economists stress changes in the amount of, and rate of return on, people's "human capital"—an amalgam of education, training, intelligence, energy, and other factors that influence individual productivity. Commentators on the left often see darker causes, citing market imperfections that allow the rich to set their own terms in a world increasingly insulated from competition and social inhibitions against greed.

Although there is some truth to most of these claims, each misses something essential. Lower tax rates on the rich undoubtedly made the distribution of income less equal than it would have been, for example, but the explosive growth of the top salaries was well under way even before the tax-reform act of 1986.

The Spread of Winner-Take-All Markets

Our claim is that growing income inequality stems from the growing importance of what we call "winner-take-all markets"—markets in which small differences in performance give rise to enormous differences in economic reward. Long familiar in entertainment, sports, and the arts, these markets have increasingly permeated law, journalism, consulting, medicine, investment banking, corporate management, publishing, design, fashion, even the hallowed halls of academe.

An economist under the influence of the human-capital metaphor might ask: Why not save money by hiring two mediocre people to fill an important position instead of paying the exorbitant salary required to attract the best?

From *Across the Board,* May 1996, pp. 44-48. Much of the material in this article is adapted from *The Winner-Take-All Society: How More and More Americans Compete for Ever Fewer and Bigger Prizes, Encouraging Economic Waste, Income Inequality, and an Impoverished Cultural Life* by Robert H. Frank and Philip J. Cook.

Although that sort of substitution might work with physical capital, it does not necessarily work with human capital. Two average surgeons or CEOs or novelists or quarterbacks are often a poor substitute for a single gifted one.

The result is that for positions for which additional talent has great value to the employer or the marketplace, there is no reason to expect that the market will compensate individuals in proportion to their human capital. For these positions—ones that confer the greatest leverage or "amplification" of human talent—small increments of talent have great value and may be greatly rewarded as a result of the normal competitive market process. This insight lies at the core of our alternative explanation of growing inequality.

When we say that people like Stephen King, Kathleen Battle, and George Will sell their services in winner-take-all markets, we don't mean that there is literally only one winner in their respective arenas. Indeed, there may be scores or even thousands of highly lucrative positions in any given winner-take-all market. More accurately, we should call them "those-near-the-top-get-a-disproportionate-share" markets, but this is a mouthful, hence our less descriptive label.

A winner-take-all market is one in which reward depends on relative, not absolute, performance. Whereas a farmer's pay depends on the absolute amount of wheat he produces and not on how that compares with the amounts produced by other farmers, a software developer's pay depends largely on her performance ranking. In the market for personal income-tax software, for instance, the market reaches quick consensus on which among the scores or even hundreds of competing programs is the most comprehensive and user-friendly. And although the best program may be only slightly better than its nearest rival, their developers' incomes may differ a thousandfold.

The dependence of economic reward on performance ranking is nothing new. What is new is how technology has greatly extended the power and reach of the planet's most gifted performers. The printing press let a relatively few gifted storytellers displace any number of minstrels and village raconteurs. Now that we listen mostly to recorded music, the world's best soprano can literally be everywhere at once. The electronic newswire allowed a small number of syndicated columnists to displace a host of local journalists. And the proliferation of personal computers enabled a handful of software developers to replace thousands of tax accountants.

Of central importance in other cases has been the emergence of the so-called information revolution. In the global village, there is unprecedented market consensus on who the top players are in each arena, and unprecedented opportunity to deal with these players. A company that made the best tire in Akron was once assured of being a player in at least the northern Ohio tire market. But today's sophisticated consumers increasingly purchase their tires from only a handful of the best tire producers worldwide.

Rules Rewritten

Before there can be large and concentrated rewards in a winner-take-all market, the top performers must not only generate high value, but there must also be effective competition for their services. Yet in many markets, a variety of formal and informal rules traditionally prevented such competition. Most major sports leagues, for example, once maintained restrictive agreements that prevented team owners from bidding for one another's most talented players. But now, however, players have won at least limited free-agency rights in all the major professional team sports.

Unlike the owners of professional sports teams, the owners of businesses were never subject to formal sanctions against bidding for one another's most talented employees. But there were often informal norms that seemed to have virtually the same effect. Under these norms, it was once the almost-universal practice to promote business executives from within, which often enabled companies to retain top executives for less than one-tenth of today's salaries.

The anti-raiding norms of business have recently begun to unravel. As recently as 1983, the business community arched its collective eyebrow when Apple hired a CEO with a background in soft-drink marketing. But since then, inter-firm and inter-industry boundaries have become increasingly permeable, and business executives today are little different from the free agents of professional sports. Thus, no one expressed surprise when RJR-Nabisco's Louis Gerstner left to head up IBM, or when Motorola's George Fisher was brought in to rescue Kodak. Firms that fail to pay standout executives their due now stand to lose them to aggressive rivals.

The Allocation of Talent

Consumers clearly gain when modern technology and increased mobility allow the most talented people to serve broader markets. Once the world's hospitals are linked by high-speed data-transmission networks, for example, the world's most gifted neurosurgeons can assist in the diagnosis and treatment of patients thousands of miles away—patients whose care would otherwise be left to less talented and experienced physicians. And we should count as a benefit that the most talented executives are now more likely to manage the most important companies.

Yet the lure of the top prizes in winner-take-all markets has also steered many of our most able graduates toward career choices that make little sense for them as individuals and still less sense for the nation as a whole. In increasing numbers, our best and brightest graduates pursue top positions in law, finance, consulting, and other overcrowded arenas, in the process forsaking careers in engineering, manufacturing, civil service, teaching, and other occupations in which an infusion of additional talent would yield greater benefit to society.

One study estimated, for example, that whereas a doubling of enrollments in engineering would cause the growth rate of the GDP to rise by half a percentage point, a doubling of enrollments in law would actually cause a decline of three-tenths of a point. Yet the number of new lawyers admitted to the bar each year more than doubled between 1970 and 1990, a period during which the average standardized test scores of new public-school teachers fell dramatically.

One might hope that such imbalances would fade as wages are bid up in underserved markets and driven down in overcrowded ones, and indeed there have been recent indications of a decline in the number of law-school applicants. For two reasons, however, such adjustments are destined to fall short.

The first is an informational problem. An intelligent decision about whether to pit one's own skills against a largely unknown field of adversaries obviously requires a well-informed estimate of the odds of winning. Yet people's assessments about these odds are notoriously inaccurate. Survey evidence shows, for example, that some 80 percent of us think we are better-than-average drivers and that more than 90 percent of workers consider themselves more productive than their average colleague. Psychologists call this the "Lake Wobegon Effect," and its importance for present purposes is that it leads people to overestimate their odds of landing a superstar position. Indeed, overconfidence is likely to be especially strong in the realm of career choice, because the biggest winners are so conspicuous. The seven-figure NBA stars appear on television several times each week, whereas the many thousands who fail to make the league attract little notice.

The second reason for persistent overcrowding in winner-take-all markets is a structural problem that economists call "the tragedy of the commons." This same problem helps explain why we see too many prospectors for gold. In the initial stages of exploiting a newly discovered field of gold, the presence of additional prospectors may significantly increase the total amount of gold that is found. Beyond some point, however, additional prospectors contribute very little. Thus, the gold found by a newcomer to a crowded gold field is largely gold that would have been found by others.

A more contemporary example: Beyond some point, an increase in the number of aspiring mergers-and-acquisitions lawyers produces much less than a proportional increase in the amount of commissions to be had from such transactions. One law student's good fortune in landing a position in a leading Wall Street firm is thus largely offset by his rival's failure to land that same position. To be sure, even those lawyers who fail to win the biggest prizes often go on to earn comfortable incomes. But career choices must be measured not in terms of absolute pay but relative to what might have been. Contestants for the top prizes in law are highly talented people who could have held interesting jobs at high pay in other fields. Those who end up processing divorce papers and real-estate closings may not starve, but neither do they realize their full potential.

Growth vs. Equality: An Agonizing Tradeoff?

In the midst of burgeoning income inequality and growing national debt, one might expect Congress to consider raising tax rates on the highest earners. Yet many Republicans in Congress are lining up behind Majority Leader Dick Armey's "flat tax," which would reduce the tax rates on the top earners by more than half. Had the Armey proposal been in effect in 1993, it would have given Disney's Michael Eisner roughly an additional $50 million to spend.

That such a proposal even receives serious consideration provides striking testimony to the staying power of trickle-down theory. Its claim that lower takes on the rich will spur economic growth has animated the Republican push for less progressive taxes for almost two decades and in recent years has even captured the attention of some prominent Democrats.

Properly qualified, the major point of trickle-down theory is well taken. Confiscatory tax rates on a nation's most productive citizens are indeed, as Britain discovered, a blueprint for economic decline. But tax rates on the highest American earners are far from confiscatory, either in absolute or in relative terms. Indeed, even with the revisions of 1993, our tax system remains the least progressive among major industrial nations.

And for tax rates like ours, the arguments for trickle-down theory were never persuasive in the first place. Thus, trickle-down theorists have always had great difficulty explaining why executives in Germany and Japan, whose pretax incomes are much smaller than those of their counterparts in the United States, continue to work hard, save, and take risks, even though they face significantly higher marginal tax rates.

To the extent that economic incentives matter at all (and it is the cornerstone of trickle-down theory that they do), the effect of higher taxes on top earners would be to cause fewer of our most talented people to compete for limited slots at the top. Moreover, the people most likely to drop out would be those whose odds of making it into the winner's circle were smallest to begin with. Thus, the value of what gets produced in winner-take-all markets would not be much reduced if higher taxes were levied on winners' incomes, and any reductions that did occur would tend to be more than offset by increased output in traditional markets. The optimistic conclusion is that a more progressive tax structure may produce not only greater equality of incomes but also higher economic growth.

A Progressive Consumption Tax

Advocates of consumption taxes have stressed their positive impact on savings, their ability to capture revenue now lost to the underground economy, and their relative simplicity. These are important advantages, to be sure. Yet they pale in comparison to the advantages of using consumption taxes to counteract the inefficiencies that arise from winner-take-all markets.

For instance, a progressive tax on consumption, like a progressive income tax, would reduce the effective rewards of landing a superstar position, and thus would reduce overcrowding in winner-take-all markets.

A progressive consumption tax would also pay another dividend—namely, it would free up hundreds of billions of dollars of resources that are largely wasted through mine-is-bigger consumption arms races. Consider, for instance, a wealthy family's decision to build an 8,000-square-foot house. It does so not just because spacious living quarters are desirable in some absolute sense, but also because houses that size have become the norm for their income bracket. To have a smaller or less well-appointed house than one's peers would entail social embarrassment. Yet if *all* wealthy families had smaller houses (as indeed most do in cities like Manhattan and Tokyo), no one would be embarrassed in the least, and several hundred thousand dollars of resources per family would be freed up for more-pressing purposes—deficit reduction, medical research, capital investment, job training, school lunches, drug-treatment programs, time for family and community, whatever. The elegant hidden feature of the progressive consumption tax is this ability to create resources virtually out of thin air.

Opponents of progressive consumption taxes will caution that such taxes will cause unemployment, citing the layoffs in the shipbuilding industry that followed imposition of a luxury tax on yachts in 1991. But that was a tax with a glaring loophole that exempted boats purchased outside the country. A general progressive consumption tax would shift employment from some activities to others, to be sure, but that is precisely the objective. Full employment for carpenters can be achieved by the construction of small number of mansions for the well-to-do or by the construction of a larger number of smaller houses for people with more modest incomes.

Another consumption-tax advantage: By eliminating certain deductions and loopholes, a consumption tax could easily be made to entail less complexity than the current income tax.

Armey's flat tax is itself a consumption tax, and that is a point in its favor. Yet the adoption of the flat tax in anything like its current form would move us in precisely the wrong direction. By reducing the tax rates on top earners by more than half, it would steer even more of our best and brightest students into winner-take-all markets that are already overcrowded. By giving the top earners much more disposable income, it would also fuel the growth of wasteful consumption expenditures, both by the rich and by others who emulate them. And most important, it would worsen the social strains of income inequality.

The irony is that the flat tax would not only exacerbate inequality, it would actually be more likely to retard economic growth than to stimulate it. The case for this tax rests on an outmoded understanding of the forces that allocate both talent and reward in a winner-take-all society. These forces suggest that equity and growth are more likely to be complements than substitutes. This is a hopeful conclusion, indeed, for it means that the very same policies that promote both fiscal integrity and equality are also likely to spur economic growth.

Working Harder, Getting Nowhere

Millions of American families hold two or three jobs but still can't afford necessities and see little relief ahead

NANCY GIBBS

The billboards around Orlando, Florida, call Kissimmee "the affordable place to live." Take I-4 south and look for the Disney World exit, then drive in the opposite direction. There, behind the souvenir shops and motels, live America's working poor. It's not a bad neighborhood. The lawns are mowed, and the kids can play safely in the street. Anybody who wants a job can find one: Orlando was one of the nation's top five cities for job growth last year, with 40,000 new positions. Only 10% of the area's residents live below the poverty line.

All this is small comfort to Terri Yates. Last year, working full time, she took home about $10,000 as a cabdriver. This year she expects to do better. But even with Terri working six-day weeks and her husband Philip driving a cab all seven days, they still can't scrape together a down payment on a house. Terri's 1985 Pontiac needs radiator and clutch work; Philip is still paying $160 a month on a 14-year-old Mazda pickup. "I'm making less money than ever in my whole life, and I'm working more," Terri says. "I have no life. The only thing that holds me together is my children. But I can't even afford to send my daughter to the dentist for a cavity." Terri is proud that she has never been on welfare, but she feels little kinship with the politicians who extol workers like her. "I feel lost," she says. "Now everything goes up but people's wages. Either you're rich or poor."

When Capitol Hill revolutionaries vow to lift millions of Americans off the dole and into the work force, this is where they would land first, among the more than 10 million Americans who hover one rung above welfare on the nation's "ladder of opportunity." These are people who tiptoe between paychecks and have no savings, who ride the bus to the discount stores, who sell their plasma until their veins scar, who don't bother to clip coupons for Cheerios because the generic version is still cheaper, and who can be wiped out by even a minor medical problem.

Last week their position looked more precarious than ever. The Labor Department reported that despite the exuberant stock market and mild inflation, real wages keep on falling. For the lowest-earning 10% of workers, the weekly paycheck averaged $225 last year, a 10% drop since the late '70s after inflation is taken into account. It means that more people than ever are working full time and still living below the poverty line of $15,141 for a family of four. Millions more are scrapping by, just one broken refrigerator away from crisis. Says Labor Secretary Robert Reich: "If we don't take steps to begin to reverse the trend for so many workers who are sinking in this new economy, we will be paying a high price as a society."

What to Do

Such figures will make it even harder for politicians to tout the virtues of work over welfare. "We've got years of stagnant wages and people who are working hard and being punished for it," President Bill Clinton told a crowd of enthusiastic New Jersey autoworkers last week. "The question is, What are we going to do about it?" He talks cheerily about enterprise zones and community block grants. Republican Bob Dole talks about patience, while the Republicans set about shrinking government, shedding regulations and flattening taxes. "We've got to take a long-term view of this," he says. "I think we're on the right track. If we're not, I assume they'll throw us out." In the meantime the working poor work, and wonder what it would be like if the politicians had to walk in their shoes, just for a day or so.

Close to the Edge

In Atlanta, a quarter of the people who call the homeless hot line are working people: schoolteachers, chefs, computer-maintenance men, airline flight attendants. The standard recommendation is that a family should budget 30% of its income for housing. Among the working poor, 70% is more typical. "It doesn't take much in the way of an unforeseen circumstance to spin these people right out of control," says Anita Beaty, director of Atlanta's Task Force for the Homeless. "You cannot pay rent and child care on minimum wage."

David Harris works from 3:30 to 11:30 p.m. at the Travelers Aid Society shelter in Salt Lake City; his wife Nancy Tillack works there from midnight to 8 a.m. This means that if they need anything more than a couple of hours of sleep, they are never together for more than an hour or so a day.

"When we're home, we have no real life," Tillack says. They take turns looking after six-year-old MacKenzie; at least they don't have to pay for day care out of their combined weekly income of $720. "It's hard on her," Tillack admits sadly. "She has our time when we're awake enough."

Soon after moving to Utah from North Carolina, they were homeless for three months. "It's payday to payday. If one of us gets sick, it would be trouble," says Tillack. "We both love our jobs. We feel useful and productive, but I don't think anybody has job security anymore. We work hard to make ourselves indispensable."

The Tax Crunch

The working poor may not expect much from the government, but at a minimum they wish the government would not make their hard lives even harder. But that is what happened in 1983, when a blue-ribbon commission headed by Alan Greenspan decided to "save" Social Security by hoisting payroll taxes to record heights. The employer and employee share of the payroll tax, which in 1960 was 3% each, for a total of $530 for people who worked full time at minimum wage, now stands at 7.65% each for a total of $1,352. Today more than three-quarters of American households pay more in payroll taxes than in income taxes. "We ought to let the working poor keep all their earnings until they reach a higher level of income," says Douglas Besharov, resident scholar at Washington's American Enterprise Institute. "But politically, it would be difficult to do, because if we lifted the payroll tax, it would become clear that Social Security is a form of intergenerational welfare." Instead, the earned income tax credit was supposed to provide relief. In 1986, when Ronald Reagan expanded the EITC, he called it "the best antipoverty, the best profamily, the best job-creation measure to come out of Congress." But now the Republican budget cutters have EITC in their sights.

Bernice Jackson can't fathom why politicians want to cut the EITC. They can't understand the hardship of a life such as hers "unless you've lived it," she says. "They can't just say we need to work harder and things will be fine."

Jackson lives in a small trailer home in Appomattox County, Virginia, with her disabled husband Virgil and their three children. For 17 years she has cooked and cleaned at a training center for mentally disabled adults; she brings home $1,000 a month, while Virgil's disability payments add $560. Credit cards are banned from the house; she is still ashamed of how they ran up a huge debt and had to declare bankruptcy. The creditors were so aggressive they came and demanded her engagement ring and Virgil's wedding band.

This year Bernice received a $1,000 refund from the EITC. Some years she needs the refund for emergency repairs. This time she indulged in a small luxury for herself; she bought a treadmill. "You'd be depressed if you didn't treat yourself sometimes," she reasons. "I work every night, and I still can't make ends meet." She knows that there are people on welfare who are collecting more than she does. "But I believe in earning my way." The family has no savings at all. Son Jodie just graduated from high school, but Bernice couldn't afford to buy his graduation pictures. The mantels are filled with trophies from the daughters' track meets; but it's a struggle to find the cash to send them to the events. "Sometimes I just have to tell my kids they just can't go," she says.

Welfare vs. Work

AN URBAN INSTITUTE STUDY CONSIDERED the case of a hypothetical Pennsylvania woman with two children who received $4,836 in AFDC, $2,701 in food stamps and $3,000 in Medicaid benefits, for a total of $10,537 in cash and benefits. If she took a full-time job at minimum wage, her family would gain $9,516 in earnings before taxes and lose Medicaid, AFDC and one-third of her food stamps. Moreover, she would have to find and pay for day care. Welfare recipients who take part in job-training and education programs are eligible for subsidized care; for the working poor, day care can consume half their take-home pay.

"They send these AFDC people back to school. How come we don't do that for the working poor?" asks Eloise Anderson, an African American who grew up in a poor neighborhood in Toledo, Ohio, and now heads the California Department of Social Services. "They made the same bad decisions, except the working poor usually live with their decisions; they go to work, they say, 'Hey, I blew it, but I'm working.' "

Heather Beck, 23, earns $1,200 a month at a machine shop in Spokane, Washington. She is the only employee left following layoffs last month. Heather handles everything from typing and filing to the actual machine work. Her rent and day-care costs alone total $1,000. Last month, she reports with great pain and embarrassment, she had to borrow $200 to pay her landlord. "It just about killed me to do it," recalls Beck. "I don't even know how I'm going to repay it."

She is on a 200-person waiting list for subsidized day care for her two children and has found little sympathy from other agencies. "They said if I wasn't living on $500 a month, they weren't going to do anything for me, and that's for a family of three!" says Beck. "My ultimate frustration that day was that their message was that it would just be easier to give up and go on welfare." Beck bristles at talk of getting people off the dole and into the work force. "The people that are really working and trying to succeed are the ones who get the least amount of help," she says.

As President Clinton likes to note, two-thirds of the nearly 40 million Americans with no health insurance live in families with full-time workers. A single illness or injury will plunge a family into crisis. Often health-care concerns override all others in determining whether someone stays on welfare or goes to work. Being poor means making choices: the phone bill or the gas bill? Cough medicine or snow boots? In hard times, health insurance is a luxury; you can't eat peace of mind. So when Briana Harris, 17, fractured her leg sliding into home in a softball game last month, her parents' pain was as real as hers. "We're going to be faced with incredible hospital costs," says Denise O'Brien, a 45-year-old mother of three. "We gave up our insurance last December. When income is low, health care is too expensive."

The family's dairy farm no longer provides enough income to support the family. Husband Larry Harris works in construction, where he makes $1,800 a month, and Denise earned $5.45 an hour last season transplanting flowers and vegetables in a greenhouse. She is searching for a full-time job and raising free-range chickens and raspberries in the meantime. She used to imagine that by this time in her life, things would be more settled. "I sat and wrote in my journal, 'Things are different in 1995,'" she says. "All of us are wondering what we're going to do. Life seems so precarious. But there are a lot of people in denial."

The Learning Gap

IF YOU HAVE GRADUATED FROM COLLEGE, the chances that you will live in poverty are less than 2%, vs. almost 20% for those with only some high school. "We are seeing a higher correlation between education, earnings and benefits than we have ever seen in the history of this country," says Labor Secretary Reich. "If you are not educated and trained, technology tends to replace you." But it is little comfort to the working poor to be told they are victims of shifts in the global labor market and of advancing technology. What are they supposed to do? they ask. Follow the jobs to Thailand? Learn to use a computer?

"If I could do it all again, I would go to college, become a doctor," says William Robinson, who works three jobs in Appomattox County, Virginia. "Just, at the time, I didn't feel like going." He works six days a week, about 80 hours, for an annual income of less than $20,000. He studies traffic for the state transportation department, he stacks shelves at the local Food Lion, and he works as a medic for the National Guard. His wife is legally blind. Before the couple got married, she received a $265 federal disability allowance, but the subsidy was cut off because William earned too much. When all the basic necessities have been covered each month, the Robinsons are left with about $14.

He stands on the wooden steps in front of his double-width trailer home and looks proudly out over the mounds of stones and gravel that one day will be a circular driveway. "Just like rich people have," he says. Mary Robinson's grandmother sold the couple the two-acre plot of land for $500. "She offered to give it to us, but I wouldn't take it for free," William, 32, explains. It took him more than two years to pay off his elderly in-law in $10 and $20 installments; he then used the equity to borrow for the trailer. "It's taken so long to get this far. But we finally did it."

He is determined that his daughter Whitney, 7, will not take her education lightly. "She is already talking about going to college," says William. "I'm encouraging her to be a doctor. When I play Barbie dolls with her, we're always playing accident, and she is the emergency technician going to the scene."

Who's to Blame?

LIBERAL CRITICS CHARGE THAT THE G.O.P. agenda is about to make the economic ladder even more slippery. "Republicans might actually be surprised to see that they are doing things that will provide less assistance to the working poor," argues Isaac Shapiro of Washington's Center on Budget and Policy Priorities, which is about to release a report detailing the impact of the budget proposals on low-income families. "They are talking of substantially reducing the EITC. They are opposed to an increase in the minimum wage. What's gotten lost in the message is that the programs that offer assistance to the working poor could be partially if not completely reversed by the new agenda."

As government training and education programs are scaled back, aspiring workers will have to rely on the private sector. Training Inc. is a national nonprofit group that helps people polish their job skills. This has given counselors like Bev Schroeder a window on people's dreams and expectations. She works at the Indianapolis branch. "When I do presentations, I'm struck by the number of people who ask, 'How much will I make at the end?' " she says. "I talk about process. They say, 'I can't take $7.50 an hour. I need at least $30,000 a year.' I say, 'What do you have now?' " The trainees talk about owning their own business someday. "You need a reality injection," Schroeder tells them.

For the past five months, Jerome Ash has spent every weekday at Training Inc. in hopes of breaking into what used to be known as the pink-collar ghetto of receptionists, typists and filing clerks. He spends game nights at Indianapolis' Market Square Arena, selling frozen drinks on a 15% commission. On his first night, during a hockey game, he made $13.05.

Along with 40 classmates, Ash spent three weeks in a simulated office, practicing how to answer the phone, send a fax, file a letter. His goal: a $7-an-hour job. "I want an office job dealing with the public." His dream? That one day, with a steady job and a nest egg, he will actually take a vacation.

The Parent Trap

HARPER'S MAGAZINE PUBLISHED THE RESPONSE of Washington State Representative Marc Boldt to a letter from a constituent asking him to fight to preserve funding for the local family-education center. "If your situation is subject to so much financial instability, then why did you have three children?" he wrote back, expressing the concern of taxpayers. "Why is your husband in a line of work that subjects him to 'frequent layoffs'? Why, in the face of your husband's ability to parent as a result of his frequent layoffs, are you refusing to work outside the home? Why should the taxpayer foot the bill [for such education programs]?"

Being single and a parent is a good way to slip into poverty. Millions of single mothers know that, but it can be even worse to be a single father. Terry Younger remembers when he tried to get help in rearing his three children. "I've gone to some of these places, said I needed some help, and they just said, 'We can't help you; you're not a lady,' " says Younger. "So it's double jeopardy, and I'm bucking the system."

Younger fixes appliances at the local college in Spokane for $1,200 a month and uses subsidized day care. The hardest part for him, he says, is saying no to his children. When they go to the supermarket, the kids put cereals and other treats in the shopping basket, and Younger has to take them back out. "You learn to shop by the sales—last week I found a sale on cream-of-mushroom soup, so instead of three cans, you're buying 10 or 12 because you know the price is going back up again." He explains, "I want to be a role model. I don't want my kids to grow up thinking that things in life are free."

There may be 10 million working poor in this country, but many of them say they are too discouraged to go out and vote, so it's easy for politicians to ignore them. Many millions more, however, are perched near the bottom of the middle class, worried about what would happen if their jobs were eliminated, or they got sick, or for some reason lost their homes. In the campaigns to come, the politician who can speak to these fears will unleash vast political energy for good or ill.

—Reported by Cathy Booth/Miami, Ann Blackman and Ann M. Simmons/Washington, Dan Cray/Los Angeles and Elizabeth Taylor/Indianapolis

Poverty's Children

Growing up in the South Bronx

JONATHAN KOZOL

The Number Six train from Manhattan to the South Bronx makes nine stops in the eighteen-minute ride between East 59th Street and Brook Avenue. When you enter the train, you are in the seventh richest Congressional district in the nation. When you leave, you are in the poorest. The 600,000 people who live here and the 450,000 people who live in Washington Heights and Harlem, across the river, make up one of the largest racially segregated concentrations of poor people in our nation.

Brook Avenue, the tenth stop on the local, lies in the center of Mott Haven, whose 48,000 people are the poorest in the South Bronx. Walking into St. Ann's Church in Mott Haven on a hot summer afternoon, one is immediately in the presence of small children. They seem to be everywhere: in the garden, in the hallways, in the kitchen, in the chapel, on the stairs. The first time I see the pastor, Martha Overall, she is carrying a newborn baby in her arms and is surrounded by three lively and excited little girls. In one of the most diseased and dangerous communities in any city of the Western world, the beautiful, old, stone church on St. Ann's Avenue is a gentle sanctuary from the terrors of the streets outside.

A seven-year-old boy named Cliffie, whose mother has come to the church to talk with the Reverend Overall, agrees to take me for a walk around the neighborhood. Reaching up to take my hand the moment we leave the church, he starts a running commentary almost instantly, interrupting now and then to say hello to men and women on the street, dozens of whom are standing just outside the gateway to St. Ann's, waiting for a soup kitchen to open.

At a tiny park in a vacant lot less than a block away, he points to a number of stuffed animals that are attached to the branches of a tree.

"Bears," he says.

"Why are there bears in the tree?" I ask.

He doesn't answer me but smiles at the bears affectionately. "I saw a boy shot in the head right over there," he says a moment later. He looks up at me pleasantly. "Would you like a chocolate-chip cookie?"

"No, thank you," I say.

He has a package of cookies and removes one. He breaks it in half, returns half to the package, and munches on the other half as we are walking. We walk a long block to a rutted street called Cypress Avenue. He gestures down a hill toward what he calls "the bad place," and asks if I want to go see it.

I say, "OK."

"They're burning bodies down there," he announces ominously.

"What kind of bodies?" I ask.

"The bodies of people!" he says in a spooky voice, as if enjoying the opportunity to terrify a grownup.

The place Cliffie is referring to turns out to be a waste incinerator that went into operation recently over the objections of the parents in the neighborhood. The incinerator, I am later reassured by the Reverend Overall, does not burn entire "bodies." What it burns are so-called red-bag products, such as amputated limbs and fetal tissue, bedding, bandages, and syringes that are transported here from New York City hospitals.

Munching another cookie as we walk, Cliffie asks me, "Do you want to go on Jackson Avenue?" Although I don't know one street from another, I agree.

"Come on," he says, "I'll take you there. We have to go around this block." He pauses, however, and pulls an asthma inhaler from his pocket, holds it to his mouth, presses it twice, and then puts it away.

As confident and grown-up as he seems in some ways, he has the round face of a baby and is scarcely more than three-and-a-half feet tall. When he has bad dreams, he tells me, "I go in my mommy's bed and crawl under the covers." At other times, when he's upset, he says, "I sleep with a picture of my mother and I dream of her."

Unlike many children I meet these days, he has an absolutely literal religious faith. When I ask him how he pictures God, he says, "He has long hair and He can walk on the deep water." To make sure I understand how unusual this is, he says, "Nobody else can."

He seems to take the lessons of religion literally also. Speaking of a time his mother sent him to the store "to get a pizza"— "three slices, one for my mom, one for my dad, and one for me"—he says he saw a homeless man who told him he was hungry. "But he was too cold to move his mouth! He couldn't talk."

"How did you know that he was hungry if he couldn't talk?"

"He pointed to my pizza."

"What did you do?"

"I gave him some!"

"Were your parents mad at you?"

He looks surprised by this. "Why would they be mad?" he asks. "God told us, 'Share!' "

When I ask him who his heroes are he first says "Michael Jackson," and then, "Oprah!"—like that, with an exclamation on the word. I try to get him to speak about "important" persons as the schools tend to define them: "Have you read about George Washington?"

"I don't even know the man," he says.

We follow Jackson Avenue past several

From *The Progressive,* October 1995, pp. 22-27. Adapted from *Amazing Grace* by Jonathan Kozol.

boarded buildings and a "flat-fix" shop, stop briefly in front of a fenced-in lot where the police of New York City bring impounded cars, and then turn left and go two blocks to a highway with an elevated road above it, where a sign says BRUCKNER BOULEVARD. Crossing beneath the elevated road, we soon arrive at Locust Avenue.

The medical waste incinerator is a new-looking building, gun-metal blue on top of cinder blocks. From one of its metal sliding doors, a sourly unpleasant odor drifts into the street. Standing in front of the building, Cliffie grumbles slightly, but does not seem terribly concerned. "You sure that you don't want a cookie?"

Again I say, "No, thank you."

"I think I'll have another one," he says, and takes one for himself.

"You want to go the hard way or the easy way back to the church?"

"Let's go the easy way," I say.

Next to another vacant lot where someone has dumped a heap of auto tires and some rusted auto parts, he points to a hypodermic needle in the tangled grass and to the bright-colored caps of crack containers, then, for no reason that I can discern, starts puffing up his cheeks and blowing out the air, a curious behavior that seems whimsical and absent-minded and disconsolate at the same time.

"The day is coming when the world will be destroyed," he finally announces. "Everyone is going to be burned to crispy cookies." He reaches into the package for another cookie, only to discover they're all gone.

Cliffie's mother is a small, wiry woman wearing blue jeans and a baseball cap, a former cocaine addict who now helps addicted women and their kids.

Inside the church, when I return Cliffie to her, she looks at me with some amusement on her face and asks, "Did this child wear you out?"

"No," I say. "I enjoyed the walk." I mention, however, that he took me to the waste incinerator and I share with her his comment about "burning bodies." She responds by giving him a half-sarcastic look, hesitating, and then saying, "Hey! You never know! Maybe this child knows something we haven't heard."

She gives him another pleasantly suspicious look and leans back in her chair. "The point is they put a *lot* of things into our neighborhood that no one wants," she says. "The waste incinerator is just one more lovely way of showing their affection."

I ask, "Does it insult you?"

"It used to," she replies. "The truth is, you get used to the offense. There's trashy things all over. There's a garbage dump three blocks away. Then there's all the trucks that come through stinking up the air, heading for the Hunts Point Market. Drivers get their drugs there and their prostitutes."

She tells me that 3,000 homeless families have been relocated by the city in this neighborhood during the past few years and asks a question I hear from many other people here. "Why do you want to put so many people with small children in a place with so much sickness? This is the *last* place in New York that they should put poor children. Clumping so many people, all with the same symptoms and same problems, in one crowded place with nothing they can grow on? Our children start to mourn themselves before their time."

Cliffie, who is listening to this while leaning on his elbow like a pensive grownup, offers his tentative approval to his mother's words. "Yes," he says. "I think that's probably true."

He says it with so much thought, and grown-up reserve, that his mother can't help smiling, even though it's not a funny statement. She looks at him hard, grabs him suddenly around the neck, and kisses him.

Alice Washington lives on a street called Boston Road, close to East Tremont Avenue, about two miles north of St. Ann's Church. Visibly fragile as a consequence of having AIDS and highly susceptible to chest infections, she lives with her son, who is a high-school senior, in a first-floor apartment with three steel locks on the door. A nurse comes once a month to take her temperature and check her heart and her blood pressure.

The nurse, says Mrs. Washington one evening when we're sitting in her kitchen, has another sixteen patients in the building. "Some are children born with AIDS. Some are older people. One is a child, twelve years old, shot in a crossfire at the bus stop on the corner. The bullet ricocheted and got her in the back. She's lost her hair. Can't go to school. She's paralyzed. I see her mother all the time. They wheel her outside in the summer.

"This happened last year, on the Fourth of July. Summer had just begun. I feel so sorry for that child."

I ask how many people in the building now have AIDS.

"In this building? Including the children, maybe twenty-seven people. That's just in this section. In the other building over there, there's maybe twenty more. Then there's lots of other people have it but don't know, afraid to know, and don't want to be tested. We're living in a bad time. What else can I say?"

She tells me that her food stamps and her welfare check have been cut off. It's a complicated story, but it seems that her food stamps and her welfare payment had been stolen from her in the street some months before. When she began the process of replacing them, there was a computer error that removed her from the rolls entirely.

She relates a story that I've heard many times from people in New York who have lost their welfare payments. "To get an emergency replacement for my check," she says, "I needed to bring three letters to the welfare office—one from my doctor, one from the hospital, and one from my social worker. . . .

"I got the doctor's letter and the social worker's letter, but the hospital's letter didn't come. So I went back and forth from welfare to the hospital—it took a week and finally I got the letter. I brought in all the letters and I waited for another week and then I went to the computer. I put my card in, but it didn't work. . . . Then the man there said, 'Your card is dead. You've been cut off.'

"My doctor says, when it comes to the poor, they can't get nothin' right. Anyway, they got me runnin' uptown, downtown, to the hospital, to 34th Street, to the welfare, with the streets so hot and everyone at welfare so impatient. I've got no choice but I don't think I can go through it anymore. I feel like somebody beat me up."

Listening to her voice, which does sound like that of someone who is feeling beaten, I find myself thinking of the words of certain politicians who believe that we have got to get much tougher with unmarried, indigent, nonworking women. How much tougher could we get with Mrs. Washington, I wonder, without settling for plain extermination?

"If poor people behaved rationally," says Lawrence Mead, a professor of political science at New York University, "they would seldom be poor for long in the first place." Many social scientists appear to hold this point of view today and argue that the largest portion of the suffering that poor people undergo has to be blamed upon their own behaviors.

But even from the most severe of academic viewpoints about "rationality" or "good" and "bad" behaviors, what has Mrs. Washington done wrong?

She was born in 1944 in New York City. She grew up in Harlem and the Bronx and went to segregated public schools, not something of her choosing, nor that of her mother and her father. She finished high school, studied bookkeeping at a secretarial college, and went to work when she was nineteen.

When she married, at the age of twenty-five, she had to choose her husband from that segregated "marriage pool," to which our social scientists sometimes quite icily refer, of frequently unemployable black men, some of whom have been involved in drugs or spent some time in prison. From her husband, after many years of what she thought to be monogamous matrimony, she contracted the AIDS virus.

She left her husband after he began to beat her. Cancer of her fallopian tubes was detected at this time, then cancer of her uterus. She had three operations. Too frail to keep on with the second of two jobs that she had held, in all, for nearly twenty years, she was forced to turn for mercy to the city of New York.

In 1983, at the age of thirty-nine, she landed with her children in a homeless shelter two blocks from Times Square, an old hotel in which the plumbing did not work and from which she and her son David and his sister had to carry buckets to a bar across the street in order to get water. After spending close to four years in three shelters in Manhattan, she was moved by the city to the neighborhood where she now lives in the South Bronx. It was at this time that she learned she carried the AIDS virus. Since the time I met Mrs. Washington, I have spent hundreds of hours talking with her in her kitchen. I have yet to figure out what she has done that was irrational.

The entire discussion of poor women and their children and their values seems to take place out of any realistic context that includes the physical surroundings of their lives.

The statement, for example, heard so often now as to assume the character of incantation, that low-income neighborhoods like the South Bronx have undergone a "breakdown of family structure" infuriates many poor women I have met, not because they think it is not true but because those who employ this phrase do so with no reference to the absolute collapse of almost every other form of life-affirming institution in the same communities.

"Nothing works here in my neighborhood," a mother named Elizabeth has told me. "Keeping a man is not the biggest problem. Keeping from being killed is bigger. Keeping your kids alive is bigger. If nothing else works, why should a marriage work? I'd rather have a peaceful little life just with my kids than live with somebody who knows that he's a failure. Men like that make everyone feel rotten."

Perhaps it is partly for this reason that so much of the debate about the breakdown of the family has a note of the unreal or incomplete to many of the poorest women I know, and to many of the priests and organizers who work with them. "Of course the family structure breaks down in a place like the South Bronx!" says a white minister who works in one of New York City's poorest neighborhoods. "Everything breaks down in a place like this. The pipes break down. The phone breaks down. The electricity and heat break down. The spirit breaks down. The body breaks down. The immune agents of the heart break down. Why wouldn't the family break down also?

"If we saw the children in these neighborhoods as part of the same human family to which we belong, we'd never put them in such places to begin with. But we do *not* think of them that way. That is one area of 'family breakdown' that the sociologists and the newspapers do not often speak of."

Mrs. Shirley Flowers, whose neighbors call her "Miss Shirley," sits for several hours every day at a table in the lobby of her building to keep out drug dealers.

When I visit, we talk for a while of some of the children I have met at St. Ann's Church, almost all of whom have relatives in prison. Mrs. Flowers speaks of one of these kids, a fourteen-year-old boy who used to live here in this building but whose mother has since died of cancer.

"The family lived upstairs. The daughter's out at Rikers Island. Been there several times. Had two of her babies there. Now a brother of hers is out there, too. Another brother's dead."

"What happens to the kids," I ask, "when mothers are in prison?"

"Some of them, their relatives take them in. Others go in foster care. Other times," she says, "a neighbor takes the baby."

She speaks of toddlers in the streets who sometimes don't know where their mothers are. "If it's dinnertime, I'll bring them in and feed them. If they're dirty, I'll give them their bath." Many of the kids, she says, have little bugs all over them. "*Piojos* is the word the Puerto Rican children use. They get into their hair and skin. I say to them, 'Stay here with me. I'll keep you safe until your mamma's home.' The children know me, so they know that they don't need to be afraid."

For the past seven years, a gang of murderers and dealers has been based four doors away from Mrs. Flowers's home on Beekman Avenue. They marketed crack in a distinctive vial with a red and orange cap, and disciplined dishonest dealers by such terrifying means as mutilations. In one mid-day mutilation, *Newsday* reported, gang enforcers punished a refractory gang member by taking him to St. Mary's Park, right at the end of Beekman Avenue, where "they hacked at him with machetes" and a serrated knife, "opening wounds so severe that some of his organs spilled out." A crowd including children from a nearby junior high school watched the killing.

In one massacre that took place on the street two years before, a man and woman were shot dead for buying crack from the wrong dealer, the dealer was shot and killed as well, and a fourth person who had no drug involvement but was walking in the alley at the wrong time was chased down the street into St. Mary's Park and shot there fourteen times.

I ask Mrs. Flowers, "Have you ever seen a shooting victim die before your eyes?"

"I've seen a *generation* die," she answers. "Some of them were killed with guns. Some lost their minds from drugs. Some from disease. Now we have AIDS, the great plague, the plague of AIDS, the plague that can't be cured. It's true. I've seen it. I've been there. I've been here in this building twenty-four years and I've seen it all."

Despite the horrors she has seen, she seems a fearless person and almost serene. I ask, "How do you keep yourself composed?"

"I pray. I talk to God. I tell him, 'Lord, it is your work. Put me to rest at night and wake me in the morning.' "

"Do your children have the same belief in God that you do?"

"Yes," she says, nodding at her daughter and her son-in-law. "They do. This family talks to God."

Before I leave, she shows me a handful of photocopied clippings from newspapers that have sent reporters here to talk with her. It occurs to me that I must be one in a long line of people who have come to ask her questions. "Do you ever get sick of all these people knocking at your door year after year to pick your brain?" I ask her.

"No," she says, "I don't get sick of it because a lot of them have been nice people. The trouble is, you answer their questions and you give them your opinions. They collect your story from you. Then you see it and you read it. You think, 'Good.' But nothing happens. It's just 'there' and then it drops. It's like they put you in a bucket, like a wishing well. Only it's a wishing well where wishes don't come true."

Anthony is waiting for me at St. Ann's, where he is playing with a furry dog who sleeps outside the front door of the church except on freezing winter nights, when the Reverend Overall brings him in to sleep inside her office. "Mother Martha does not have a heart of stone. She has a heart of gold," says this remarkable boy.

We walk down the street to Children's Park, and then, because there are too many addicts there and it is growing very cold, we walk another block and find a sandwich shop where we can sit inside and talk. When I first met Anthony, the Reverend Overall had told me that he was "unusual." Nothing, however, has prepared me for the fascination and intelligence this thirteen-year-old displays as he describes the things that interest him and those that sadden him. He speaks, moreover, in a frequently inverted syntax, which I take at first to be the consequences of his bilingualism, but soon discover also has some other, more literary explanations.

When I ask him, "Anthony, do you have a happy life?" he answers, "Mr. Jonathan, my life is like the life of Edgar Allan Poe."

I ask him how he knows of Edgar Allan Poe.

"Because I have read his books," he says. "Did you know he lived here in the Bronx?"

"No," I say. "I didn't."

"Yes. It's true. He lived here in a cottage with his wife."

I ask him what he's read of Poe, and he replies, "*The Masque of the Red Death*—and many other stories."

"Why is your life like his?"

"Because he had not a very happy life. He always began a job but for some reason never finished it, which is my problem, too." He adds, "His wife had tuberculosis, but he loved her anyway. After she died, he had a breakdown he could never get out of."

I ask about *The Masque of the Red Death*, which I have never read.

"It's about a plague that stalks the Earth," he says. "For many, many days has it been on the Earth. But a man decides to hold a party because he is not afraid. He thinks the plague will never come to him if he can make things very safe. So he closes all the windows, all the gates, and all the doors, even the little peepholes in the doors. 'Seal them!' he said. And they sealed them. This was because he didn't want the plague to get inside."

"What was this plague like?" I ask.

"Little, sharp pins, like tuberculosis," he replies. "Or else like AIDS, because of the disease that gets into the blood, but maybe more like cancer. There was not AIDS in those days. I know that there was cancer."

I press him a bit and ask, "What is the meaning of this 'plague'?"

"A plague is an evil in one way," he says, "but not an evil in another way, because it could have a purpose." He then launches into a brief lecture on the history of plagues. "Now there was also a plague of Egypt where the firstborn died. The plague of Egypt is, of course, not over. It's over in Egypt but it could have gone to other places. Plagues are never really over. They can move from place to place.

"Sadness is one plague today. Desperate would be a plague. Drugs are a plague also but the one who gets it does not have to be the firstborn. It can be the second son. It could be the youngest."

"Anthony, what should we do to end this plague?"

"Mr. Jonathan," he answers, "only God can do that. I cannot be God."

I ask him when he thinks this plague will end "or else go someplace else."

"Mr. Jonathan," he says again, "I don't know when. I think it will only happen in the Kingdom of Heaven, but even the angels do not know when that will come. I only know that this is not His kingdom."

"How can you be sure of that?" I ask.

"This," he says with a gesture out the window that seems to take in many things beyond the dealers on the sidewalk and the tawdry-looking storefront medical office and the "*Farmacia*" sign across the street, "this out here is not God's kingdom. A kingdom is a place of glory. This is a place of pain."

Anthony meets me in the garden of St. Ann's and takes me for a walk to see the building where he lives, a few blocks to the west and north, and the building where his grandmother lives, which is close to the same neighborhood and which he says he likes to visit because "my grandma feeds me."

His grandmother, he tells me, is "the happiest person that I know."

I ask him why he thinks that she is happy.

"I don't know why," he says. "I think that feeding people makes her happy." Children from the neighborhood, he says, come to her house and she makes ices for them and bakes cookies. "I think that she likes children more than grownups."

His uncle, however, who lives with his grandmother, is, he says, "not happy. He has many troubles."

His eyes look worried when he says this.

"Anthony," I ask. "What troubles does your uncle have?"

"Mr. Jonathan, my uncle is a sick man. He has AIDS."

"What does he do during the day?" I ask. "Is he well enough to go outside?"

"Yes, he goes out. . . ." Then, in a grown-up voice, he adds, "How can I say this? He goes out but he stays in. He stays inside himself. He does not look at people. He looks down. The man looks at the ground. I don't know why. I think that he's afraid to look up at the world."

"Anthony, is your uncle a drug-user?"

"That," he answers, "is something that I do not want to know."

"Do you cry for your uncle?"

"Yes, I cry. It's not a sin to cry."

"Do you know other children who cry?"

"Many cry."

"Do you know children who are happy?"

"Truly happy? No."

"Happy at all?"

"Not many. . . . Well, to tell the truth, not any who are happy for more than one day." Then he corrects himself. "No! Not for one day. For fifteen minutes." He thinks this over, as if to check that he is being accurate, then reports, "Not any. That's no lie."

I wonder at times if a sense of the dramatic might lead Anthony to overstate his answers to my questions, so I challenge him by telling him that I've met children in the schoolyard who seem cheerful.

"Cheerful? Yes. Happy is not the same as cheerful," he replies.

"I think there are certain children who are happy anywhere," I tell him. But he holds his ground.

"Whenever you see a child who enjoys life in this neighborhood, come and see me right away. I'll have to go and see a doctor."

He stops at that moment and waves his hand around him at the neighborhood. "Would you be happy if you had to live here?"

"No, Anthony. I wouldn't," I reply.

We walk as far as Alexander Avenue, then circle back. As we walk, we pass a painted memorial to a victim of gunfire that has been partly whited over in one of the periodic cleanups by the city. A name and date can still be read, however. Sometimes, the Reverend Overall has told me, the city needs to use sandblasters to remove these tributes to the dead.

"How old would you like to live to be?" I ask Anthony.

"That's easy," he replies. "One hundred and thirteen."

"That number's quite exact," I say. "How did you decide on that?"

"I'm thirteen. I'd like to live another 100 years."

"Why *exactly* 100 years?"

"I would like to live to see the human race grow up."

Upside-Down Welfare

AMERICANS SPEND VAST AMOUNTS ON "WELFARE" EACH YEAR, BUT LESS THAN 10 PERCENT GOES TO POOR PEOPLE.

Daniel D. Huff

Daniel D. Huff is a professor of social work at Boise State University in Idaho.

At this writing George Bush's "kinder, gentler nation" is still rather callous. One in four of our nation's children is now born into poverty, up from one in five a decade ago. One in six has no health insurance. Following years of progress in preventing infant deaths, improvements in infant mortality have stopped; our rate is now worse than in nineteen other nations (a black baby born in Boston or Washington, D.C., is more likely to die before his or her first birthday than a baby born in Jamaica). Twenty million Americans remain hungry; a half million children malnourished. The average American's real wages have declined since 1980. For the poor and the near poor, the drop has been more severe. Meanwhile, the share of household income of the richest fifth of the American population continues to rise. The gap between the two groups is now wider than at any time in the last fifty years.

—*Robert B. Reich*
The Resurgent Liberal[1]

The decade of the eighties was hard. For many of us who came of age in the sixties and early seventies, watching low-income individuals and families lose most of the meager gains won during that now dim and distant past has been frustrating. Not only have low-income people grown both in numbers and proportion of population, but also those who now find themselves economically disadvantaged are predominantly women and children, with strong representation from mentally ill and developmentally disabled people.[2] Ironically, those who should be the highest on any rational list of priorities for government assistance are receiving the least.

Our conservative friends tell us that we are now spending more of our national budget on welfare programs than at any time in our past. But how can that be? How can poverty be growing—particularly among our most vulnerable populations—at the same time that we are spending hundreds of billions of dollars for welfare?

The answer lies in our understanding of what social welfare is and who it should help. In fact, we have developed a set of extremely elaborate programs that are designed to shift income from one group of citizens to another, but need is seldom a criterion for receiving benefits. Presently, we have three types of welfare programs. First are the poverty programs, such as Food Stamps and Aid to Families with Dependent Children (AFDC), which represent a relatively small amount of money and are designed to serve only low-income people.

Second are those programs, such as Social Security and Medicare, whose benefits serve mostly middle- and upper-income individuals. The third category of welfare is a newer set of programs designed to redistribute wealth to American businesses. Unlike the traditional poverty programs, these "upside-down welfare" programs are not customarily called "welfare." Many of their benefits are funneled through obscure and "off-budget" devices that avoid the scrutiny and debate that normally accompany implementation of more conventional welfare designs. Upside-down welfare represents an immense redistribution of our national wealth and explains why so little has been done for low-income people over the past decade. While we have been redistributing our nation's wealth through a variety of benefits, most of this money has not gone to help poor people.

The upside-down welfare state is extensive and breaks naturally into two different categories. First are those schemes primarily benefiting middle- and upper-income persons. That system represents a "gilded" welfare state, which provides the nonpoor with such benefits as low-cost government insurance for their oceanside homes, tax breaks for their investments, subsidized

From *Public Welfare*, Winter 1992, pp. 36-40, 46, 47. © 1992 by The American Public Welfare Association. Reprinted by permission.

medical care, and supplemental retirement benefits. The second is reserved for corporations, rather than individuals, and provides a redistribution system that annually transfers billions of dollars from ordinary taxpayers to the richest and largest corporations in America—a welfare program for Wall Street.

Wall Street Welfare and Business Subsidies

I live in a Northwestern city that is a hub for a large agricultural area. My rural neighbors are among the first to shout about too much government spending and are particularly angered when some big-city liberal advocates increased budgets for poverty programs. As a rule, my neighbors are anti-government in attitude and intolerant of what they call "government handouts." In spite of these attitudes, my farmer friends receive immense benefits from a wide variety of government programs.

In this area of the country, almost all the farming is done under irrigation. Water is provided to farmers by the government at a rate so low that the farmers' water bills represent only a fraction of the cost of water usage. For every dollar spent by the Bureau of Reclamation to provide water, the farmer pays the government 10 cents.[3] During the course of the year, my farmer friends typically take advantage of subsidized loans to purchase various necessities and subsidized insurance to protect them from the perils of pests and weather. When harvest season arrives, the local agricultural community lines up for government loans on their crops—loans they have to repay only if the crop values are greater than the so-called target price set by the U.S. Department of Agriculture (USDA).

Many of the farms in this area are devoted to growing the sugar beet, an ugly plant that only reluctantly surrenders its sugar after extensive and expensive processing. The sole reason there is any market at all for these beets rests on a government trade policy that does not allow imports of cheaper cane sugar. This indirect subsidy to the sugar beet growers and processors costs consumers $3 billion a year.[4]

Government programs designed to increase the income of the nation's farms costs approximately $25 billion a year.[5] USDA estimated that fewer than five million Americans lived on farms in 1990.[6] We disburse $25 billion a year for five million people on farms, while spending only $15 billion a year to support 11 million women and children on AFDC. Paradoxically, most of these funds never even reach the poorer segments of rural America. Clifton Luttrell, agricultural economist for the CATO Institute, estimates that less than 20 percent of farm subsidies trickle down to the poorest farmers. He explains that if the real purpose of farm subsidies is to eliminate farm poverty, the government could send every low-income farmer in America a check of sufficient size to pull him or her out of poverty for a cost of $2 billion to $3 billion.[7] Clearly, our current farm programs are not a very efficient means of helping the country's poorer farmers.

Unfortunately, the upside-down welfare that benefits farmers is only a small illustration of an extensive system. Local manufacturers wishing to sell their wares abroad to sometimes unstable and unreliable governments arrange for government loans at rates far below those available at commercial banks. The source of these loans, the Export-Import Bank (Ex-Im Bank), has been accused of falsifying its books in an attempt to make this subsidy appear smaller than it really is. An example of what one observer termed "creative bookkeeping" was listing loans to prerevolutionary China as fully collectible. In 1989, House Banking Committee Chair Henry Gonzales suggested that the Ex-Im Bank was so awash in red ink that even if it were liquidated, it would leave a shortfall of $4 billion to $6 billion—

Table 1. Welfare Spending by Program

Welfare Programs for Individuals	Amount (in billions)
Means-Tested Poverty Programs	
Medicaid	$ 49
AFDC	17
SSI	13
Food Stamps	15
Other (loans, etc.)	24
Total	$ 117
Gilded Welfare Middle-Class Programs	
Social Security	247
Medicare	104
Other retirement programs	62
Miscellaneous benefits	50
Tax expenditures	300
Fringe benefits (health & retirement)	385
Total	$1,148
Grand Total	$1,265

Source: U.S. Budget, 1990; Statistical Abstracts of the United States, 1990

Figure 1. Welfare Spending For Individuals
Total = $1.265 trillion

Source: U.S. Budget, 1990; Statistical Abstracts of the United States, 1990

a loss that would be passed directly to American taxpayers.[8]

Much larger subsidies are received by the defense industry, which since World War II has enjoyed the free use of more than $100 billion worth of plants and machinery owned by the U.S. Department of Defense and has accepted gifts of more than $10 billion worth of shops and equipment.[9]

The savings and loan (S&L) debacle is a more current example of upside-down welfare. The General Accounting Office now estimates that the financial rescue of the S&L industry will cost the American taxpayer $500 billion.[10] This gigantic public relief program for the S&L industry is necessary because of deregulation policies that enriched a small number of investors and developers in a handful of states. Taxpayers from all over the country will end up paying the bill for what amounts to one of the largest transfer-of-wealth programs in American history.

Clearly, our current farm programs are not very efficient means of helping the country's poorer farmers.

Corporate welfare programs represent a huge income redistribution system, providing American business interests with more than $200 billion a year.

Gilded Welfare: Public Assistance for the Middle and Upper Classes

I have a wealthy friend who is adamant in his denunciation of government handouts. Programs for low-income people earn his special disdain, for he believes such policies sap incentive and subsidize immoral behavior. He seems to harbor no such fears for his own set of personal subsidies, however, which are substantial. My comrade enjoys flying his own airplane at our local airport, which happily provides him services at fees representing only a fraction of their true costs. His beachfront vacation home is located in an area that is so frequently exposed to heavy weather that he cannot obtain private insurance and instead is forced to use subsidized government insurance. My friend spends many of his summer weekends camping in a neighboring national park where the fees he pays represent about one-third the actual costs of maintaining the park and its facilities. His considerable use of electricity, both at home and at his place of business, is subsidized through our regional power supplier at rates approximately one-half of those paid by consumers on the East Coast.[11]

Needless to say, my friend is hardly alone. Most of us view our personal subsidies as important benefits and just compensation for the taxes we pay. Unfortunately, while the middle-and upper-income groups have become sophisticated in lobbying and advocating for their welfare benefits, low-income Americans have proven to be less capable at seizing their share of government benefits. This fact is evident if we examine two of the largest categories of the current federal budget, Social Security and tax expenditures.

Most of us view our personal subsidies as important benefits and just compensation for the taxes we pay.

Social Security represents annual allocations of more than $240 billion.[12] Although one might imagine that the bulk of those dollars are divided among relatively low-income retirees, in fact in 1988 only 20 percent of Social Security benefits went to recipients with incomes less than 200 percent of the poverty level.[13] Although Social Security is commonly seen as an insurance program, current Social Security benefits are heavily subsidized by current workers and are two to five times greater than those of comparable private retirement programs.[14]

Social Security, therefore, is an inelegant solution to poverty, since 96 percent of the approximately 35 million low-income Americans are not elderly. The total spending for all the programs specifically designated for low-income people of all ages is only about $100 billion, while we spend $250 billion on retirees. According to Alice Munnell of the Brookings Institute, sending every American in poverty a check large enough to raise his or her income above the poverty level would cost only $130 billion to $160 billion.[15]

In spite of Social Security's beneficence, it is not very effective as an antipoverty program. Even staunch advocates of the system, like policy analysts Merton and Joan Bernstein, admit that although more than two-thirds of Social Security recipients had incomes higher than 125 percent of the poverty level, 27 percent—almost one-third—of all beneficiaries collect so little in benefits that they are forced to live in poverty.[16] The record is even more disheartening for minorities. In 1990, more than 40 percent of all black women currently receiving Social Security benefits were living below the poverty line.[17]

The system of tax breaks for individuals highly favors the rich. Tax breaks—or, in current budgetary parlance, "tax expenditures"—represent uncollected revenues and are every bit as real as actual expenditures. Not collecting taxes has the same fiscal impact as a grant or subsidy, with the obvious advantage that tax expenditures are "off budget" and not subject to the same scrutiny as direct spending. Uncollected personal taxes made possible by the assorted tax breaks written into the 1986 tax reform bill came to a staggering $281 billion per year in 1990.[18] Obviously, the bulk of savings generated by those tax breaks is distributed among the affluent. In 1990, *Wall Street Journal* reporters David Wessel and Jeffrey Birnbaum estimated that three-quarters of all itemized deductions are taken by individuals earning $50,000 or more per year.[19] Half the benefits from the interest deductions on mortgages went to the wealthiest 10 percent of the population, and 80 percent of the benefits claimed through the deductibility of state and local taxes goes to individuals with annual incomes of more than $30,000.[20]

The $90 billion a year in tax breaks represented by the exemption granted to employer-paid health and retirement plans represents a double inequity.[21] Not only do lower-income workers benefit far less from the tax savings, but they enjoy coverage from these programs at proportionally lower rates than do upper-income workers. According to a study conducted by the U.S. Agency for Health Care Policy and Research, lower-income workers are twice as unlikely to have health insurance than their more affluent colleagues.[22] Kevin Phillips in his book, *The Politics of Rich and Poor,* notes that only four in 10 American workers even have retirement plans.[23]

Even the ostensibly benign private charity system has been accused of unfairly disbursing its largess. Most of the more than $100 billion collected every year by private nonprofits is spent on activities that only remotely affect lower-income Americans. According to Lester Solomon's Urban Institute study on private nonprofits, less than $22 billion annually is earmarked for social services, with the bulk of philanthropic spending directed to such enterprises as private hospitals, preparatory schools, universities, and a variety of cultural activities.[24] Even money spent on social services is, upon closer inspection, suspect. The highly visible United Way programs have come under attack from such organizations as the United Black Fund for underwriting mostly middle class and white "charities," such as the Boy Scouts, the Young Men's Christian Association, and the Red Cross.

Altogether, Americans spend more than a trillion dollars every year on welfare; but less than 10 percent of that is specifically earmarked for low-income people.[25] The rest, such as Social Security, private charity, tax breaks, fringe benefits, and assorted subsidies for American businesses, benefit primarily those who are relatively well-to-do. Although nothing is inherently wrong or immoral about any individual component of this upside-down welfare, spending so much on the nonpoor while cutting poverty programs seems shamefully hypocritical at best, and at worst inimical to the nation's economic and social stability.

Spending so much on the nonpoor while cutting poverty programs seems shamefully hypocritical.

> *So there is the fundamental paradox of the welfare state: that it is not built for the desperate, but for those who are already capable of helping themselves. As long as the illusion persists that the poor are merrily freeloading on the public dole, so long will the other America continue unthreatened. The truth, it must be understood, is the exact opposite. The poor get less out of the welfare state than any group in America.*
>
> —*Michael Harrington*
> The Other America[26]

PW

1. Robert Reich, *The Resurgent Liberal* (New York: Time Books, 1989), 235.
2. Robert Greenstein and Herman Leonard, *The Bush Administration Budget: Rhetoric and Reality* (Washington, D.C.: Center on Budget and Policy Practices, 1990), 1–5.
3. General Accounting Office, *Federal Charges for Irrigation Projects* (Washington, D.C.: Government Printing Office, 1981), 26–27.
4. Consumers for World Trade, "The Economic Effects of Significant Import Restraints" (Statement before the International Trade Commission: Investigation 232-263, April 5, 1989).
5. James Bovard, "Farm Policy Follies," *The Public Interest* (Spring 1989): 75–88.
6. Richard D. Hylton, "Wall Street's Latest Diversification Strategy: Down on the Farm," *New York Times,* Sept. 23, 1990, sec. F, p. 11.
7. Clifton Luttrell, *The High Cost of Farm Welfare* (Washington, D.C.: The CATO Institute, 1989), 125.
8. James Gannon, "Lawmakers' View: Export-Import Bank as a Red Ink Gusher," *The Idaho Statesman* (Boise), May 21, 1989, sec. F, p. 5.
9. Richard Stubbing and Richard Mindel, *The Defense Game* (New York: Harper & Row, 1986), 55.
10. Michael Galtner, "Biggest Robbery in History—You're the Victim," *Wall Street Journal,* Aug. 9, 1990, sec. A, p. 9.
11. Congressional Budget Office, *Charging for Federal Services* (Washington, D.C.: Government Printing Office, 1983), 66; Edward Flattan, "Federal Subsidies the Height of Folly," *The Idaho Statesman* (Boise), August 5, 1989, sec. F, p. 2; J. Peter Grace, *Burning Money* (New York: MacMillan Publishing, 1984), 174, 182.
12. Congressional Budget Office, *Special Analysis of the Budget* (Washington, D.C.: Government Printing Office, 1989), A38.
13. Pete Peterson and Neil Howe, *On Borrowing Time* (San Francisco: Institute for Contemporary Studies, KS Press, 1988), 102.
14. Howard Karger and David Stoesz, *American Social Welfare Policy* (White Plains, N.Y.: Longman, 1990), 169–172.
15. Alicia Munnell, "Lessons From the Income Maintenance Experiment: An Overview," in *Lessons From the Income Maintenance Experiment,* ed. Alicia Munnell (Washington, D.C.: Federal Reserve Board and The Brookings Institution, 1986), 4.
16. Merton C. and Joan Bernstein, *Social Security: The System That Works* (New York: Basic Books, 1987), 169–172.
17. Regina O'Grady-LeShane, "Old Women and Poverty," *Journal of Social Work* 35 (September 1990): 422–424.
18. Robert Reischauer, "The Federal Budget: Economics and Subsidies for the Rich," in *The Federal Budget: Economics and Politics,* ed. Aaron Woldauski and Michael Baskin (San Francisco: Institute for Contemporary Studies, 1989), 247.
19. Jeffrey Birnbaum and David Wessel, "Tax Breaks: Who Gets What," *Wall Street Journal,* July 20, 1990, sec. A, p. 3.
20. Reischauer.
21. Joseph Peachman, *Federal Tax Policy* (Washington, D.C.: The Brookings Institution, 1987), 361.
22. Ron Winslow, "Health Costs: Insurance Net For the Poor Frayed Badly in the 1980s," *Wall Street Journal,* Sept. 21, 1990, sec. B, p. 1.
23. Kevin Phillips, *The Politics of Rich and Poor* (New York: Random House, 1990), 42.
24. Lester Solomon, *The Non Profit Sector and the New Federal Budget* (Washington, D.C.: The Urban Institute Press, 1986), 19.
25. Congressional Budget Office, *Special Analysis of the Budget* (Washington, D.C.: Government Printing Office, 1990), A37–A39.
26. Michael Harrington, *The Other America* (New York: MacMillan Publishing, 1962), 161.

Dismantling The Welfare State

IS IT THE ANSWER TO AMERICA'S SOCIAL PROBLEMS?

Address by D. STANLEY EITZEN, *Professor Emeritus of Sociology, Colorado State University*
Delivered at the College of Liberal Arts and Sciences Alumni Week Lecture, University of Kansas, Lawrence, Kansas, April 25, 1996

The welfare state is under attack by Republicans, the New Democrats (including President Clinton), and various pressure groups, most notably corporate America and the Christian Coalition. The emerging ideological consensus among these groups has three related propositions. First, government subsidies exacerbate social problems rather than solving them. Second, individuals must rely on their own resources and motivations if they are to succeed. And, third, individuals who fail are to blame for their failure. In a most telling irony, these propositions are assumed for individuals but not for corporations. My task this evening is to present the facts of two representative social problems, to assess the current public policy debate surrounding them, to examine the consequences of dismantling the welfare state, and to propose an alternative vision.

REPRESENTATIVE SOCIAL PROBLEMS AND PUBLIC POLICY

Poverty. Using the government's official definition of poverty, which understates the actual numbers, 15.1 percent of the population (39.3 million people) were officially poor in 1993. If the cost of basic necessities, especially housing, were included in the formula instead of just multiplying food costs by three, the number in poverty would be approximately 60 million.

Some social categories are over represented among the poor. In terms of race, 9.9 percent of whites are officially poor, compared to 30.6 percent of Latinos and about one-third of African Americans and Native Americans. Two out of three impoverished adults are women. Slightly more than one-fourth of children under age six are poor. Comparatively, poor American children don't do very well. In a study of 18 industrialized nations, our poor children ranked 16th in living standards.

The outlook for these poor children is grim. Because they had the misfortune to be born to disadvantaged parents they are denied, for the most part, adequate nutrition, decent medical and dental care, and a safe and secure environment. As a result, by age 5 or so, they are less alert, less curious, and less effective at interacting with their peers than more privileged children. Thus, they begin school already behind. They, most likely, attend the most poorly staffed, overcrowded, and ill-equipped schools. Poor children, in short, are more likely than more advantaged children to have health problems, to not do well in school, to be in trouble with the law, and, as adults to be unemployed and on welfare. Progressives argue that unless society intervenes with meaningful programs and adequate support, many of these poor children will fail. But the current plan sponsored by Republicans and supported for the most part by Democrats is to reduce programs aimed at helping the poor and their children. This reduction process has been underway since 1981 when President Reagan took office.

Related to poverty is the issue of welfare. We have a social safety net of AFDC, WIC, food stamps, Head Start, subsidized housing, to help those in need. Conservatives see this safety net as the problem since they believe it destroys incentives to work and encourages illegitimacy. They argue that welfare is the problem. Thus, social problems will get worse if we are more generous to the poor; they will get better if less is spent on the poor. This logic is a trifecta for the conservatives — by spending less on the poor, we save money, government is reduced, and the lot of the poor improves. Such a deal!

Using this logic, subsidies for compensatory education programs for poor children such as Head Start for preschool children and Chapter 1 for elementary school children have been slashed. Various welfare programs have been reduced by more than a quarter in the past 20 years. The Contract With America proposes to reduce welfare further by eliminating the earned income tax credit for the near poor, to deny AFDC to unmarried mothers under 18, and reduce Supplemental Security Income, food stamps, subsidized housing and other forms of public assistance to non-citizens; and cap the number of years to receive welfare benefits without providing job training, jobs, decent minimum wage, or child care.

Foremost among the social pathologies that the conservatives are most concerned with is illegitimacy, which they feel is the result of young poor women choosing babies to get welfare benefits. This concern is a bit misplaced, however, because only 8 percent of welfare-dependent households are currently headed by teen mothers.

For Progressives, rising illegitimacy is mostly the result of epochal changes in sexual and family mores, and not the result of AFDC. The esteemed social scientist, Frances Fox Piven points out that out-of-wedlock births are increasing in all strata of society, not just among welfare recipients or potential welfare recipients. Moreover, illegitimate births are increasing in all Western countries. Illegitimacy among the poor is also a consequence of the changing economy. When the mother or the father has a good job, the couple tends to get married. But with jobs leaving the inner cities for the suburbs, leaving for the nonunionized states, and leaving for the low-wage economies overseas during the last two decades, illegitimate births have risen dramatically. This argument runs counter to the conservative one which says that unmarried women are having additional children to increase their AFDC payments. The progressives challenge this prevailing myth with the following empirical facts:

- Since 1972, the value of the average AFDC check has withered by 40 percent, yet the ratio of out-of-wedlock births has risen in the same period by 140 percent.
- States that have lower welfare benefits usually have more out-of-wedlock births than states with higher benefits.
- The teen out-of-wedlock birth rate in the United States is much higher than the rates in countries where welfare benefits are much more generous.

These data surely squelch the argument that generous welfare encourages out-of-wedlock births.

Rising economic inequality. The U. S. has the most unfair distribution of wealth and income in the industrialized world. Moreover, the rate of growth in inequality is faster than in any other industrialized country.

The facts concerning inequality include:

- The richest 1 percent own more wealth ($3.6 trillion in 1992) than the bottom 90 percent ($3.4 trillion).
- Between 1983 and 1989 (the Reagan years) the nation's net worth increased from $13.5 trillion to $20.2 trillion, and 58 percent of that $6.7 trillion increase went to the fortunate top one-half of 1 percent. That works out to a $3.9 million bonanza per wealthy household.
- In 1960, the average CEO earned about as much as 41 factory workers. In 1992 that CEO makes as much as 157 factory workers. In 1995 the average compensation of CEOs (salary, bonus, and stock options) increased by 26.9 percent compared to the 2.8 percent increase in wages for the average worker.
- In a 15-year period ending in 1993, the richest 1 percent almost doubled their income and had their tax rates cut by 23 percent. In sharp contrast, the poorest one-fifth saw their tax rates go up and their incomes go down.
- The real value (adjusted for inflation) of a standard welfare benefit package has declined by some 26 percent since 1972.
- From 1967 to 1979 a full-time, year-round worker paid the minimum wage earned above the official poverty line for a family of three. Now a worker earning the minimum wage of $4.25 earns $8,840 annually, which is $3,427 below the 1995 poverty line for a family of three.

These data point to two related problems — wage decline and wealth stratification. Let's put these problems in historical perspective. After World War II, the United States entered a 30 year period of unprecedented economic growth where each segment of the population, from the top 20 percent to the poorest 20 percent and everyone in between saw their incomes double. But following the Vietnam War, the economy began slowing down. In short order, global competition began heating up and technological innovations changed the workplace forever. In the ensuing years, we have seen accelerated changes involving deindustrialization, capital flight as companies merged, moved their operations to low wage economies overseas or to low wage regions within the U. S., corporate downsizing, a decline in union membership and clout, and the replacement of permanent workers by temporaries or independent contractors. With the changing economy, wages stagnated, fewer workers were covered by adequate health care benefits and pensions, and the numbers of the underemployed rose to unprecedented heights.

While the masses lost ground or stagnated because of the economic transformation, the fortunate few have done very well with higher salaries, greater profits, and a rising stock market. A major reduction in taxes under President Reagan increased the gulf between the "haves" and the "have-nots."

This rising inequality gap has enormous consequences.

Economist Lester Thurow is very concerned about this trend:

> These are unchartered waters for American democracy. Since accurate data have been kept, beginning in 1929, America has never experienced falling real wages for a majority of its work force while its per-capita GDP was rising. In effect, we are conducting an enormous social and political experiment — something like putting a pressure cooker on the stove over a full flame and waiting to see how long it takes to explode.

Given the threats to individuals, families, and society that are part of the income/wealth inequality package, what are our policy makers doing to reduce it? Actually, rather than limiting this trend, current policies and proposals are increasing it. I refer to "trickle down" policies such as efforts to cut capital gains taxes and reduce inheritance taxes. Also the various proposals for a flat tax or a consumption tax to replace the income tax clearly will benefit the wealthy at the expense of the less affluent. While the Republicans are more inclined than Democrats toward measures that exacerbate inequality, the leadership in both political parties support policies that lead to a larger gap between the rich and the poor and that increase the constriction of the working class. Rather than fight the conservatives, the Democratic leadership has aimed for some right-of-center compromise. This submissive strategy reminds me of the quip by columnist Mark Shields: "If the Biblical prediction is true that the meek shall inherit the earth, then the Democrats will be land barons."

THE CONSEQUENCES OF DISMANTLING THE WELFARE STATE

What will be the consequences for individuals, families, and society of dismantling the welfare state? The American welfare state, modest in comparison to Canada and the European welfare states, emerged in the 1930s as a reaction to the instability of the Great Depression and capitalism run amuck. Motivated by a fear of radical unrest by the disadvantaged and disaffected and the need to save capitalism from its own self-destructive tendencies (economic instability, rape of the environment, worker exploitation, lack of worker and consumer safety), the creators of the New Deal under Roosevelt and the Great Society under Johnson instituted Social Security, the minimum wage, federal aid to education, health programs, nutrition, subsidized housing, and other services. Although these programs are not as generous as found in the social democracies of Canada and Europe, they have worked. Since Lyndon Johnson's War on Poverty (because it was not an all out war, some have called it Johnson's Skirmish on Poverty) began, for example, the poverty rate has been reduced by 20 percent and the rate of elderly poverty cut by one-half.

The prevailing hostility of conservatives to the New Deal legacy of "big government" is based on two fundamental beliefs. First, they believe that the unequal distribution of economic rewards is none of the government's business. They value individualism and a market economy. Inequality is not evil; rather it is good because it motivates people to compete and it weeds out the weak. This laissez-faire approach, consistent with Social Darwinism, guarantees an exaggerated inequality and leads to what economists Robert Frank and Philip Cook have called a "winner-take-all" society.

The second fundamental belief of the conservative creed is that government efforts to reduce poverty and class inequality actually cause the very problems they seek to solve. Welfare dependency, in this view, is the source of poverty, illegitimacy, laziness, crime, unemployment, and other social pathologies. They agree with Charles Murray that only when poor people are confronted with a "sink or swim" world will they ever really develop the will and the skill to stay afloat.

These beliefs lead to the obvious solution — do away with the welfare state and the quicker the better. This leads to the following set of questions: Will dismantling the welfare state be beneficial or will it create chaos? Will reducing or eliminating the safety net to the economically disadvantaged save them or hurt them? Will it make society safer or more dangerous? Are we on the right track when we eliminate all AFDC payments to mothers under age 18? Will poverty be reduced when the pool of unskilled workers is expanded as welfare recipients lose their benefits after a specified period yet there is no effort to create jobs for them or to provide a reasonable minimum wage?

I believe the answers to these questions are self-evident. Society will be worse off rather than better off, as evidenced by a comparison of our society now with the more generous welfare states. The number of people on the economic margins will rise. Homelessness will increase. Family disruption will escalate. Crime rates will swell. Public safety will become much more problematic.

This phenomenon of economic inequality has implications for democracy, crime, and civil unrest. As economist Lester Thurow has asked: "How much inequality can a democracy take? The income gap in America is eroding the social contract. If the promise of a higher standard of living is limited to a few at the top, the rest of the citizenry, as history shows, is likely to grow disaffected, or worse."

Criminologists have shown that poverty, unemployment, and economic inequality are powerful determinants of street crime. Compare, for example, the homicide rate in the United States (9.4 per 100,000) to that of Sweden (1.3), Germany (1.2), and France (1.1).Or contrast the homicide rate in cities of comparable size:

— Chicago with 2.8 million residents had 930 homicides compared to Paris with 2.2 million, which had 88 murders.

— Baltimore with 736,000 inhabitants had 321 murders, compared to Amsterdam with 700,000 residents but only 47 homicides.

Or consider the differences in rape rates, where the number of rapes reported per 100,000 women for the latest period available was 118 for the United States, a rate nearly three times higher than found in Sweden and Denmark.

Another comparison between the United States and the generous welfare states is the incarceration rate. In 1993 the U. S. rate was 519 per 100,000 population, compared to Canada's 116, France's 84, Germany's 80, Sweden's 69, and the Netherlands' 49. Imprisoning so many, as we do, is very expensive, costing about $80,000 to build each cell, and about $25,000 per prisoner per year.

How do the welfare states of Canada and Europe differ from the United States on other dimensions? Most significantly, they are more generous. They each have a comprehensive, universal health care insurance system. They have a much more ample minimum wage. On average, they mandate that workers have a four week paid vacation. These social democracies provide pensions and nursing home care for the elderly. They have paid maternity (and in some cases paternity) leave. Education is free through college. Unemployment benefits are significantly higher than in the United States. These benefits are costly with income, inheritance, and sales taxes considerably higher than in the United States. The trade-off is that poverty is rare, street crime is relatively insignificant, the population feels relatively safe from crime and from the insecurities over income, illness, and old age. Most important, there is large middle and working class with a much stronger feeling of community, of social solidarity than found in the United States.

The United States, in sharp contrast, has the highest poverty rate by far among the industrialized countries, a withering bond among those of different social classes, a growing racial divide, and is moving rapidly toward a two-tiered society. The consequences of an extreme bi-polar society are seen in the following description by journalist James Fallows:

> If you had a million dollars, where would you want to live, Switzerland or the Philippines? Think about all the extra costs, monetary and otherwise, if you chose a vastly unequal country like the Philippines. Maybe you'd pay less in taxes, but you'd wind up shuttling between little fenced-in enclaves. You'd have private security guards. You'd socialize only in private clubs. You'd visit only private parks and beaches. Your kids would go to private schools. They'd study in private libraries.

The United States is not the Philippines but we are already seeing a dramatic rise in private schooling, home schooling, and in the number of walled and gated affluent neighborhood enclaves on the one hand and ever greater segregation of the poor and especially poor racial minorities in segregated and deteriorating neighborhoods and inferior schools, on the other. Personal safety is more and more problematic as the violent crime rates increase among the young. Finally, democracy is on the wane as more and more people opt out of the electoral process, presumably because, among other things, they are alienated and their votes do not count (consistently, the U. S. has the lowest voter turnout among the industrialized nations).

A PROGRESSIVE PLAN TO SOLVE SOCIETY'S SOCIAL PROBLEMS

What would be the planks of a progressive platform? Let me suggest some possibilities.

Eliminating Poverty:

1. Raise the minimum wage to an amount that would keep the worker above the poverty line. This would mean $7.60 for an urban family of four. Index this wage according to the rate of inflation and the methods used to establish the poverty line.

2. Secure the earned income tax credit for the near poor, which would allow them to keep more of their income and thus escape poverty.

3. Institute a universal health care insurance plan for all residents.
4. Provide a national, universal child care subsidy.
5. Provide job training and a job. The government can generate jobs in child care, road building and maintenance, building mass transit, waste recycling, parks and national forests maintenance, cleaning up pollution, converting to renewable energy, environmental protection, assisting in schools and hospitals, building and renovating affordable housing, and the like.
6. Furnish compensatory education for preschool and school children from high risk situations.
7. End all redlining and other forms of discrimination by banks and insurance companies that artificially limit the opportunities for the poor and racial minorities.

Family Protection:

1. Paid maternity leave with guaranteed job retention.
2. Access to family planning information and technology, and subsidize birth control.
3. Subsidized child care for working parents, including school and recreation programs for older children to provide adult supervision after school for otherwise latchkey children.
4. Universal health care insurance that provides for prevention of as well as treatment of health problems.
5. Cost-free public education through college.
6. A social security system that provides adequately for retirement and a health care program that supplies the special health needs of the elderly, including home care and nursing home care.

Dealing with Corporate Downsizing: We are in a transition phase where the global economy and the computer chip revolution are causing severe dislocations in businesses, the communities where they are located, and for their workers. Government needs to help ensure the survival and enhance the dignity of downsized workers.

1. Insist that corporations give a sufficient advanced warning to communities and workers affected by downsizing or other corporate moves that have a negative impact.
2. In an economy of shifting employment, benefits (health care, pensions, job training) must be universal.
3. A government-corporate partnership to retrain workers. This happens in Germany, Japan, and Switzerland. Why not here?
4. A full employment policy. The displacement that has become endemic to American industry will be bearable only when other jobs are plentiful. Education and training are lauded as a big part of the solution, but education alone does not produce jobs. Thus, we need a government jobs program that provides work and decent pay while building the physical and social infrastructure of society.

Reduction of Extreme Income/Wealth Disparities. The aim here is not the elimination of wealth inequality but to reduce it. The high taxes in the generous welfare states, for example, still allow for very wealthy individuals and families. What they have done, contrary to our system, is to bring the wealthy down a bit and raise those at the bottom significantly.

1. Restore real progressivity to the tax system by closing loopholes for corporations and the wealthy, and with lower rates for the bottom and higher rates for the top.
2. Increase inheritance taxes.
3. End the discriminatory financing of public schools through private property taxes.
4. Take the power of money out of politics by public financing of local, state, and federal campaigns, free media for debate among candidates, and eliminating all political contributions.

IMPLEMENTATION OF THE PROPOSALS

Three questions remain: (1) Why should we adopt a progressive plan? (2) How do we pay for these programs? (3) Is there any hope to enact progressive solutions to social problems?

Rationale for Adopting a Progressive Plan

Why should we adopt a progressive plan to deal with our social problems? Foremost, these are serious problems and market solutions will not alleviate them. I am convinced that public policy based on abandoning the powerless will exacerbate their problems and societal problems.

A second reason to favor progressive solutions has to do with domestic security. We ignore the problems of poverty and wealth inequality at our own peril. If we continue on the present path of ignoring these problems or reducing or eliminating programs to deal with them, we will be less secure, we will have more problem people that require greater control, and, at an ever greater social and economic cost.

The final argument for a progressive attack on social problems is an ethical one. We need, in my view, to have a moral obligation to others, to our neighbors (broadly defined) and their children, to those unlike us as well as those similar to us, and to future generations. We need to restore a moral commitment to the safety net. As Jonathan Kozol, an expert on poverty, has argued: "There is something ethically embarrassing about resting a national agenda on the basis of sheer greed. It's more important in the long run, more true to the American character at its best, to lodge the argument in terms of simple justice."

Or consider this moral warning from an unlikely source, the very conservative British Chancellor of the Exchequer Kenneth Clarke, who explained his resistance to calls for a minimalist state: "This is a modern state. It is not the fifties, not southeast Asia. I believe in North American free-market economics, but I do not wish to see [here] the dereliction and decay of American cities and the absolute poverty of the American poor."

Financing the Progressive Plan

There are several sources of additional funds. The first is to reduce defense spending. We are the world's mightiest nation by far and there no longer is a Soviet threat, yet we maintain a defense budget of $260 billion that is more than is spent by all of our allies combined. Put another way, we are second in per capita military spending only to the nations of the Mideast. Our per capita defense spending is $1,110, compared to $750 for England, $427 for Canada, and $263 for Japan. We could easily decrease our annual defense spending by $100 billion or so without affecting our safety.

A second source of funds would be to reduce or eliminate corporate welfare and subsidies to the wealthy. At the moment, corporations receive $51 billion in direct subsidies and $53 billion in tax breaks. The wealthiest Americans pay lower tax rates and have more tax loopholes than found in any other modern nation. Annually we have about $400 billion in "tax expenditures" (i. e., money that is legally allowed to escape taxation). The economically advantaged receive most of these tax advantages. Representative George Miller of California has said that if corporations paid today the taxes that they did forty years ago, the deficit would vanish in a year.

Third, we should increase tax revenues. Totalling federal,

state, and local taxes equals 30 percent of the Gross Domestic Product. This is the lowest rate of any industrialized nations. In comparison, the English pay 36 percent, Germans and Canadians pay 37 percent, the French pay 44 percent, and at the high end the Swedes pay 56 percent.

There are also long-term savings that would accrue if a progressive plan were implemented. Research has shown, for example, that for every $1 invested in comprehensive childhood programs, $7 is saved in social costs, unemployment insurance, incarceration, and welfare. Just regarding crime, the costs, including incarceration, amount to $500 billion a year. Consider how much less on a per capita basis the Dutch, Swedes, or Germans pay for crime because of their progressive policies.

The Potential for a New Progressive Era

The crucial question — is there any hope of mounting a successful progressive program? At first the negative side seems overwhelming. Foremost is the fundamental belief in individualism. We celebrate individualism, which is the antithesis of cooperation, social solidarity, and acceptance of the redistribution of resources.

A second barrier to progressive policies is political. To begin, the majorities in the federal and state legislatures are political conservatives, which means they will opt for reducing or eliminating the welfare state rather than expanding these government programs. The political debates in these assemblies are from the right to the political center, with little, if any, voice from the political left. Another political barrier is that the two major parties are financed by big business and wealthy individuals. Also, the two-party system that has evolved in this country makes it structurally difficult for third parties to emerge as viable alternatives.

The social conservatives, who are about 30 percent of those who vote, are very organized. The conservative economic agenda seems assured by the generous contributions to both parties from the business community and wealthy individuals. At present, there are no countervailing pressures from the left on political parties and candidates.

The $5 trillion debt, much of which was generated during the Reagan and Bush presidencies, and the annual deficits are viewed by politicians of both parties as a giant weight on government that makes it difficult to fund existing programs, let alone institute new ones.

One of the necessary ingredients for a generous welfare state is the existence of a heavily unionized workforce. This condition is not present in the United States, as unions have declined significantly in membership since the 1960s.

The provisions of the Contract With America are exactly opposite position from the progressive view. The Republicans, fresh from their victories in the 1994 Congressional elections, where they became the majority in each house of Congress and won a number of governorships, claimed a mandate for their Contract With America. Most political pundits agree that they had such a mandate, but do they? Only 38 percent of the electorate voted in 1994 and only 52 percent of that three-eights or 19.8 percent of the public voted for the Republicans. Those who voted were disproportionately white, relatively affluent, and suburban. What about the poor and the near poor, racial minorities, bluecollar laborers, and city dwellers who chose not to vote? Why didn't they vote? What is the source of their apathy? Are they alienated from the political process because neither political party was speaking to their needs?

These questions lead to the possibility that the progressives might eventually prevail. Although this is unlikely, at least in the near term, there are some plausible arguments for this optimistic view. First, the current crop of Republicans may take what they consider a mandate and go too far with it, thereby alienating substantial portions of the citizenry. We must remind ourselves, however, that there is no guarantee that if the Republicans go too far, that things will only get better. This can occur, but so, too, can the situation steadily worsen.

Second, looking at the lessons from history, exactly 100 years ago the Progressive movement began as a reaction to unchecked capitalism, the robber barons, economic exploitation, and political corruption. Out of the Progressive Era came an activist government that addressed labor problems by instituting workplace safety regulations, prohibiting child labor, mandating the eight-hour workday, and providing disability compensation. The government broke up business monopolies, established a national parks system, and gave women the vote.

If the Republicans and the New Democrats today go too far and the marketplace replaces the welfare state completely, then it may lead to a further unraveling of social solidarity and a less secure society. In short, it may lead to a search for new answers — perhaps a new progressive era, just as it did 100 years ago.

Two necessary conditions for a successful welfare state are a strong union movement and a class-based labor party or Social Democratic party. While such a movement has been moribund for two decades or so in the United States, there is some evidence that unions are undergoing a shift toward more organizing and more ambitious and aggressive approaches against hostile employers and unfriendly laws. If the momentum accelerates, then there is hope for a labor renaissance and an organized push for progressive social policies.

A new progressive era will work only if a class-based political party emerges that addresses the needs of the masses. In particular, progressive leaders need to articulate a vision, a sense of direction, that builds a sense of community. Our future, we must recognize, depends on the welfare of all in the community in which we live, and the society of which we are a part, not just our personal accumulated wealth.

The alternative vision that I have proposed may seem very radical. I do not think so. Most of the suggestions are found in one form or another in each of the Western social democracies except the United States. Can we learn from these progressive societies? Should we learn from them? Can we afford a more generous welfare state? Can we afford not to adopt a more progressive plan?

Whites' myths about blacks

Though some white views have softened, mistaken beliefs persist

After the riots last spring, the *Los Angeles Times* asked city residents a simple, open-ended question. What did Angelenos think was "the most important action that must be taken" to begin a citywide healing process? Poll results from nearly 900 residents showed that the two most *unpopular* antidotes were the standard solutions favored by liberals or conservatives: "more government financial aid" and a "crackdown on gangs, drugs and lawlessness." Slightly more in favor were the human-capital remedies of the economists to "improve education" and "improve the economy." But the No. 1 solution was the psychologists' remedy. What the city most needed, residents concluded, was to "renew efforts among groups to communicate [with] and understand each other."

That prescription sounds squishy, yet in all the post-riot analysis there may have been too much talk about the ostensibly "tangible" roots of the riots (such as cutbacks in urban aid or weak job markets) and too little discussion of racial misunderstanding and ethnic stereotyping. In 1964, Martin Luther King Jr. warned that "we must learn to live together as brothers or perish together as fools," and the famed 1968 Kerner Commission report emphasized that narrowing racial inequalities would require "new attitudes, new understanding and, above all, new will." In the intervening years, however, those new attitudes and understanding have been all too lacking. And continuing strife in cities may only further stoke racial prejudice and hostilities.

Changing times. At first glance, recent trends in white attitudes toward blacks are deceptively upbeat. Fifty years ago, a *majority* of white Americans supported segregation and discrimination against blacks; just 25 years ago, 71 percent of whites felt blacks were moving too fast in their drive for equality. Today, by contrast, overwhelming majorities of whites support the principle of equal treatment for the races in schools, jobs, housing and other public spheres. Moreover, several national surveys taken after the riots contain encouraging signs of interracial accord. Most whites and blacks agree the Rodney King verdict and the violence that followed it were both unjust. The same polls show little evidence the riots initially made whites less sympathetic to the plight of poor blacks. For instance, both whites and blacks agree by large margins that jobs and training are more effective ways to prevent future unrest than strengthening police forces.

Yet much of the black-white convergence may be misleading for two reasons. For starters, blacks are still far more likely than whites to identify race discrimination as a pervasive problem in American society, especially when it comes to police and the criminal-justice system. At the same time, it seems probable that the riots at best only temporarily shifted whites' views of blacks. Annual polls taken by the National Opinion Research Center have found consistently since 1973 that most whites believe the government spends enough or too much to improve the condition of blacks (65 percent of whites thought so in 1991). Yet a *New York Times*/CBS survey taken *after* the riots found that a hefty majority of the American public (61 percent) now believes that the government spends too little to improve the condition of blacks.

In all likelihood, white prejudices are now evolving a bit like a virus. While the most virulent forms have been largely stamped out, new and more resistant strains continue to emerge. In the old racist formula, the innate "inferiority" of blacks accounted for their plight; in the modern-day cultural version, a lack of ambition and laziness do. Some modern-day stereotypes are simply false; others contain a kernel of truth but are vastly overblown. Regrettably, a large group of whites continues to harbor core myths about blacks based almost solely on their impressions of the most disadvantaged. Some examples:

■ **The work ethic.** The white myth: Blacks lack motivation. A 1990 NORC poll found that 62 percent of whites rated blacks as lazier than whites, and 78 percent thought them more likely to prefer welfare to being self-supporting.

Fact: For most of this century, blacks were actually *more* likely to work than whites. A greater percentage of black men than white men were in the work force from 1890 until after World War II, and black women outpaced white women until mid-1990. As late as 1970, black males ages 20 to 24 had higher labor-force participation rates than their white counterparts.

Today, the labor-force participation of the races is closer. After a 25-year influx of white women into the job market, white and black women participate in the labor force at nearly identical rates. Black men are slightly less likely than white men to be in the work force—69.5 percent vs. 76.4 percent. The only large gap between the two races occurs among teenagers: Last year, 55.8 percent of white teens were in the labor force, compared with 35.4 percent of black teens.

Blacks, who make up 12 percent of the population, are disproportionately represented on the welfare rolls; 40 percent of recipients of Aid to Families with Dependent Children are black, while 55 percent are white or Hispanic. However, numerous surveys have failed to find any evidence that most blacks prefer welfare to work.

In 1987, under the supervision of University of Chicago sociologist William Julius Wilson, the NORC surveyed 2,490 residents of Chicago's inner-city poverty tracts, including 1,200 blacks, 500 Puerto Ricans and 400 Mexicans. Roughly 80 percent of black parents surveyed said they preferred working to welfare, even when public aid provided the same money and medical coverage.

The most that can be said for white suspicions about black motivation is that a small segment of blacks has a more casual attitude toward welfare than do their low-income ethnic peers. In Wilson's survey, black parents were about twice as likely as Mexican parents to believe people have a right to welfare without working. And inner-city black fathers who did not finish high school and lacked a car were, in practice, twice as likely to be unemployed as were similarly situated Mexicans.

■ **Crime and the police.** The white myth: Blacks are given to violence and resent tough law enforcement. The 1990 NORC survey found that half of whites rated blacks as more violence-prone than whites. An 11-city survey of police in ghetto precincts taken after the 1960s riots showed 30 percent of white officers believed "most" blacks "regard the police as enemies."

Fact: The vast majority of blacks have long held favorable attitudes toward the police. As Samuel Walker reports in the 1992 edition of "The Police in America," 85 percent of blacks rate the crime-fighting performance of police as either good or fair, just below the 90 percent approval rating given by whites. Some blacks, especially young males, tend to hold hostile views toward the police, and ugly encounters with young blacks often stand out in the minds of cops. Yet studies consistently show that white officers have "seriously overestimated the degree of public hostility among blacks," says Walker. Even *after* the recent riots, a *Los Angeles Times* poll found that 60 percent of local blacks felt the police did a good job of holding down crime, not much below the white figure of 72 percent.

Blacks, or at least young black males, do commit a disproportionate share of crime; blacks account for roughly 45 percent of all arrests for violent crime. Still, the disparity between black and white arrest rates results partly from the fact that blacks *ask* police to arrest juveniles and other suspects more often than whites do. The vast majority of victims of black crime are themselves black, and it is blacks, more than whites, who are likely to be afraid to walk alone at night or to feel unsafe at home. In fact, one of the gripes blacks have with cops is underpolicing. Walker writes: "Black Americans are nearly as likely as whites to ask for more, not less, police protection."

■ **Job and housing bias.** The white myth: Blacks no longer face widespread job and housing discrimination. Three of four respondents in a 1990 Gallup Poll said that "blacks have as good a chance as white people in my community" to get any job for which they are qualified, and a survey last year found that 53 percent of Americans believed that blacks actually got "too much consideration" in job hiring. In June, a national survey by the Federal National Mortgage Association reported that most whites also believe blacks have as good a chance as whites in their community to get housing they can afford.

Fact: Researchers have documented the persistence of discrimination by testing what happens when pairs of whites and blacks with identical housing needs and credentials—apart from their race—apply for housing. The most recent national study, funded by the Department of Housing and Urban Development, found that in 1989, real-estate agents discriminated against black applicants slightly over half the time, showing them fewer rental apartments than they showed whites, steering them to minority neighborhoods, providing them with less assistance in finding a mortgage and so on. According to University of Chicago sociologist Douglas Massey, 60 to 90 percent of the housing units presented to whites were not made available to blacks. Even more disappointing, the evaluators found no evidence that discrimination had declined since HUD's last national study in 1977. And last week, the Federal Reserve Board reported that black applicants are currently twice as likely to be rejected for mortgages as economically comparable whites.

In the workplace, discrimination seems slightly less pervasive. Still, a 1991 Urban Institute analysis of matched white and black male college students who applied for 476 entry-level jobs in Chicago and Washington found "entrenched and widespread" discrimination in the hiring process; 1 white applicant in 5 advanced further than his equally qualified black counterpart.

■ **Taking responsibility.** The white myth: Blacks blame everyone but themselves for their problems. Since 1977, a majority of whites have agreed that the main reason blacks tend to "have worse jobs, income and housing than whites" is that they "just don't have the motivation or willpower to pull themselves up out of poverty." Fifty-seven percent of whites ascribed to that belief when NORC last asked the question, in 1991.

Fact: When it comes to apportioning blame, blacks neither presume that big government is the answer to their problems nor shy away from self-criticism. A 1992 Gallup Poll of 511 blacks found that just 1 in 4 blacks believed the most important way they could improve conditions in their communities was to "put more pressure on government to address their problems"; 2 of 3 opted for trying harder either to "solve their communities' problems themselves" or to "better themselves personally and their families."

In fact, blacks are almost as likely as whites to "blame the victim" and invoke the virtues of individual responsibility. In a 1988 Gallup Poll asking, "Why do you think poor blacks have not been able to rise out of poverty—is it mainly the fault of blacks themselves or is it the fault of society?" 30 percent of blacks responded that black poverty was the fault of blacks themselves; 29 percent of whites said the same. A 1992 poll for the *Washington Post* found that 52 percent of blacks—and 38 percent of whites—agreed that "if blacks would try harder, they could be just as well off as whites." Often, the status of race relations is a secondary concern for black voters. A poll released just last week by the Joint Center for Political and Economic Studies found that black and white voters both ranked the economy, public education and health care as their "most important" issues. Only 14 percent of blacks and 5 percent of whites cited the state of race relations.

A development with uncertain consequences is that both whites and blacks exaggerate the extent of white stereotyping. Both groups display a classic polling phenomenon—the "I'm OK, but you're not" syndrome. Whites are likely to overestimate other whites' support for racial segregation; blacks are likely to exaggerate whites' beliefs that blacks have no self-discipline or are prone to violent crime. Moreover, blacks and whites are far more sanguine about race relations and police fairness in their own communities than they are about other areas or the nation at large. A *New York Times*/CBS News poll after the L.A. riots found that just 1 in 4 Americans thought race relations were good nationwide, but 3 out of 4 believed race relations were generally good in their communities.

The downside to this syndrome is that it could make it easier for whites and blacks in suburban and upscale neighborhoods to write off blacks in poorer areas. A *Los Angeles Times* poll taken days after the riots found that nearly 80 percent of city residents felt they would suffer few if any hardships because of the riots' aftereffects, and 2 out of 3 respondents said their lives were already back to normal.

On the other hand, the fact that whites and blacks mix more at work, at home and socially than in previous decades suggests that increases in interracial contact could eventually help diminish stereotyping by both races. More tolerance will not solve the nation's race problem by itself. But it sure wouldn't hurt if, one day, "them" became "us."

BY JEANNYE THORNTON AND DAVID WHITMAN WITH DORIAN FRIEDMAN

Affirmative

It Benefits Everyone

Jesse L. Jackson

Jesse L. Jackson is currently head of the National Rainbow Coalition. He has been active in civil rights issues since the late '60s.

In the tradition of its predecessors, busing and law and order, the issue of affirmative action has become the operative buzzword for racial politics in the 1996 presidential campaign season. While we know that most Americans have benefited from affirmative action programs—Latinos, Asian Americans, Native Americans, African Americans, veterans, the disabled, and women of all races and ethnic backgrounds—current political rhetoric has forced a black face on the issue. This is not only inaccurate but also intellectually dishonest and manipulative.

From statehouses to the halls of Congress, politicians who until very recently lauded the benefits of affirmative action have now commenced a full frontal assault on such programs, the only mechanism proved truly effective for achieving equal opportunity in American workplaces and universities. Senate Majority Leader Bob Dole was right in 1986 when he led the bipartisan fight to maintain the Nixon administration's policy of goals and timetables in the face of Ronald Reagan's attempts to dismantle it.

California Gov. Pete Wilson, as mayor of San Diego, himself championed the city's affirmative action programs as a necessary means to achieve equity. In an eloquent appeal to the city council, he once stated that "it must come from the heart, but we must have goals to do it."

Americans are anxious. Our fears, while real and justified, are

(Continued on page 140)

■ ***Virginia Beach, Florida, 1959:*** **Affirmative action is needed to prevent America from returning to its racist past.**

This article originally appeared in *The World & I*, November 1995, pp. 74-83. Reprinted by permission of *The World & I*, a publication of The Washington Times Corporation.

Action

Let's Get Rid of It

Armstrong Williams

***Armstrong Williams is a Washington, D.C., business executive, talk show host, and writer; his new book is* Beyond Blame: How We Can Succeed by Breaking the Dependency Barrier.**

Imagine, if you will, a time when advertisements for jobs might say "whites encouraged to apply." Imagine the time when simply being white could get you into college outside all considerations of intellectual achievement or athletic ability or any other talent or merit you might possess. Imagine the time of the old-boy network, when you could get a city contract because you were a part of it, even though someone else might offer a lower bid.

■ ***Civil rights demonstration, 1963:* Affirmative action is not necessary because of strides made in the last 30 years.**

Those were the days of Jim Crow. It is also the story of today's "affirmative action." All of the above-mentioned instances have occurred many times over the past 20 years, and there is no dispute about that fact. The controversy is over whether a few years of affirmative action alleviates the many years of Jim Crow laws and the centuries of slavery before that. The answer to that question is no, it does not compensate for those injustices. No amount of either retribution or special assistance will eradicate those injustices. They are indelibly etched in America's past. The question remains: Will they remain a part of our future?

I want to make clear that I am not saying that the magnitude of racial discrimination inherent in affirmative action is comparable to the days of Jim Crow. The effect of affirmative action is minor compared to the discrimination that blacks faced in racially segregated America

(Continued on page 142)

■ ***Easy target for modern conservatives:*** **Some blacks believe Republican leaders, like George Bush and Bob Dole, are attempting to turn the clock back on civil rights by attacking affirmative action.**

(Continued from page 138)
being dangerously misdirected. Politicians who once supported affirmative action as an effective way to level the tilted playing field now would have America believe that affirmative action is at the root of our economic distress. Our jobs have not gone from white to black and brown, from men to women. What workers really feel is the pain of the globalization of the economy.

Where we once exported products, we now export plants and jobs. The real culprits are NAFTA and GATT—destructive economic policies that have sent our jobs across the border and overseas to cheaper labor markets, leaving our plants closed, without a plan to retrain our workers and invest in our economic future.

Affirmative action was never intended to be an antipoverty program. Opponents are using fear, scapegoating, and hysteria to shift the debate from the real issue: jobs. If we had a full-employment economy today, we would not be fighting over minutiae. The affirmative action fight is over what is left, not over what is needed.

HISTORY IS UNBROKEN CONTINUITY

Opponents of affirmative action would have us turn a blind eye to the past, opting for a scorched-earth approach to history. After 250 years of slavery, 100 years of apartheid, and 40 years of discrimination, we cannot burn the books and start anew at this point by instituting a "color-blind" code of justice.

Race- and gender-conscious programs were crafted precisely because individuals were discriminated against, historically and currently, because of their race or gender. We must not strive to be race neutral. We must work toward a race-inclusive and race-caring society. This summer, while the *Adarand* Court was imposing more stringent requirements for upholding federal affirmative action programs, seven out of nine justices on the most conservative Supreme Court we have had in generations rejected the notion of a color-blind Constitution.

The unbroken record of race and gender discrimination warranted the legal remedy of affirmative action. When we consider what true reparations for past discrimination entail, merely equalizing the laws of competition by leveling the playing field is indeed a conservative form of redress.

Our legal history is replete with the cancer of racism—from the 1857 *Dred Scott* decision, maintaining that blacks at three-fifths human had no rights whites were bound to respect, to the 1896 *Plessy v. Ferguson*'s "separate but equal" mandate of apartheid. The 1954 *Brown v. Board of Education* decision was an effort to heal race cancer with "race cure," to fight exclusion with race inclusion. Title VII of the 1964 Civil Rights Act outlawed discrimination by mandating negative action to offset negative behavior. In 1965, President Lyndon B. Johnson recognized that positive, or affirmative, action was necessary to overcome the vestiges of a discriminatory past.

DISPELLING THE MYTHS

The current political debate over affirmative action has been based on myth and anecdote rather than data and facts. The American public deserves an honest and informed discussion of the issue rather than a steady stream of divisive and misleading sound bites. Despite typical claims, affirmative action is not quotas or preferential treatment of the unqualified over the qualified. It does not demean merit and is not reverse discrimination.

Contrary to popular opinion, affirmative action does not require quotas. Unless a court imposes them, quotas are illegal. Quotas are used only as a

Why Affirmative Action Is Needed

- ✓ After 250 years of slavery, a clean slate cannot be made by instituting a "color-blind" code of justice.
- ✓ It helps to bring about a race-inclusive and race-caring society.
- ✓ It requires women and people of color to be included in the applicant pools of universities, workplaces, and unions.

last resort to remedy a manifest imbalance in a company's work force or to compensate for a widespread and persistent pattern of discrimination. All a company must do is prove that it has made a "good-faith effort" to meet flexible goals, targets, and timetables that have been established to diversify its pool of applicants, ensuring that women and people of color are included in the hiring or promotion pool.

To say that affirmative action is no longer necessary is to ignore the clear evidence of present-day racism and sexism.

Courts may also act affirmatively to root out more subtle forms of discrimination. If a court finds that qualified women and people of color are not sufficiently chosen from the wider pool of applicants, it may impose a quota to bring the employer up to the level of a nondiscriminating employer.

In university admissions, race may be one of several factors that admissions officers use in creating a diverse student body—a benefit that serves all of its students. The Supreme Court has definitively outlawed the use of rigid numerical quotas, while it affirmed the consideration of race and gender along with special talent, geographic origin, athletic ability, or legacy status.

Furthermore, we must look at the way we define merit. We have yet to see proof of any correlation between standardized test scores and a student's success in his academic or postacademic careers. As I understand it, motivation is the primary predictor of future success.

Can we genuinely believe that the child who was educated in an impoverished farming community without an SAT-preparation course, worked two jobs after school, and still graduated with an "A" average is any less qualified to attend a university than her affluent counterpart who had the benefit of extracurricular activities, honors courses, and training to earn a "competitive" score on the SAT? I think not.

Affirmative action does not mandate reverse discrimination. These policies merely require women and people of color to be included in the applicant pools of universities, workplaces, and unions. If an unqualified applicant is hired or promoted over a qualified one in the name of diversity, this is discrimination, and it is actionable in court just as it is under all other circumstances.

A recent Rutgers University study commissioned by the Department of Labor found that a majority of claims of reverse discrimination were brought by disgruntled job applicants who were determined by courts to be less qualified than the successful woman or person of color who received the job or promotion. Reverse discrimination is not only illegal; it is rare. Less than 2 percent of the 90,000 employment discrimination cases before the Equal Employment Opportunity Commission are reverse discrimination cases.

White men form 33 percent of the population and 48 percent of the college-educated work force. Yet, they constitute 80 percent of the tenured professors, 80 percent of the House of Representatives, 86 percent of the partners in major law firms, 88 percent of the management-level jobs in advertising, 90 percent of the top positions in media, 90 percent of the officers of major corporations, 90 percent of senators, 92 percent of the *Forbes* 400, 97 percent of school superintendents, 99.9 percent of professional athletic team owners, and 100 percent of U.S. presidents. It is clear that the notion of the "angry white male" is not grounded in reality. Rather, it is an error in perception that has been wrongfully validated by divisive political tactics.

IT WORKS WHEN ENFORCED

Many opponents concede that our discriminatory past necessitated the need for positive affirmative steps to root out this pervasive evil, yet today they have come to the irrational conclusion that after 30 years on the books, these policies are no longer necessary. History demonstrates that when these policies were enforced as they were during the 1970s, the employment of women and people of color increased dramatically. These gains, however, were offset by the assaults on affirmative action during the Reagan-Bush era.

To say that affirmative action is no longer necessary is to ignore the clear evidence of present-day racism and sexism. The Department of Labor's *Glass Ceiling Report* found that women in the largest corporations hold less than 5 percent of top management posts, while African Americans, Latinos, and Asian Americans hold less than 1 percent of these positions. White males hold 95 percent of these jobs.

The unemployment rates of African Americans and Latinos are twice that of whites. Women are 53 percent of the population, African Americans 13 percent, and Latinos 10 percent. Yet, in the 1994 labor market, 22 percent of all doctors were women, 4 percent African American, and 5 percent Latino. Twenty-four percent of all lawyers were women, 3 percent African American, and 3 percent Latino. Thirty-one percent of all scientists were women, 4 percent African American, and 1 percent Latino.

Affirmative action has not only benefited those who have been historically locked out; it has benefited our nation as a whole.

The pay differential between white men, white women, and people of color persists. In 1993, for every dollar a white man earned, an African American man made 74 cents, a white woman 70 cents, a Latino man 64 cents, an African American woman 63 cents, and a Latina 54 cents.

The Urban Institute has documented the rampant nature of discrimination in the workplace. Sending equally qualified African Americans and whites to apply for the same jobs, they found that in nearly a quarter of the cases, whites moved further through the hiring process than blacks. The institute likewise found that whites received 33 percent more of the interviews and 52 percent more job offers than equally qualified Latinos. Even when African Americans and Latinos are hired, they are promoted and paid less.

We cannot fall prey to the destructive tactic of "divide and conquer" for the sake of political expediency. Affirmative action has not only benefited those who have been historically locked out; it has benefited our nation as a whole. Two-income-earner households have enabled American families to provide for their children. Race- and gender-inclusive policies turn tax consumers into tax producers. A diversified corporate America is better able to compete in this increasingly globalized economy. Let us not be misled: Increasing the educational and employment opportunities for a majority of Americans is good for the nation and good for our future.

(Continued from page 139)

prior to the 1960s. Blacks were treated as inferior and as second-class citizens. They were deprived of basic civil rights and often of human dignity.

RACISM IN BLACK AND WHITE

What I am saying, though, is that in principle the legal incorporation of race into decisions regarding jobs, promotions, contracts, or school enrollment is the same. Making judgments based on race is racism, any way you want to try to cover it over. I disagree with affirmative action on principle for the same reason I disagree with Jim Crow laws, or, for that matter, slavery.

Race is a false basis for legal distinctions. Racial differences are accidental. Our human nature, which we share regardless of race, is essential. We are rational, moral, and spiritual men and women, created in the image and likeness of God, regardless of the hue of our outer covering. Just as you can't judge a book by its cover, you can't judge a man by his color.

One could perhaps overlook the race consciousness of affirmative action programs if they were working as a temporary help for a downtrodden minority, but affirmative action programs have been neither temporary nor helpful. What was designed to be a temporary breach in a racially hostile society devolved into a system of permanent bureaucratic racial set-asides.

As a permanent system, it undermines the earned successes of hardworking ethnic minori-

■ ***Speaking at a fund-raising event in 1963:*** **Dr. Martin Luther King's dream was that Americans would be recognized not by the color of their skin but the content of their character.**

ties or women, while its presumptions offer a ready excuse for those who do not wish to earn their way through effort. As permanent public policy, affirmative action affirms the stereotype that ethnic and racial minorities can never compete on an equal playing field with other Americans. The special rights the black leadership perpetually tries to legislate are admissions that we are not equal—an admission, quite frankly, that I am not prepared to make.

Further, affirmative action has not advanced the interests of black Americans as a class. It certainly has benefited some: usually, white, middle-class women and elite blacks who need it least. Studies indicate that affirmative action has moved employed blacks around in the labor pool but has not lowered the black unemployment rate.

Further, affirmative action in university admissions has resulted in graduation rates among black students that are almost half those of other students. Universities are interested in having black bodies on campus, but those students are often not completing their education. Some analysts believe that the racial double standard for admitting minority students does more harm than good.

■

Making judgments based on race is racism, any way you want to try to cover it over.

■

So we have with affirmative action the worst of all possible worlds: a morally abhorrent policy that is utterly failing to achieve its objective. It fails the test of both principle and pragmatism. We are creating special racial tribes that destroy any sense we have of a unified American nation. The ultimate irony: The more race-based laws we enact, the less real justice we will enjoy and the less real progress we will make, because all our achievements will appear unearned.

Some argue that many people have benefited from affirmative action. Perhaps, but many also benefited from slavery and its effects. Their descendants have to live with that guilt. Can we tolerate a policy that confers guilt upon our children for sins that are not their own? The answer is, unequivocally, no.

IT IS NOT CIVIL RIGHTS

Nevertheless, affirmative action originated from a legitimate need. Supporters of affirmative action point to a history of truly grotesque treatment of our African American forebears and to the lagging earnings and participation in top professions among minorities today. Opponents cite the double standards and inherent unfairness of the set-aside programs, as well as the regressive usage of race as a criterion for higher education enrollment, hiring, promotions, and contracts and grants. Both sides have some legitimate concerns.

People often speak of affirmative action as though it were a natural part of the civil rights movement and consistent with the movement's aims. To avoid confusion and unnecessary friction, I think it is important to draw some crucial distinctions here. Many people confuse affirmative action with civil rights; they are afraid that legal civil rights protections won during the 1960s would be lost by rejecting affirmative action. Others equate affirmative action with the quotas and preferential treatment it has become today.

Affirmative action originally was not necessarily either of these things. It was never supposed to embody reverse discrimination. The original civil rights legislation in 1964 contained a section known as Title VII, which is the basic employment statute prohibiting discrimination on the basis of race, color, religious origin, or sex. The bill's principal sponsor in the Senate, Hubert Humphrey, famously said upon its passage that he would physically eat the bill if it was ever used to discriminate against anyone on the basis of color. These laws are not affirmative action but basic civil rights protections against discrimination.

Over the years the courts began to make exceptions for what could be called "remedial" affirmative action, compensating individuals for specific acts of discrimination, especially racial discrimination, which was then still a common occurrence. They offered what is termed "make whole" relief, placing the party discriminated against in the position he would have otherwise been in had there not been discrimination. The courts could restore back pay, promotions, or other benefits that had actually been denied. Today's debate does not address this kind of affirmative action, which amounts to eminently fair restitution.

Today's debate about affirmative action centers on something beyond that. In the 1970s, courts began ruling that statistical disparities in enrollment or the workplace could be used to help determine whether discrimination had occurred. Businesses then could be presumed guilty by race-based head count and forced to prove their innocence of discriminatory practice. Also, courts ruled that tests or requirements with a "disparate impact" on applicants by race could be considered discriminatory. This applied even to hiring criteria that although not designed to discriminate might have that effect, like requiring a high-school diploma.

REAL AFFIRMATIVE ACTION

Though the Supreme Court undercut some of these later developments in subsequent decisions, they were fully restored in the 1991 Civil Rights Act, which was first called a "quota bill" and then signed by President George Bush.

Real affirmative action was to open opportunities where doors had been shut—and make no mistake, doors have been shut in the faces of blacks throughout America's history. There should be proper redress of grievances. There should be active recruiting of qualified minorities. I wholeheartedly support a ban on discrimination, as well as legal recourse to "make whole" those who have been hurt by actual instances of discrimination. That sort of affirmative action is designed to eliminate racial considerations and punish discrimination. Those laws were very necessary to protect minorities, especially blacks in the 1960s.

Yet I just as wholeheartedly oppose quota-oriented affirmative action based on color-conscious head counts, which reinstate race-based judgments for hiring and admissions. There is a world of difference between the two—in fact, the two approaches are opposite in spirit. Preferring minori-

Why Affirmative Action Has to Go

✗ It is a revival of the "Jim Crow" laws that subjugated blacks for 100 years.

✗ It undermines the earned successes of hardworking ethnic minorities and women.

✗ It offers a ready excuse for those who do not wish to earn their way through effort.

✗ It affirms the stereotype that ethnic and racial minorities can never compete on an equal playing field with other Americans.

ties, strictly on the basis of race, over other Americans equally or better qualified for jobs, advancements, and schools is the antithesis of the American ideal of equality and Martin Luther King's "dream."

As a short-term remedy for racism, it was argued, affirmative action had some merit. It was supposed to be a temporary fix to help the first group of disadvantaged blacks get a shot at equal opportunity with a little boost. As a permanent institution, affirmative action is patronizing, degrading, and self-defeating. Federally legislated "goals and timetables," (that is, quotas) will do far less for advancing minorities than they can do by proving themselves in open competition. Excellence would eradicate all the stigmas that attach to minorities or women through racism or sexism and that are furthered by affirmative action.

As a permanent institution, affirmative action is patronizing, degrading, and self-defeating.

While recognizing the realities of a racist history, we must address it without merely instituting a different shade of the same race-based laws and judgments. We must put the injustices of our past behind us without erecting new injustices in their place.

Americans need policies that promise more progress than affirmative action has delivered. Affirmative action helps a few, but its overall effect is to hurt the groups it is designed to aid. It helps some in the short run through unjustly hurting others. That inherent unjustness, in turn, aggravates already tense race relationships. Worst of all, affirmative action subverts the only really functional and morally acceptable criterion for judging anything or anyone: merit.

The Longest Climb

Only a handful of women have made it to the top of corporate America. A ground-breaking study of those who have made it shows that they are rewriting the rules of success.

Lisa Mainiero, Ph.D.

Lisa A. Mainiero, Ph.D., received her doctorate in organizational behavior from Yale University in 1983. She is an associate professor of management at the School of Business at Fairfield University in Fairfield, Connecticut.

By no means have women shattered the invisible barrier to top offices in the symbolic heart of corporate America—the Fortune 500. Women still hold a paltry seven percent of positions within three levels of chief executive officer in America's leading corporations. But slowly progress is being made. A few women have managed to pierce isolated holes through the glass ceiling.

Just who are these women? How did they manage to get there? And can we all learn something from their experience?

We've seen their pictures in all the right glossy business magazines—*Fortune, Business Week, Money.* But information about these women—about the career paths they followed, the obstacles they surmounted—has been limited. The popular news media has given us pictures of individual women who bravely forged their way to the top. No one has studied these women in the aggregate to dissect recurring patterns in their career paths. Until now.

Through referral from one to the next—what's technically known as a snowball sample—I found 55 respected, credible executive women who hold powerful positions in major U.S. corporations. Their mean age is 48 and their industries run the gamut: two-thirds are in banking, finance, or manufacturing, 15 percent in communications, television, or publishing, 4 percent in health or public service, and the remaining 12 percent in a variety of industries. I interviewed them at length, searching for common benchmarks in their careers and what they learned at each. I expected to hear of major feats of corporate politics, but, ironically, the vast majority saw themselves as apolitical. They told me that hard work and being in the right place at the right time had made the difference. Still, I beg to differ.

From their resounding denial of office politicking on, the same stories emerged again and again, no matter what the context. So I didn't have to work too terribly hard to trace the path they took to the top. Their careers sorted easily into four stages, taking a total of about 20 years, that I labeled according to key seasoning lessons the women learned in each: Political Naiveté, Building Credibility, Refining a Style, and Shouldering Responsibility.

In stage one—their first or second jobs—these women came to realize that their level of candor and directness set them apart from the rest. Over 60 percent of them told me that, purely out of naiveté, they took an unpopular stand early in their careers. One woman from the home-appliance industry characterized herself this way: "I would be the person who would come out of the crowd and say, 'Gee, Mr. Vice President, this is what is going on and guess what, you are screwing up. Do you know you are being laughed at?' " The comment typifies the mindset of these women. They didn't mince words or pull punches, they simply said, "The emperor has no clothes."

Their penchant for the truth and outspokenness clashes with the commonly held definition of politics—conforming to what upper management expects. Because they were women, they had escaped the traditional socialization process that men go through that might have "cured" them of their directness. No one took them aside to tell them the do's and don'ts: when to talk, when to shut up, who to ignore, and who to pay attention to.

But their early political blunders paid off. Their directness made them valuable in the eyes of senior management because no one else was willing to tell them what was going on. And senior management may have been more willing to take candor from a woman because they didn't know what to expect from them.

Looking back, the women believe that their political mishaps also helped them define the boundaries of corporate culture. By going slightly beyond the range of appropriate in their comments and personal style, the women became acutely aware of the norms driving their companies. That same woman in the home-appliance firm told me, "A big eye-opener for me was when these two guys sat me down and said, 'We don't always want to know what you think. You are usually right and you're too logical and we don't know how to argue with you.' This turned out to be the biggest piece of advice I ever got in my career. I learned to get people to come to me to ask for my opinion rather than telling them straight out." As another woman put it, "I learned how to say the right thing to the right person at the right time."

As the burgeoning executives moved into middle management, they had to make sure they were perceived as hard workers. For them, stage two was all about building credibility. They often were one of two or three women within a department and, given the times, expectations for them were low. They had to work against the stereotype of the woman who came to the office each day solely to snag a husband. The women said they worked inordinately hard, trading nights and weekends for meetings and pencil pushing to prove their mettle.

For a third of the women, building credibility meant performing remarkable feats of management while working within the system. One woman told me of her request to be placed into a commodities trading group. At the time there were no women in any such group. Still her boss agreed to do it. "The gentleman who ran the group was not pleased, although today he is one of my good friends. My attitude at that point was key. If I had taken the view that this man would have to just take it, I could have been heading for trouble," she relayed. But she was grateful and sympathetic. "I wasn't pushy or masculine; I had definite ideas about what I wanted to accomplish, but I did it in a more subtle way."

Others told me they took serious risks or were involved with corporate innovations to build credibility and jump start their careers. One woman from an electronics firm was given the opportunity to run a division in charge of developing a new device. There were people around her who had much more seniority—she was 26 years old at the time—and who knew a lot more about managing a factory than she did, but she was offered the job. Her manager took a risk on her, knowing she had significant people, organizational, and management skills. She accepted the job. Though she woke up every morning dodging fears of failure, she managed to pull it off. They took a risk on her and she made it happen. That device has since become a household name.

I also heard a lot about gaining credibility through managing people. Some women focused on developing a team-oriented participative style among their peers. They made sure their work was done and done well. As a result each developed a network of people who would say readily that she worked hard and deserved a promotion.

In the third stage, which I call refining a style, my interviewees had reached Division Manager level. That meant they were dealing with multimillion-dollar accounts, with several hundred employees under them. In describing this stage, they got philosophical on me because the nature of their job at that point gave them a new lens on their careers.

Eighty percent of the women told me they were very concerned with honing the team-building style they developed in the previous stage. These women were not tokens who felt territorial. Though persistent and tough, they had learned how to create a management vision, share power, and give others responsibility. The highest ranking woman in a clothing-manufacturing corporation describes her style this way: "Define clear objectives, give your people lots of leeway, stay informed on their progress, and especially, delegate and empower others to do what they need to do. My job is to get them the resources they need so they can be empowered to do the work."

Knowing these women were managers willing to give people a chance, employees vied to get into their area. Suddenly these women found a ground swell of support from below, as well as the support engendered from above and with peers from the previous stages. Here they sealed their credibility and management style.

Apart from empowering, a major component of that style was what Judith Rosener, in a classic *Harvard Business Review* article on the ways women lead, termed personal influence. Because women historically have not been in positions of power in the workplace, they were forced to step behind the scenes and manage through sheer personal influence. For example, one woman told me how she once drove an hour and a half to another manager's home at night, sat him down, and had a heart-to-heart about a project he was obstructing. The next day, everything was fine. So, in addition to learning how to be direct early on, they learned how to influence others on a very personal, covert level, to make things happen.

Some women described the level of responsibilities at this stage as overwhelming. To be effective, they had to make tough decisions. They faced the fine line between aggressive and pushy and managed to create a unique brand of assertiveness. A senior vice president of a beverage company captured it well: "There are some differences in the women who succeed. They tend to be better listeners, to have more insight into people, to come right out and say what they think, and not pull any punches. And very often they are right."

Enter stage four, the top floor, aptly called shouldering responsibility. Now in charge of huge subsidiaries and divisions, the executives' decisions touch thousands upon thousands of lives. And they feel the responsibility keenly. They put in long hours and when they do come home, work comes with them. They can't shake the thought that if, for example, they decided to stop one product line, 5,000 people might lose their jobs. And they have the added responsibility of being the sole woman (or one of two) at the top. Many admitted straight out that they thought their careers had been jump started because they were women, so they took on the role of mentor, publicly and personally, for other women on their way up.

As the resident female executive, they were the ones trotted out for dinner speeches and other public relations events. And they often had open-door policies for men and women who needed career advice. They were very accessible and willing to give of their time. One woman told me that on some days she and her secretary joked about putting a "Now serving..." sign on her door because so many people were lining up for just five minutes of her insight.

Bear in mind that the women were taking the time to be mentors while shouldering an enormous workload. Managing a sense of balance in their lives as executive, mentor, and family member was a vexing issue for them. In general though, most were happy with their lives and the choices they made. They had been ambitious, worked long, hard hours and "made it" in their companies. But there were personal trade-offs that they had difficulty articulating.

The compromises they made may have been in personal relationships, but, as a group, these women didn't come close to the stereotype of lonely, barren woman at the top. About a third were married to their original mates; another third had been divorced or were remarried; a little over 10 percent were single; and the remainder wouldn't comment. Of those who were married, more than 60 percent had children, though a third made reference to stepchildren.

Family life, for those who chose to have kids, is a tug of war. As one woman put it: "Recently I had a son. Now I am more emotionally torn than ever before. We have a great nanny, but it's hard to leave him." She had recently been to France for a week on a business trip. "There I was, sitting on the French Riviera in one of the best hotels in the world, and all I wanted to do was go home."

They all had the resources for child care, and some had husbands with flexible jobs who picked up the slack at home. So they structured their lives to allow for their careers, but they weren't happy having to spend time away from their families. A few told me they wouldn't want the same for their daughters.

I was the one who said, "Mr. Vice President, you're screwing up."

For some of the single women, their major regret was not taking the time out to get married and have children. But I heard others who echoed the woman who said, "I'm glad I didn't have kids. I wasn't cut out to be a mother. I was cut out to do this."

Now at the top, reflecting back on choices they made behind their office and bedroom doors, these women still resoundingly deny that politics had anything to do with their ascension. But all along, they were performing political acts, even though they didn't realize it or wouldn't label them that way. Take the woman who drove an hour and a half to talk to her colleague at home—that was nothing short of a political act. Yet if she were to read this, she would flatly deny it and say she was simply communicating.

Maybe what these women have done is redefine politics as another word for simply dealing with others in the workplace. In American corporate culture, politics implies back stabbing and game playing. But with the recent infusion of women, politics has come to mean the art of communication—what women do best. They have mastered communication overtly by being open, direct, and honest and covertly by slipping behind the scenes and developing a personal rapport. Creating relationships with people who work for them and with them proved just as important as interacting with those above.

In the words of one woman, "When I hear the term, 'corporate politics,' I think about human interrelations." This was how many of the women really felt. They would agree with an assertion made by organizational psychologist Abraham Zaleznik—that superior business performance requires managers to overcome their concern about politicking so they can realize and speak the truth about their firms.

But what sealed their success was an awareness of the culture surrounding them. These women achieved prominence by learning a very subtle, but key, business lesson—how to walk a tightrope between the norms of a culture and a critique of it. Corporate culture has slowly folded the criticisms into its seams and is struggling to create an evolving management vanguard—one that is more open, honest, and blind to gender.

VIOLENCE AGAINST WOMEN

It may be the biggest human rights issue in the world—and it is certainly one of the least discussed. Yet increasingly, women are finding ways to fight the mutilation, rape, beating, and murder that have been their lot.

Toni Nelson

Toni Nelson is a staff researcher at the Worldwatch Institute.

A GIRL IS MUTILATED IN EGYPT

It is not a ritual that many people would expect—much less want—to witness. Yet in the fall of 1994, the television network CNN brought the practice of female genital mutilation (FGM) into living rooms around the world, by broadcasting the amputation of a young Egyptian girl's clitoris. Coinciding with the United Nations International Conference on Population and Development in Cairo, the broadcast was one of several recent events that have galvanized efforts to combat the various forms of violence that threaten women and girls throughout the world. The experience suffered by 10-year-old Nagla Hamza focused international attention on the plight of the more than 100 million women and girls in Africa victimized by FGM. In doing so, it helped spur conference delegates into formulating an official "Programme of Action" that condemned FGM and outlined measures to eliminate the practice.

Euphemistically referred to as female circumcision, FGM encompasses a variety of practices ranging from excision, the partial or total removal of the clitoris and labia minora, to infibulation, in which all the external genitals are cut away and the area is restitched, leaving only a small opening for the passage of urine and menstrual blood. Nagla's mutilation, performed by a local barber without anesthesia or sanitary precautions, was typical. Although the physical and psychological consequences of FGM are severe and often life-threatening, the practice persists due to beliefs that emerged from ancient tribal customs but which have now come to be associated with certain major religions. In Israel, for instance, FGM is practiced by Jewish migrants from the Ethiopian Falasha community; elsewhere in Africa, it is found among Christian and Islamic populations. But FGM has no inherent association with any of these religions. Although some Islamic scholars consider it an important part of that religion, FGM actually predates Islam, and neither the Qur'an, the primary source for Islamic law, nor the Hadith, collections of the Prophet Mohammed's lessons, explicitly require the practice.

Justifications for FGM vary among the societies where it occurs (FGM is practiced in 28 African nations, as well as in scattered tribal communities in the Arabian Peninsula and various parts of South Asia). But most explanations relate in some way to male interest in controlling women's emotions and sexual behavior. One of the most common explanations is the need to lessen desire so women will preserve their virginity until marriage. The late Gad-Alhaq Ali Gad-Alhaq, Sheik of Cairo's al-Azhar Islamic University at the time of the CNN broadcast, explained it this way: the purpose of FGM is "to

From *World Watch,* July/August 1996, pp. 33-38.

moderate sexual desire while saving womanly pleasures in order that women may enjoy their husbands." For Mimi Ramsey, an anti-FGM activist in the United States who was mutilated in her native Ethiopia at age six, FGM is meant to reinforce the power men have over women: "the reason for my mutilation is for a man to be able to control me, to make me a good wife." Today, migrants are bringing FGM out of its traditional societies and into Europe, North America, and Australia. Approximately 2 million girls are at risk each year.

As in other countries where the practice is commonplace, Egypt's official policy on FGM has been ambiguous. Although a Ministry of Health decree in 1959 prohibited health professionals and public hospitals from performing the procedure, and national law makes it a crime to permanently mutilate anyone, clitoridectomies and other forms of FGM are not explicitly prohibited. An estimated 80 percent of Egyptian women and girls, or more than 18 million people, have undergone some form of FGM, which is often carried out by barbers in street booths on the main squares of both small towns and large cities.

Before the CNN broadcast, Egyptian public opinion seemed to be turning against the practice. In early 1994, activists founded the Egyptian Task Force Against Female Genital Mutilation. Later that year, during the population conference, Population and Family Welfare Minister Maher Mahran vowed to delegates that "Egypt is going to work on the elimination of female genital mutilation." Plans were even laid for legislation that would outlaw FGM. But some members of Egypt's religious community saw the broadcast as a form of Western imperialism and used it to challenge both the secular government of Hosni Mubarak and the conference itself.

In October 1994, Sheik Gad-Alhaq ruled that FGM is a religious obligation for Muslims. The same month, Minister of Health Dr. Ali Abdel Fattah issued a decree permitting the practice in selected government hospitals. The Minister's directive came just 10 days after a committee of experts convened by him condemned FGM and denied that it had any religious justification. Fattah affirmed his personal opposition, but insisted that the decree was necessary to "save those victimized girls from being 'slaughtered' by unprofessionals."

In the wake of the Minister's decision, plans for the bill outlawing FGM were postponed. Contending that Fattah had effectively legalized the procedure, national and international nongovernmental organizations sought to reverse the decision through petition drives, public education initiatives, and lawsuits. And on October 17, 1995, Fattah reversed his decision, and the Ministry of Health once again banned FGM in public hospitals. The anti-FGM legislation, however, remains on hold.

VIOLENCE IS A UNIVERSAL THREAT

Egypt's confused and ambivalent response to FGM mirrors in many ways the intensifying international debate on all forms of violence against women. And even though FGM itself may seem just a grotesque anomaly to people brought up in cultures where it isn't practiced, FGM is grounded in attitudes and assumptions that are, unfortunately, all too common. Throughout the world, women's inferior social status makes them vulnerable to abuse and denies them the financial and legal means necessary to improve their situations. Over the past decade, women's groups around the world have succeeded in showing how prevalent this problem is and how much violence it is causing—a major accomplishment, given the fact that the issue was not even mentioned during the first UN Women's Conference in 1975 or in the 1979 UN Convention on All Forms of Discrimination Against Women. But as the situation in Egypt demonstrates, effective policy responses remain elusive.

Violence stalks women throughout their lives, "from cradle to grave"—in the judgment of *Human Development Report 1995,* the UN's annual assessment of social and economic progress around the world. Gender-specific violence is almost a cultural constant, both emerging from and reinforcing the social relationships that give men power over women. This is most obvious in the implicit acceptance, across cultures, of domestic violence—of a man's prerogative to beat his wife. Large-scale surveys in 10 countries, including Colombia, Canada, and the United States, estimate that as many as one-third of women have been physically assaulted by an intimate male partner. More limited studies report that rates of physical abuse among some groups in Latin America, Asia, and Africa may reach 60 percent or more.

Belying the oft-cried cliché about "family values," studies have shown that the biggest threat to women is domestic violence. In 1992, the *Journal of the American Medical Association* published a study that found that women in the United States are more likely to be assaulted, injured, raped, or murdered by a current or former male partner than by all other types of attackers combines. In Canada, a 1987 study showed that 62 percent of the women murdered in that year were killed by an intimate male partner. And in India, the husband or in-laws of a newly married woman may think it justified to murder her if they consider her dowry inadequate, so that a more lucrative match can be made. One popular method is to pour kerosene on the woman and set her on fire—hence the term "bride burning." One in four deaths among women aged 16 to 24 in the urban areas of Maharashtra state (including Bombay) is attributed to "accidental burns." About 5,000 "dowry deaths" occur in India every year, according to government estimates, and some observers think the number is actually much higher. Subhadra Chaturvedi, one of India's leading attorneys, puts the death toll at a minimum of 12,000 a year.

The preference for sons, common in many cultures, can lead to violence against female infants—and even against female fetuses. In India, for example, a 1990 study of amniocentesis in a large Bombay hospital found that 95.5 percent of fetuses identified as female were aborted, compared with only a small percentage of male fetuses. (Amniocentesis involves the removal of a sample of amniotic fluid from the womb; this can be used to determine the baby's sex and the presence of certain inherited diseases.) Female infanticide is still practiced in rural areas of India; a 1992 study by Cornell University demographer Sabu George found that 58 percent of female infant deaths (19 of 33) within a 12-village region of Tamil Nadu state were due to infanticide. The problem is especially pronounced in China, where the imposition of the one-child-per-family rule has led to a precipitous decline in the number of girls: studies in 1987 and 1994 found a half-million fewer female infants in each of those years than would be expected, given the typical biological ratio of male to female births.

Women are also the primary victims of sexual crimes, which include sexual abuse, rape, and forced prostitution. Girls are the overwhelming target of child sexual assaults; in the United States, 78 percent of substantiated child sexual abuse cases involve girls. According to a 1994 World Bank study, *Violence Against Women: The Hidden Health Burden*, national surveys suggest that up to one-third of women in Norway, the United States, Canada, New Zealand, Barbados, and the Netherlands are sexually abused during childhood. Often very young children are the victims: a national study in the United States and studies in several Latin American cities indicate that 13 to 32 percent of abused girls are age 10 and under.

Rape haunts women throughout their lives, exposing them to unwanted pregnancy, disease, social stigma, and psychological trauma. In the United States, which has some of the best data on the problem, a 1993 review of rape studies suggests that between 14 and 20 percent of women will be victims of completed rapes during their lifetimes. In some cultures, a woman who has been raped is perceived as having violated the family honor, and she may be forced to marry her attacker or even killed. One study of female homicide in Alexandria, Egypt, for example, found that 47 percent of women murdered were killed by a family member following a rape.

In war, rape is often used as both a physical and psychological weapon. An investigation of recent conflicts in the former Yugoslavia, Peru, Kashmir, and Somalia by the international human rights group, Human Rights Watch, found that "rape of women civilians has been deployed as a tactical weapon to terrorize civilian communities or to achieve 'ethnic cleansing'." Studies suggest that tens of thousands of Muslim and Serbian women in Bosnia have been raped during the conflict there.

A growing number of women and girls, particularly in developing countries, are being forced into prostitution. Typically, girls from poor, remote villages are purchased outright from their families or lured away with promises of jobs or false marriage proposals. They are then taken to brothels, often in other countries, and forced to work there until they pay off their "debts"—a task that becomes almost impossible as the brothel owner charges them for clothes, food, medicine, and often even their own purchase price. According to Human Rights Watch, an estimated 20,000 to 30,000 Burmese girls and women currently work in brothels in Thailand; their ranks are now expanding by as many as 10,000 new recruits each year. Some 20,000 to 50,000 Nepalese girls are working in Indian brothels. As the fear of AIDS intensifies, customers are demanding ever younger prostitutes, and the age at which girls are being forced into prostitution is dropping; the average age of the Nepalese recruits, for example, declined from 14-16 years in the 1980s, to 10-14 years by 1994.

THE HIDDEN COSTS OF VIOLENCE

Whether it takes the form of enforced prostitution, rape, genital mutilation, or domestic abuse, gender-based violence is doing enormous damage—both to the women who experience it, and to societies as a whole. Yet activists, health officials, and development agencies have only recently begun to quantify the problem's full costs. Currently, they are focusing on two particularly burdensome aspects of the violence: the health care costs, and the effects on economic productivity.

The most visible effects of violence are those associated with physical injuries that require medical care. FGM, for example, often causes severe health problems. Typically performed in unsterile environments by untrained midwives or barbers working without anesthesia, the procedure causes intense pain and can result in infection or death. Long-term effects include chronic pain, urine retention, abscesses, lack of sexual sensitivity, and depression. For the approximately 15 percent of mutilated women who have been infibulated, the health-related consequences are even worse. Not only must these women be cut and stitched repeatedly, on their wedding night and again with each childbirth, but sexual dysfunction and pain during intercourse are common.

Infibulated women are also much more likely to have difficulties giving birth. Their labor often results, for instance, in vesico-vaginal fistulas—holes in the vaginal and rectal areas that cause continuous leakage of urine and feces. An estimated 1.5 to 2 million African women have fistulas, with some 50,000 to 100,000 new cases occurring annually. Infibulation also greatly increases the danger to the child during labor. A study of 33 infibulated women in delivery at Somalia's Benadir Hospital found that five of their babies died and 21 suffered oxygen deprivation.

Other forms of violence are taking a heavy toll as well. A 1994 national survey in Canada, for example, found that broken bones occurred in 12 percent of spousal assaults, and internal injuries and miscarriages in 10 percent. Long-term effects may be less obvious but they are often just as serious. In the United States, battered women are four to five times more likely than non-battered women to require psychiatric treatment and five times more likely to attempt suicide. And even these effects are just one part of a much broader legacy of misery. A large body of psychological literature has documented the erosion of self esteem, of social abilities, and of mental health in general, that often follows in the wake of violence. And the problem is compounded because violence tends to be cyclical: people who are abused tend to become abusers themselves. Whether it's through such direct abuse or indirectly, through the destruction of family life, violence against women tends to spill over into the next generation as violence against children.

Only a few studies have attempted to assign an actual dollar value to gender-based violence, but their findings suggest that the problem constitutes a substantial health care burden. In the United States, a 1991 study at a major health maintenance organization (a type of group medical practice) found that women who had been raped or beaten at any point in their lifetimes had medical costs two-and-a-half times higher during that year than women who had not been victimized. In the state of Pennsylvania, a health insurer study estimated that violence against women cost the health care system approximately $326.6 million in 1992. And in Canada, a 1995 study of violence against women, which examined not only medical costs, but also the value of community support services and lost work, put the annual cost to the country at Cdn $1.5 billion (US $1.1 billion).

One important consequence of violence is its effect on women's productivity. In its *World Development Report 1993*, the World Bank estimated that in advanced market economies, 19 percent of the total disease burden of women aged 15 to 44—nearly one out of every five healthy days of life lost—can be linked to domestic violence or rape. (Violence against women is just as pervasive in developing countries, but because the incidence of disease is higher in those regions, it represents only 5 percent of their total disease burden.) Similarly, a 1993 study in the United States showed a correlation between violence and lower earnings. After controlling for other factors that affect income, the study found that women who have been abused earn 3 to 20 percent less each year than women who have not been abused, with the discrepancy depending on the type of sexual abuse experienced and the number of perpetrators.

Violence can also prevent women from participating in public life—a form of oppression that can cripple Third World development projects. Fear may keep women at home; for example, health workers in India have identified fear of rape as an impediment to their outreach efforts in rural sites. The general problem was acknowledged plainly in a UN report published in 1992, *Battered Dreams: Violence Against Women as an Obstacle to Development*: "Where violence keeps a woman from participating in a development project, force is used to deprive her of earnings, or fear of sexual assault prevents her from taking a job or attending a public function, development does not occur." Development efforts aimed at reducing fertility levels may also be affected, since gender-based violence, or the threat of it, may limit women's use of contraception. According to the 1994 World Bank study, a woman's contraceptive use often depends in large part on her partner's approval.

A recurrent motive in much of this violence is an interest in preventing women from gaining autonomy outside the home. Husbands may physically prevent their wives from attending development meetings, or they may intimidate them into not seeking employment or accepting promotions at work. The World Bank study relates a chilling example of the way in which violence can be used to control women's behavior: "In a particularly gruesome example of male backlash, a female leader of the highly successful government sponsored Women's Development Programme in Rajasthan, India, was recently gang raped [in her home in front of her husband] by male community members because they disapproved of her organizing efforts against child marriage." The men succeeded in disrupting the project by instilling fear in the local organizers.

Women Break the Silence

"These women are holding back a silent scream so strong it could shake the earth." That is how Dr. Nahid Toubia, Executive Director of the U.S.-based anti-FGM organization RAINBO, described FGM victims when she testified at the 1993 Global Tribunal on Violations of Women's Human Rights.

Yet her statement would apply just as well to the millions of women all over the world who have been victims of other forms of violence. Until recently, the problem of gender-based violence has remained largely invisible. Because the stigma attached to many forms of violence makes them difficult to discuss openly, and because violence typically occurs inside the home, accurate information on the magnitude of the problem has been extremely scarce. Governments, by claiming jurisdiction only over human rights abuses perpetrated in the public sphere by agents of the state, have reinforced this invisibility. Even human rights work has traditionally confined itself to the public sphere and largely ignored many of the abuses to which women are most vulnerable.

But today, the victims of violence are beginning to find their voices. Women's groups have won a place for "private sphere" violence on human rights agendas, and they are achieving important changes in both national laws and international conventions. The first major reform came in June 1993, at the UN Second World Conference on Human Rights in Vienna. In a drive leading up to the conference, activists collected almost half a million signatures from 124 countries on a petition insisting that the conference address gender violence. The result: for the first time, violence against women was recognized as an abuse of women's human rights, and nine paragraphs on "The equal status and human rights of women" were incorporated into the Vienna Declaration and Programme of Action.

More recently, 18 members of the Organization of American States have ratified the Inter-American Convention on the Prevention, Punishment and Eradication of Violence Against Women. Many activists consider this convention, which went into effect on March 5, 1995, the strongest existing piece of international legislation in the field. And the Pan American Health Organization (PAHO) has become the first development agency to make a significant financial commitment to the issue. PAHO has received $4 million from Sweden, Norway, and the Netherlands, with the possibility of an additional $2.5 million from the Inter-American Development Bank, to conduct research on violence and establish support services for women in Latin America.

National governments are also drawing up legislation to combat various forms of gender violence. A growing number of countries, including South Africa, Israel, Argentina, the Bahamas, Australia, and the United States have all passed special domestic violence laws. Typically, these clarify the definition of domestic violence and strengthen protections available to the victims. In September 1994, India passed its "Pre-natal Diagnostic Techniques (Regulation and Prevention of Misuse) Act," which outlaws the use of prenatal testing for sex-selection. India is also developing a program to eradicate female infanticide. FGM is being banned in a growing number of countries, too. At least nine European countries now prohibit the practice, as does Australia. In the United States, a bill criminalizing FGM was passed by the Senate in May, but had yet to become law. More significant, perhaps, is the African legislation: FGM is now illegal in both Ghana and Kenya.

It is true, of course, that laws don't necessarily translate into real-life changes. But it is possible that the movement to stop FGM will yield the first solid success in the struggle to make human rights a reality for women. Over the past decade, the Inter-African Committee on Traditional Practices Affecting the Health of Women and Children, an NGO dedicated to abolishing FGM, has set up committees in 25 African countries. And in March 1995, Ghana used its anti-FGM statute to arrest the parents and circumciser of an eight-year-old girl who was rushed to the hospital with excessive bleeding. In Burkina Faso, some circumcising midwives have been convicted under more general legislation. These are modest steps, perhaps, but legal precedent can be a powerful tool for reform.

In the United States, an important precedent is currently being set by a 19-year-old woman from the nation of Togo, in west Africa. Fleeing an arranged marriage and the ritual FGM that would accompany it, Fauziya Kasinga arrived in the United States seeking asylum in December 1994. She has spent much of the time since then in prison, and her request for asylum, denied by a lower court, is at the time of writing under appeal. People are eligible for asylum in the United States if they are judged to have a reasonable fear of persecution due to their race, religion, nationality, political opinions, or membership in a social group. However, U.S. asylum law makes no explicit provision for gender-based violence. In 1993, Canada became the world's first country to make the threat of FGM grounds for granting refugee status.

Whichever way the decision on Kasinga's case goes, it will be adopted as a binding general precedent in U.S. immigration cases (barring the passage of federal legislation that reverses it). But even while her fate remains in doubt, Kasinga has already won an important moral victory. Her insistence on her right *not* to be mutilated—and on the moral obligation of others to shield her from violence if they can—has made the threat she faces a matter of conscience, of politics, and of policy. Given the accumulating evidence of how deeply gender-based violence infects our societies, in both the developing and the industrialized countries, we have little choice but to recognize it as the fundamental moral and economic challenge that it is.

Social Institutions: Issues, Crises, and Changes

- The Political Sphere (Articles 30–32)
- The Economic Sphere (Articles 33 and 34)
- The Social Sphere: Abortion, Health, Education, and Religion (Articles 35–38)

Social institutions are the building blocks of social structure. They accomplish the important tasks of society—the regulation of reproduction, socialization of children, production and distribution of economic goods, law enforcement and social control, and organization of religion and other value systems, for example.

Social institutions are not rigid arrangements; they reflect changing social conditions. Institutions generally change slowly. At the present time, however, many of the social institutions in the United States and many other parts of the world are in crisis and are undergoing rapid change. Eastern European countries are literally transforming their political and economic institutions. Economic institutions such as stock markets are becoming truly international, and when a major country experiences a recession, many other countries experience the effects. In the United States, major reform movements are active in political, economic, family, medical, and educational institutions.

The first section examines American political institutions. The first selection presents a sophisticated analysis of the way special interests influence government. Votes on specific bills are seldom bought by political action committees (PACs), but access is bought, and access is often converted to changes in the details of legislation that determine its real effects on the special interests. Next, Robert Wright agrees that the government is ruled by special interests, but argues that the politicians are too accessible to the people. We have "hyper democracy,"

UNIT 5

which hampers effective governance. Then, Philip Howard identifies another impediment to effective government—excessive bureaucratic regulations. He tells stories of outrageously ridiculous regulations and argues that lawyers are gumming up the arteries of society.

The next section examines the considerable changes currently taking place in corporate America. Jonathan Rowe reviews the history of the corporation and points out that corporations until a century ago were accountable to the governments that gave them charters of incorporation. He describes how almost overnight the states began to compete with each other for corporations by relaxing their laws to the point where corporations now are not accountable to any government beyond simply obeying the laws. This history provides a background for a discussion of what forms of accountability are needed for today's corporations. David Pearce Snyder examines the recent dramatic changes in the job market and forecasts the future trends and their consequences. The picture is not rosy.

The social sphere is also in turmoil, as illustrated by the unit's last section. One of the most contested issues is abortion, and Frederica Mathewes-Green tries to find areas of agreement and compromise in this field of battle. She finds that no one thinks that abortion is a positive event, but for some people it is better than the alternatives. Why not improve the alternatives and make abortion unnecessary, she asks. Health care and education are other troubled institutions. Willard Gaylin discusses its economic crisis. He concludes that the United States will be forced to develop some plan of health care rationing. The next entry examines a variety of recent initiatives for transforming public education systems from failing bureaucracies to deregulated systems with competition, choice, and an emphasis on performance. The last essay assesses whether the government could turn over its programs for the needy to religious groups that already have effective helping programs. Its conclusion is no.

Looking Ahead: Challenge Questions

Discuss whether or not it is important to preserve some continuity in institutions.

How can institutions outlive their usefulness?

Why are institutions so difficult to change? Cite examples where changes are instituted from the top down, and others where they are instituted from the bottom up. Do you see a similar pattern of development for these types of change?

Discuss whether or not it is possible to reform the political system to greatly reduce the corrupting role of money in politics.

What basic changes in the economic system are evident in things that you observe daily?

How should issues like abortion be decided?

MONEY CHANGES EVERYTHING

Daniel Clawson, Alan Neustadtl, and Denise Scott

In the past twenty years political action committees, or PACs, have transformed campaign finance. The chair of the PAC at one of the twenty-five largest manufacturing companies in the United States explained to us why his corporation has a PAC:

> The PAC gives you access. It makes you a player. These congressmen, in particular, are constantly fundraising. Their elections are very expensive and getting increasingly expensive each year. So they have an on-going need for funds.
>
> It profits us in a sense to be able to provide some funds because in the provision of it you get to know people, you help them out. There's no real quid pro quo. There is nobody whose vote you can count on, not with the kind of money we are talking about here. But the PAC gives you access, puts you in the game.
>
> You know, some congressman has got X number of ergs of energy, and here's a person or a company who wants to come see him and give him a thousand dollars, and here's another one who wants to just stop by and say hello. And he only has time to see one. Which one? So the PAC's an attention getter.

Most analyses of campaign finance focus on the candidates who receive the money, not on the people and political action committees that give it. PACs are entities that collect money from many contributors, pool it, and then make donations to candidates. Donors may give to a PAC because they are in basic agreement with its aims, but once they have donated they lose direct control over their money, trusting the PAC to decide which candidates should receive contributions.

Corporate PACs have unusual power that has been largely unexamined. In this book we begin the process of giving corporate PACs, and business-government relations in general, the scrutiny they deserve. By far the most important source for our analysis is a set of in-depth interviews we conducted with corporate executives who direct and control their corporations' political activity. The insight these interviews provide into the way corporate executives think, the goals they pursue, and the methods they use to achieve those goals is far more revealing than most analyses made by outside critics. We think most readers will be troubled, as we are, by the world view and activities of corporate PAC directors. . . .

WHY DOES THE AIR STINK?

Everybody wants clean air. Who could oppose it? "I spent seven years of my life trying to stop the Clean Air Act," explained the PAC director for a major corporation that is a heavy-duty polluter. Nonetheless, he was perfectly willing to use his corporation's PAC to contribute to members of Congress who voted for the act:

> How a person votes on the final piece of legislation often is not representative of what they have done. Somebody will do a lot of things during the process. How many guys voted against the Clean Air Act? But during the process some of them were very sympathetic to some of our concerns.

In the world of Congress and political action committees things are not always what they seem. Members of Congress want to vote for clean air, but they also want to receive campaign contributions from corporate PACs and pass a law that business accepts as "reasonable." The compromise solution to this dilemma is to gut the bill by crafting dozens of loopholes inserted in private meetings or in subcommittee hearings that don't receive much (if any) attention in the press. Then the public vote on the final bill can be nearly unanimous: members of Congress can assure their constituents that they voted for the final bill and their corporate PAC contributors that they helped weaken the bill in private. We can use the Clean Air Act of 1990 to introduce and explain this process.

The public strongly supports clean air and is unimpressed when corporate officials and apologists trot out their normal arguments: "corporations are already doing all they reasonably can to improve environmental quality"; "we need to balance the costs against the benefits"; "people will lose their jobs if we make controls any stricter." The original Clean Air Act was passed in 1970, revised in 1977, and not revised again until 1990. Although the initial goal of its supporters was to have us breathing clean air by 1975, the deadline for compliance has been repeatedly extended—and the 1990 legislation provides a new set of deadlines to be reached sometime far in the future.

Because corporations control the production process unless the government specifically intervenes, any delay in government action leaves corporations free to do as they choose. Not only have laws been slow to come, but corporations have fought to delay or subvert implementation. The 1970 law ordered the Environmental Protection Agency (EPA) to regulate the hundreds of poisonous chemicals that are emitted by corporations, but as William Greider notes, "in twenty years of stalling, dodging, and fighting off court orders, the EPA has managed to issue regulatory standards for a total of seven toxics."

Corporations have done exceptionally well politically, given the problem they face: the interests of business often are diametrically opposed to those of the public. Clean air laws and amendments have been few and far between, enforcement is ineffective, and the penalties for infractions are minimal. On the one hand, corporations have had to pay billions; on the other hand, the costs to date are a small fraction of what would be needed to clean up the environment.

This corporate struggle for the right to pollute takes place on many fronts. One front is public relations: the Chemical Manufacturers Association took out a two-page Earth Day ad in the Washington Post to demonstrate its concern for the environment; coincidentally many of the corporate signers are also on the EPA's list of high-risk producers. Another front is research: expert studies delay action while more information is gathered. The federally funded National Acid Precipitation Assessment Program (NAPAP) took ten years and $600 million to figure out whether acid rain was a problem. Both business and the Reagan administration argued that no action should be taken until the study was completed. The study was discredited when its summary of findings minimized the impact of acid rain—even though this did not accurately represent the expert research in the report. But the key site of struggle has been Congress, where for years corporations have succeeded in defeating environmental legislation. In 1987 utility companies were offered a compromise bill on acid rain, but they "were very adamant that they had beat the thing since 1981 and they could always beat it," according to Representative Edward Madigan (R-Ill.). Throughout the 1980s the utilities defeated all efforts at change, but their intransigence probably hurt them when revisions finally were made.

The stage was set for a revision of the Clean Air Act when George Bush was elected as "the environmental president" and George Mitchell, a strong supporter of environmentalism, became the Senate majority leader. But what sort of clean air bill would it be? "What we wanted," said Richard Ayres, head of the environmentalists' Clean Air Coalition, "is a health-based standard—one-in-1-million cancer risk." Such a standard would require corporations to clean up their plants until the cancer risk from their operations was reduced to one in a million. "The Senate bill still has the requirement," Ayres said, "but there are forty pages of extensions and exceptions and qualifications and loopholes that largely render the health standard a nullity." Greider reports, for example, that "according to the EPA, there are now twenty-six coke ovens that pose a cancer risk greater than 1 in 1000 and six where the risk is greater than 1 in 100. Yet the new clean-air bill will give the steel industry another thirty years to deal with the problem."

This change from what the bill was supposed to do to what it did do came about through what corporate executives like to call the "access" process. The main aim of most corporate political action committee contributions is to help corporate executives attain "access" to key members of Congress and their staffs. Corporate executives (and corporate PAC money) work to persuade the member of Congress to accept a carefully predesigned loophole that sounds innocent but effectively undercuts the stated intention of the bill. Representative Dingell (D-Mich.), chair of the House Committee on Energy and Commerce, is a strong industry supporter; one of the people we interviewed called him "the point man for the Business Roundtable on clean air." Representative Waxman (D-Calif.), chair of the Subcommittee on Health and the Environment, is an environmentalist. Observers of the Clean Air Act legislative process expected a confrontation and contested votes on the floor of the Congress.

The problem for corporations was that, as one Republican staff aide said, "If any bill has the blessing of Waxman and the environmental groups, unless it is totally in outer space, who's going to vote against it?" But corporations successfully minimized public votes. Somehow Waxman was persuaded to make behind-the-scenes compromises with Dingell so members didn't have to publicly side with business against the environment during an election year. Often the access process leads to loopholes that protect a single corporation, but for "clean" air most special deals targeted entire industries, not specific companies. The initial bill, for example, required cars to be able to use strictly specified cleaner fuels. But the auto industry wanted the rules loosened, and Congress eventually modified the bill by incorporating a variant of a formula suggested by the head of General Motors' fuels and lubricants department.

Nor did corporations stop fighting after they gutted the bill through amendments. Business pressed the EPA for favorable regulations to implement the law: "The cost of this legislation could vary dramatically, depending on how EPA interprets it," said William D. Fay, vice president of the National Coal Association, who headed the hilariously misnamed Clean Air Working Group, an industry coalition that fought to weaken the legislation. An EPA aide working on acid rain regulations reported, "We're having a hard time getting our work done because of the number of phone calls we're getting" from corporations and their lawyers.

Corporations trying to convince federal regulators to adopt the "right" regulations don't rely exclusively on the cogency of their arguments. They often exert pressure on a member of Congress to intervene for them at the EPA or other agency. Senators and representatives regularly intervene on behalf of constituents and contributors by doing everything from straightening out a social security problem to asking a regulatory agency to explain why it is pressuring a company. This process—like campaign finance—-usually follows accepted etiquette. In addressing a regulatory agency the senator does not say, "Lay off my campaign contributors, or I'll cut your budget." One standard phrasing for letters asks regulators to resolve the problem "as quickly as possible within applicable rules and regulations." No matter how mild and careful the inquiry, the agency receiving the request is certain to give it extra attention; only after careful consideration will they refuse to make any accommodation.

The power disparity between business and environmentalists is enormous during the legislative process but even larger thereafter. When the Clean Air Act passed, corporations and industry groups offered positions, typically with large pay increases, to congressional staff members who wrote the law. The former congressional staff members who work for corporations know how to evade the law and can persuasively claim to EPA that they know what Congress intended. Environmental organizations pay substantially less than Congress and can't afford large staffs. They are rarely able to become involved in the details of the administrative process or influence implementation and enforcement.

Having pushed Congress for a law, and the Environmental Protection Agency for regulations, allowing as much pollution as possible, business then went to the Quayle Council for rules allowing even more pollution. Vice President J. Danforth Quayle's Council, technically the Council on Competitiveness, was created by President Bush specifically to help reduce regulations on business. Quayle told the *Boston Globe* "that his council has an 'open door' to business groups and that he has a bias against regulations." The Council reviews, and can override, all federal regulations, including those by the EPA setting the limits at which a chemical is subject to regulation. The council also recommended that corporations be allowed to increase their polluting emissions if a state did not object within seven days of the proposed increase. Corporations thus have multiple opportunities to win. If they lose in Congress, they can win at the regulatory agency; if they lose there, they can try again at the Quayle Council. If they lose there, they can try to reduce the money available to enforce regulations, tie up the issue in the courts, or accept a minimal fine.

The operation of the Quayle Council probably would have received little publicity, but reporters discovered that the executive director of the Council, Allan Hubbard, had a clear conflict of interest. Hubbard chaired the biweekly White House meetings on the Clean Air Act. He owns half of World Wide Chemical, received an average of more than a million dollars a year in profits from it while directing the Council, and continues to attend quarterly stockholder meetings. According to the *Boston Globe,* "Records on file with the Indianapolis Air Pollution Control Board show that World Wide Chemical emitted 17,000 to 19,000 pounds of chemicals into the air last year." The company "does not have the permit required to release the emissions," "is putting out nearly four times the allowable emissions without a permit, and could be subject to a $2,500-a-day penalty," according to David Jordan, director of the Indianapolis Air Pollution Board.

In business-government relations attention focuses on scandal. It is outrageous that Hubbard will personally benefit by eliminating regulations that his own company is violating, but the key issue here is not this obvious conflict of interest. The real issue is the *system* of business-government relations, and especially of campaign finance, that offers business so many opportunities to craft loopholes, undermine regulations, and subvert enforcement. Still worse, many of these actions take place outside of public scrutiny. If the Quayle Council were headed by a Boy Scout we'd still object to giving business yet another way to use backroom deals to increase our risk of getting cancer. In *Money Talks* we try to analyze not just the exceptional cases, but the day-to-day reality of corporate-government relations. . . .

MYTH ONE: KEY VOTES ARE THE ISSUE

Many critics of PACs and campaign finance seem to feel that a corporate PAC officer walks into a member's office and says, "Senator, I want you to vote against the Clean Air Act. Here's $5,000 to do so." This view, in this crude form, is simply wrong. The (liberal) critics who hold this view seem to reason as follows: (1) we know that PAC money gives corporations power in relation to Congress; (2) power is the ability to make someone do something against their will; (3) therefore campaign money must force members to switch their votes on key issues. We come to the same conclusion about the outcome—corporate power in relation to Congress—but differ from conventional critics on both the understanding of power and the nature of the process through which campaign money exercises its influence.

The debate over campaign finance is frequently posed as, "Did special interests buy the member's vote on a key issue?" Media accounts as well as most academic analyses in practice adopt this approach. With the question framed in this way, we have to agree with the corporate political action committee directors we interviewed, who answered, "No, they didn't." But they believed it followed that they have no power and maybe not even any influence, and we certainly don't agree with that. If power means the ability to force a member of Congress to vote a certain way on a major bill, corporate PACs rarely have power. However, corporations and their PACs have a great deal of power if power means the ability to exercise a field of influence that shapes the behavior of other social actors. In fact, corporations have effective hegemony: some alternatives are never seriously considered, and others seem natural and inevitable; some alternatives generate enormous controversy and costs, and others are minor and involve noncontroversial favors. Members of Congress meet regularly with some people, share trust, discuss the issues honestly off the record, and become friends, while other people have a hard time getting in the door much less getting any help. Members don't have to be forced; most of them are eager to do favors for corporations and do so without the public's knowledge. If citizens did understand what was happening their outrage might put an end to the behavior, but even if the favors are brought to light the media will probably present them as at least arguably good public policy.

High-Visibility Issues

Corporate PAC officers could stress two key facts: First, on important highly visible issues they cannot determine the way a member of Congress votes; second, even for low-visibility issues the entire process is loose and uncertain. The more visible an issue, the less likely that a member's vote will be determined by campaign contributions. If the whole world is watching, a member from an environmentally conscious district can't vote against the Clean Air Act because it is simply too popular. An April 1990 poll by Louis Harris and Associates reported that when asked, "Should Congress make the 1970 Clean Air Act stricter than it is now, keep it about the same, or make it less strict?" 73 percent of respondents answered, "Make it stricter"; 23 percent, "Keep it about the same"; and only 2 percent, "Make it less strict" (with 2 percent not sure). Few members could risk openly voting against such sentiments. To oppose the bill they'd have to have a very good reason—perhaps that it would cost their district several hundred jobs, perhaps that the bill was fatally flawed, but never, never, never that they had been promised $5,000, $10,000, or $50,000 for doing so.

The PAC officers we interviewed understood this point, although they weren't always careful to distinguish between high- and low-visibility issues. (As we discuss below, we believe low-visibility issues are an entirely different story.) Virtually all access-oriented PACs went out of their way at some point in the interview to make it clear that they do not and could not buy a member's vote on any significant issue. No corporate official felt otherwise; moreover, these opinions seemed genuine and not merely for public consumption. They pointed out that the maximum legal donation by a PAC is $5,000 per candidate per election. Given that in 1988 the cost of an average winning House campaign was $388,000 and for the Senate $3,745,000, no individual company can provide the financial margin of victory in any but the closest of races. A member of Congress would be a fool to trade 5 percent of the district's votes for the maximum donation an individual PAC can make ($5,000) or even for ten times that amount. Most PACs therefore feel they have little influence. Even the one person who conceded possible influence in some rare circumstances considered it unlikely:

> You certainly aren't going to be able to buy anybody for $500 or $1,000 or $10,000. It's a joke. Occasionally something will happen where everybody in one industry will be for one specific solution to a problem, and they may then also pour money to one guy. And he suddenly looks out and says, "I haven't got $7,000 coming in from this group, I've got $70,000." That might get his attention: "I've got to support what they want." But that's a rarity, and it doesn't happen too often. Most likely, after the election he's going to rationalize that it wasn't that important and they would have supported him anyway. I just don't think that PACs are that important.

This statement by a senior vice president at a large *Fortune* 500 company probably reflects one part of the reality: most of the time members' votes can't be bought; occasionally a group of corporations support the same position and combine resources to influence a member's vote even on a major contested issue. Even if that happens, the member's behavior is far from certain.

Low-Visibility Issues and Nonissues

This is true only if we limit our attention to highly visible, publicly contested issues. Most corporate PACs, and most government relations units, focus only a small fraction of their time, money, and energy on the final votes on such issues. So-called access-oriented PACs have a different purpose and style. Their aim is not to influence the member's public vote on the final piece of legislation, but rather to be sure that the bill's wording exempts their company from the bill's most costly or damaging provisions. If tax law is going to be changed, the aim of the company's government relations unit, and its associated PAC, is to be sure that the law has built-in loopholes that protect the company. The law may say that corporate tax rates are increased, and that's what the media and the public think, but section 739, subsection J, paragraph iii, contains a hard-to-decipher phrase. No ordinary mortal can figure out what it means or to whom it applies, but the consequence is that the company doesn't pay the taxes you'd think it would. For example, the 1986 Tax "Reform" Act contained a provision limited to a single company, identified as a "corporation incorporated on June 13, 1917, which has its principal place of business in Bartlesville, Oklahoma." With that provision in the bill, Philips Petroleum didn't mind at all if Congress wanted to "reform" the tax laws.

Two characteristics of such provisions structure the way they are produced. First, by their very nature such provisions, targeted at one (or at most a few) corporations or industries, are unlikely to mobilize widespread business support. Other businesses may not want to oppose these provisions, but neither are they likely to make them a priority, though the broader the scope the broader the support. Business as a whole is somewhat uneasy about very narrow provisions, although most corporations and industry trade associations feel they must fight for their own. Peak business associations such as the Business Roundtable generally prefer a "clean" bill with clear provisions favoring business in general rather than a "Christmas tree" with thousands of special-interest provisions. Most corporations play the game, however, and part of playing the game is not to object to or publicize what other corporations are doing. But they don't feel good about what they do, and if general-interest business associations took a stand they would probably speak against, rather than in favor of, these provisions.

Second, however, these are low-visibility issues; in fact, most of them are not "issues" at all in that they are never examined or contested. The corporation's field of power both makes the member willing to cooperate and gets the media and public to in practice accept these loopholes as noncontroversial. Members don't usually have to take a stand on these matters or be willing to face public scrutiny. If the proposal does become contested, the member probably can back off and drop the issue with few consequences, and the corporation probably can go down the hall and try again with another member. . . .

What Is Power?

Our analysis is based on an understanding of power that differs from that usually articulated by both business and politicians. The corporate PAC directors we interviewed insisted that they have no power:

> If you were to ask me what kind of access and influence do we have, being roughly the 150th largest PAC, I would have to tell you that on the basis of our money we have zero. . . . If you look at the level of our contributions, we know we're not going to buy anybody's vote, we're not going to rent anybody, or whatever the cliches have been over the years. We know that.

The executives who expressed these views used the word *power* in roughly the same sense that it is usually used within political science, which is also the way the term was defined by Max Weber, the classical sociological theorist. Power, according to this common conception, is the ability to make someone do something against his or her will. If that is what power means, then corporations rarely have power in relation to members of Congress. As one corporate senior vice president said to us, "You certainly aren't going to be able to buy anybody for $500 or $1,000 or $10,000. It's a joke." In this regard we agree with the corporate officials we interviewed: a PAC is not in a position to say to a member of Congress, "Either you vote for this bill, or we will defeat your bid for reelection." Rarely do they even say, "Vote for this bill, or you won't get any money from us." Therefore, if power is the ability to make someone do something against his or her will, then PAC donations rarely give corporations power over members of Congress.

This definition of power as the ability to make someone do something against his or her will is what Steven Lukes calls a *one-dimensional view of power.* A *two-dimensional view* recognizes the existence of nondecisions: a potential issue never gets articulated or, if articulated by someone somewhere, never receives serious consideration. In 1989 and 1990 one of the major political battles, and a focus of great effort by corporate PACs, was the Clean Air Act. Yet twenty or thirty years earlier, before the rise of the environmental movement, pollution was a nonissue: it simply didn't get considered, although its effects were, in retrospect, of great importance. In one Sherlock Holmes story the key clue is that the dog didn't bark. A two-dimensional view of power makes the same point: in some situations no one notices power is being exercised—because there is no overt conflict.

Even this model of power is too restrictive, however, because it still focuses on discrete decisions and nondecisions. Tom Wartenberg . . . argues instead for a *field theory of power* that analyzes social power as similar to a magnetic field. A magnetic field alters the motion of objects susceptible to magnetism. Similarly, the mere presence of a powerful social agent alters social space for others and causes them to orient to the powerful agent. One of the executives we interviewed took it for granted that "if we go see the congressman who represents [a city where the company has a major plant], where 10,000 of our employees are also his constituents, we don't need a PAC to go see him." The corporation is so important in that area that the member has to orient himself or herself in relation to the corporation and its concerns. In a different sense, the mere act of accepting a campaign contribution changes the way a member relates to a PAC, creating a sense of obligation and need to reciprocate. The PAC contribution has altered the member's social space, his or her awareness of the company and wish to help it, even if no explicit commitments have been made.

Business Is Different

Power therefore is not just the ability to force people to do something against their will; it is most effective (and least recognized) when it shapes the field of action. Moreover, business's vast resources, influence on the economy, and general legitimacy place it on a different footing from other so-called special interests. Business donors are often treated differently from other campaign contributors. When a member of Congress accepts a $1,000 donation from a corporate PAC, goes to a committee hearing, and proposes "minor" changes in a bill's wording, those changes are often accepted without discussion or examination. The changes "clarify" the language of the bill, perhaps legalizing higher levels of pollution for a specific pollutant or exempting the company from some tax. The media do not report on this change, and no one speaks against it. . . .

Even groups with great social legitimacy encounter more opposition and controversy than business faces for proposals that are virtually without public support. Contrast the largely unopposed commitment of more than $500 billion for the bailout of savings and loan associations with the sharp debate, close votes, and defeats for the rights of men and women to take *unpaid* parental leaves. Although the classic phrase for something noncontroversial that everyone must support is to call it a "motherhood" issue, and it would cost little to guarantee every woman the right to an unpaid parental leave, nonetheless this measure generated intense scrutiny and controversy, ultimately going down to defeat. Few people are prepared to publicly defend pollution or tax evasion, but business is routinely able to win pollution exemptions and tax loopholes. Although cumulatively these provisions may trouble people, individually most are allowed to pass without scrutiny. *No* analysis of corporate political activity makes sense unless it begins with a recognition that the PAC is a vital element of corporate power, but it does not operate by itself. The PAC donation is always backed by the wider range of business power and influence.

Corporations are different from other special-interest groups not only because business has far more resources, but also because of this acceptance and legitimacy. When people feel that "the system" is screwing them, they tend to blame politicians, the government, the media—but rarely business. Although much of the public is outraged at the way money influences elections and public policy, the issue is almost always posed in terms of what politicians do or don't do. This pervasive double standard largely exempts business from criticism. We, on the other hand, believe it is vital to scrutinize business as well. . . .

The Limits to Business Power

We have argued that power is more than winning an open conflict, and business is different from other groups because of its pervasive influence on our society—the way it shapes the social space for all other actors. These two arguments, however, are joined with a third: a recognition of, in fact an insistence on, the limits to business power. We stress the power of business, but business does not feel powerful. As one executive said to us,

> I really wish that our PAC in particular, and our lobbyists, had the influence that is generally perceived by the general population. If you see it written in the press, and you talk to people, they tell you about all that influence that you've got, and frankly I think that's far overplayed, as far as the influence goes. Certainly you can get access to a candidate, and certainly you can get your position known; but as far as influencing that decision, the only way you influence it is by the providing of information.

Executives believe that corporations are constantly under attack, primarily because government simply doesn't understand that business is crucial to everything society does but can easily be crippled by well-intentioned but unrealistic government policies. A widespread view among the people we interviewed is that "far and away the vast majority of things that we do are literally to protect ourselves from public policy that is poorly crafted and nonresponsive to the needs and realities and circumstances of our company." These misguided policies, they feel, can come from many sources—labor unions, environmentalists, the pressure of unrealistic public-interest groups, the government's constant need for money, or the weight of its oppressive bureaucracy. Simply maintaining equilibrium requires a pervasive effort: if attention slips for even a minute, an onerous regulation will be imposed or a precious resource taken away. To some extent such a view is an obvious consequence of the position of the people we interviewed: if business could be sure of always winning, the government relations unit (and thus their jobs) would be unnecessary; if it is easy to win, they deserve little credit for their victories and much blame for defeats. But evidently the corporation agrees with them, since it devotes significant resources to political action of many kinds, including the awareness and involvement of top officials. Chief executive officers and members of the board of directors repeatedly express similar views. . . .

Like the rest of us, business executives can usually think of other things they'd like to have but know they can't get at this time or that they could win but wouldn't consider worth the price that would have to be paid. More important, the odds may be very much in their favor, their opponents may be hobbled with one hand tied behind their back, but it is still a contest requiring pervasive effort. Perhaps once upon a time business could simply make its wishes known and receive what it wanted; today corporations must form PACs, lobby actively, make their case to the public, run advocacy ads, and engage in a multitude of behaviors that they wish were unnecessary. From the outside we are impressed with the high success rates over a wide range of issues and with the lack of a credible challenge to the general authority of business. From the inside they are impressed with the serious consequences of occasional losses and with the continuing effort needed to maintain their privileged position.

HYPER DEMOCRACY

Washington isn't dangerously disconnected from the people; the trouble may be it's too plugged in

ROBERT WRIGHT

THE MOST VILIFIED EXPANSE OF asphalt in the history of the universe is, almost certainly, the Washington Beltway. Whereas most municipal freeways are associated with fairly mundane evils—potholes, rush-hour traffic—the Beltway has come to symbolize nothing less than a looming threat to American democracy. It is the great invisible buffer, impermeable to communication, that separates the nation's capital from the nation. It is what keeps many politicians—the ones with an "inside the Beltway" mentality—out of touch with the needs of the citizenry. It is the reason Washington's "media élites" are so clueless as to what's really on America's mind. It is why voters get congressional gridlock when they want action, and congressional action when they want nothing in particular. In a typical indictment, one columnist recently called some piece of Washington policymaking "too secret, too expert, too Beltway."

The solution, some observers say, is simple: use information technology to break through the Beltway barrier. Ross Perot champions an "electronic town hall," a kind of cyberdemocracy that, via push-button voting, would let people make the wise policy decisions their so-called representatives are failing to make for them. And now, vaguely similar noises are coming from someone with real power—inside-the-Beltway power, no less. Speaker of the House Newt Gingrich, who last week spoke at a Washington conference called Democracy in Virtual America, is trying to move Congress toward a "virtual Congress." He envisions a House committee holding "a hearing in five cities by television while the actual committee is sitting here." He's also letting C-SPAN's cameras, the electorate's virtual eyeballs, peer into more congressional hearings. And under a new program called "Thomas," after Thomas Jefferson, all House documents are being put on the Internet for mass perusal by modem. Thomas, says Gingrich, will shift power "toward the citizens out of the Beltway." It will get "legislative materials beyond the cynicism of the élite." And as this online material sparks online debate, Americans can "begin to have electronic town-hall meetings."

This may sound visionary, but it's nothing compared with the vision sketched by Gingrich's favorite futurists, Alvin and Heidi Toffler, in their book *Creating a New Civilization.* The Tofflers view the old-fashioned, physical Congress as suffering from a progressive erosion of relevance that calls for a wholesale rethinking of the Constitution. "Today's spectacular advances in communications technology open, for the first time, a mind-boggling array of possibilities for direct citizen participation in political decision-making." And since our "pseudo-representatives" are so "unresponsive," we the people must begin to "shift from depending on representatives to representing ourselves."

One problem with all this enthusiasm about electronically wiring the citizenry to the Washington policymaking machine is that in a sense, it's already happened. Politicians are quite in touch with opinion polls and have learned not to ignore the Rush Limbaughs of the world, with their ability to marshal rage over topics ranging from Hillary to the House post office. Public feedback fills Washington fax machines, phones and E-mail boxes. From C-SPAN's studios just off Capitol Hill, lawmakers chat with callers live—including callers who have been monitoring their work via C-SPAN cameras *on* Capitol Hill. More messages from the real world pass through the Beltway barrier than ever before. And contrary to popular belief, politicians pay attention. What we have today is much more of a cyberdemocracy than the visionaries may realize.

The other problem with all the plans for a new cyberdemocracy is that judging by the one we already have, it wouldn't be a smashing success. Some of the information technologies that so pervade Washington life have not only failed to cure our ills but actually seem to have made them worse. Intensely felt public opinion leads to the impulsive passage of dubious laws; and meanwhile, the same force fosters the gridlock that keeps the nation from balancing its budget, among other things, as a host of groups clamor to protect their benefits. In both cases, the problem is that the emerging cyberdemocracy amounts to a kind of "hyperdemocracy": a nation that, contrary to all Beltway-related stereotypes, is thoroughly plugged in to Washington—too plugged in for its own good.

The worst may be yet to come. The trend toward hyperdemocracy has happened without anyone planning it, and there is no clear reason for it to stop now. With or without a new Tofflerian constitution, there is cause to worry that the nation's inevitable immersion in cyberspace, its descent into a wired world of ultra-narrowcasting and online discourse, may render democracy more hyper and in some ways less functional. We have seen the future, and it doesn't entirely work.

"ELECTRONIC TOWN HALLS" featuring push-button voting have always faced one major rhetorical handicap: the long shadow of the Founding Fathers. The Founders explicitly took lawmaking power out of the people's hands, opting for a representative democracy and not a direct democracy. What concerned them, especially James Madison, was the specter of popular "passions" unleashed. Their ideal was cool deliberation by elected representatives, buffered from the often shifting winds of opinion—inside-the-Beltway deliberation. Madison insisted in the *Federalist Papers* on the need to "refine and enlarge the public views by passing them through the medium of a chosen body of citizens, whose wisdom may best discern the true interest of their country and whose patriotism and love of justice will be least likely to sacrifice it to temporary or partial considerations."

Madison would not have enjoyed watching how the "three strikes and you're out" provision wound up in last year's crime bill. The idea first took shape in California, where 18-year-old Kimber Reynolds had been murdered by a career felon. It was electronic from its very inception: the legislation was co-authored by talk-radio host Ray Appleton from Fresno who knew the victim's father and had fielded outraged calls after the killer's lengthy criminal record came to light. As the idea gained ground in California, it spread east. Its popularity was electronically catalyzed—on talk radio, especially—and electronically expressed in telephone polls, on the airwaves, by fax. President Clinton, with the support of Congress, complied promptly and cheerfully with the people's will. A push-button referendum would not have worked more effectively.

And as Madison might have guessed, the result was more gratifying viscerally than intellectually. "Three strikes" was notable not only for the shortage of politicians eager to loudly denounce it but also for the shortage of policy analysts who enthusiastically embraced it. While liberals deemed it draconian, many conservatives found it a constitutionally dubious exertion of federal power, as well as a sloppy form of draconianism. The law does nothing to raise the cost of the first two strikes, and meanwhile spends precious money imprisoning men past middle age, after most of them have been pacified by ebbing testosterone, free of charge. Of course, on the positive side, the law does have a catchy title. (How would the crime bill read if baseball allowed each batter five strikes?)

That policy "élites" aren't wild about something does not mean it's a mistake. But whatever the merits, the process that produced "three strikes and you're out" reflects a shift in American governance since the republic's founding—the growing porousness of the supposedly impregnable buffer around Washington. This was outside-the-Beltway politics, and is typical of our era.

This constant canvassing of public sentiment, one of two basic kinds of hyperdemocracy, is a straightforward outgrowth of information technology. The second basic kind—the one more specifically linked to gridlock and to the budget deficit—is a bit more subtle and more pernicious. And like the first one, it ultimately gets back to Madison. In addition to his dread of mass "passions," Madison had a second nightmare about "pure democracy": it "can admit of no cure for the mischiefs of faction."

He was mostly worried about oppressive majority factions. The modern special-interest group was a species unknown to him. Still, he had a fundamental insight that explains the subsequent origin of that species and its growth. The beauty of a large country, he noted, is the damper it places on factionalism. For when people are dispersed far and wide, even if some of them have "a common motive," the distance among them will make it hard for them to organize—"to discover their own strength and to act in unison with each other." The history of communications technology over the past 200 years is the history of those words becoming less true.

When town halls become call-in shows, deliberation loses and slogans become laws

Technologies ranging from the telegraph to the telephone, from typewriter to carbon paper have all made mass organization easier and cheaper. And since the 1960s, the technologies have unfolded relentlessly: computerized mass mailing, the personal computer and printer, the fax, the modem and increasingly supple software for keeping tabs on members or prospective members. The number of associations, both political and apolitical, has grown in lockstep with these advances. One bellwether—the size of the American Society of Association Executives—went from 2,000 in 1965 to 20,000 in 1990. As for sheerly political organizations: no one knows exactly how many lobbyists there are in Washington, but the *Congressional Quarterly* estimates that between 1975 and 1985 alone the number more than doubled and may even have quadrupled.

There was a second impetus to interest-group growth: in the 1960s, just as the technology of computerized direct mail was emerging, a proliferation of government programs created fresh issues to get interested in. Combined, the two factors were explosive. The American Association of Retired Persons, founded in 1958, did its first lobbying in 1965 with the arrival of Medicare. Over the next 25 years, its membership grew from a million to more than 30 million. Today it sends out 50 million pieces of mail a year. And when its members talk—especially about Medicare or Social Security—Congress listens.

Information technology has also revolutionized the form such talk can take. Meet Jack Bonner, voice for hire. On behalf of an interest group, Bonner and Associates can spew 10,000 faxes a night. But Bonner is better known for applying a more personal touch. When he works on a piece of legislation, he first isolates the likely swing votes, then has his software scan a database of the corresponding congressional districts, seeking residents whose profiles suggest sympathy with his cause. When influence is in order—after, say, a sudden and threatening development at a committee hearing—his people call these sympathizers, describe the looming peril and offer to "patch" them directly through to a congressional office to voice their protest. "But only in their own words," stresses Bonner, mindful that congressional staffs are getting better at spotting pseudo-grass-roots ("Astroturf") lobbying. Bonner charges $350 to $500 per call generated.

The striking thing about many modern special interests is how unspecial they are. Whereas a century ago lobbying was done on behalf of titans of industry, the members of, for example, AARP are no one in particular—just a bunch of people with an average income of $28,000 who happen to have gray hair. Indeed, they're so common that they account for one in six American adults—maybe you, maybe your mother, certainly someone you know. And if you're not in AARP, perhaps you are in the National Taxpayers' Union, the National Rifle Association or, less probably, the Possum Growers and Breeders Association. Or the American Association of Sex Educators, Counselors and Therapists. Or the Beer Drinkers of America—190,000 members strong and devoted to low beer taxes. "Almost every American who reads these words is a member of a lobby," writes Jonathan Rauch in his recent book *Demosclerosis*. "We have met the special interests, and they are us."

That lobbying has embraced the middle classes hardly means it's now an equal-opportunity enterprise. Wealthy people can still afford more of it, and the poor are still on the sidelines. Housing projects aren't leading targets for direct-mail solicitations. Still, lobbying has gotten *more* egalitarian, more democratic, as technology has made mobilizing groups cheaper.

On its face, that seems fine. If we must have lobbyists, they might as well represent regular people, not just oil barons. The trouble is that regular people, like oil barons, are usually asking for money, whether in the form of crop subsidies for farmers, tax breaks for shopkeepers, Medicare or Social Security payments, or various other benefits. So the increasingly "democratic" face of interest groups means the American government is asked to pay more, which means finally Americans of all classes are too. And the ultimate cost could be larger still. The budget deficit is not only a grave problem in itself, a theft of resources from the next generation, but also one reason politicians feel too strapped for cash to earnestly confront the other leading contender for gravest problem: the existence of an urban underclass. This sort of predicament is what the Founders designed representative democracy to solve. "They saw the public interest as a transcendent thing that enlightened people would be able to see and promote. It wasn't just a question of adding up all the interests," says historian Gordon Wood, author of *The Radicalism of the American Revolution.*

AMERICAN UNIVERSITY POLITICAL scientist James Thurber, author of the forthcoming book *Remaking Congress,* calls politics in the information age "hyperpluralism." He remembers sitting in congressional hearings for the 1986 tax-reform law as lobbyists watched the proceedings with cellular phones at the ready. "They started dialing the instant anyone in that room even *thought* about changing a tax break." Their calls alerted interested parties and brought a deluge of protest borne by phone, letter or fax. "There is no buffer allowing Representatives to think about what's going on," Thurber says. "In the old days you had a few months or weeks, at least a few days. Now you may have a few seconds before the wave hits."

The firms that orchestrate those waves from special interests often describe themselves as nonideological. But it is inherent in special-interest work that they will time and again be employed to defend the budget deficit against brutal assault at the hands of fiscal responsibility. When in February 1993 President Clinton proposed an energy tax that was hailed by economists and environmentalists, something called the Energy Tax Policy Alliance paid for a fatal multimedia campaign. When he suggested in the same budget plan cutting the business-lunch deduction from 80% to 50%, it was the National Restaurant Association that stirred to action, sending local TV stations satellite feeds of busboys and waitresses fretting about their imperiled jobs. And the restaurateurs hired Jack Bonner to roll out the Astroturf. "I see it as the triumph of democracy," Bonner said of his livelihood in a Washington *Post* interview. "In a democracy, the more groups taking their message to the people outside the Beltway and the more people taking their message to Congress, the better off the system is."

Special interests are legendary for distorting facts and preying on fear. The letter from the National Committee to Preserve Social Security and Medicare that helped trigger the rapid-fire repeal of a 1988 law to ensure catastrophic coverage under Medicare began with the words, "Your Federal Taxes for 1989 May Increase by Up to $1,600 . . . Just Because You Are Over the Age of 65"—even though 60% of all seniors wouldn't have paid a dime more in taxes. The tone of cool reason favored by the Founding Fathers is similarly lacking from this Jerry Falwell mailing: "American troops are again facing madman Saddam Hussein in the Persian Gulf—but the enemy here at home may be much more dangerous! . . . *Homosexuals are Bill Clinton's #1 allies.*"

Still, special interests often do traffic in facts. Their stock in trade is sounding alarms about legislative threats to people's interests, and often they can do that honestly. This, in a sense, is more disturbing than the cases of dishonesty or demagoguery. It means that the corruption of the public interest by special interests is no easily cured pathology, but a stubbornly rational pattern of behavior. The costs of each group's selfishness are spread diffusely across the whole nation, after all, while the benefits are captured by the group. Though every group might prosper in the long run if all groups surrendered just enough to balance the budget, it makes no sense for any of them to surrender unilaterally.

Given that accurate information, rationally processed, often leads people to undermine the public good, how excited should we be about Gingrich's Thomas, the online data base of congressional documents? Granted, there may not be a lobbyist manipulating the data flow. But that does not mean interest-group politics won't result. In cyberspace, technology may have finally reached a point where groups form spontaneously; on the Internet, passing information to a neighbor of like interest is a push-button exercise and can easily trigger a chain reaction. The result is a mass mailing that requires neither a centralized mass mailer nor the cost of postage and paper. And the next step can be a genuine, unrehearsed protest—grass roots, not Astroturf—that rolls into Congress or the White House via E-mail. Gingrich promises that Thomas will take power away from lobbyists, but if so, that may just mean Thomas has taken over their dirty work. (And after all, why should lobbyists be exempt from technological unemployment in the information age?)

Already the spontaneous formation of a single-issue interest group has been seen on the Net. In 1993 the Federal Government announced plans to promote the Clipper chip, which would have ensured the government's ability to decipher messages sent over phone lines by modem. The circulation of an anti-Clipper petition turned into a kind of impromptu online civil-liberties demonstration, boosting the number of signatures from 40 to 47,500.

> Protest has gone from the street to the fax, further hamstringing the capital's lawmakers

The oft-expressed hope for cyberspace is that any tendency toward fragmentation into contending groups will be offset by a capacity for edifying deliberation. And decorous dialogue has indeed been seen there. But cyberspace is also notorious for bursts of hostility that face-to-face contact would have suppressed. And a perusal of the Internet's newsgroups suggests that any tendencies toward convergence will have some real gaps to bridge. There's *alt.-politics.greens, alt.politics.libertarian, alt.-politics.radical-left, alt.fan.dan-quayle, alt.politics.nationalism.white, alt.fan.g-gordon-liddy, alt.rush-limbaugh.die.a.-flaming.death.* In a nation that has trouble fixing its attention on the public good and is facing increasingly bitter cultural wars, this is not a wholly encouraging glimpse of the future. There's no *alt.transcendent-.public.interest* in sight.

Not to worry. In the Gingrich camp, optimism runs rampant. Alvin Toffler and a few other seers prepared a "Magna Carta for the Knowledge Age" for the Progress and Freedom Foundation, which supports Gingrich. The authors dismiss in Tofflerian language those who fret about social balkanization in cyberspace as "Second Wave ideologues" (that is, Industrial Revolution dinosaurs, not clued in to the "Third Wave," the knowledge revolution). "Rather than being a centrifugal force helping to tear society apart, cyberspace can be one of the main forms of glue holding together an increasingly free and diverse society." The key to a "secure and stable civilization" is to make "appropriate social arrangements." Unfortunately, they never get around to specifying the social arrangements.

IF THERE ARE "ARRANGEMENTS" THAT would indeed bring stability to a cyberdemocratic society, they might be found by first dispelling all residues of election-year rhetoric and acknowledging that Washington, far from being out of touch, is too plugged in, and that if history is any guide, the problem will only grow as technology advances. The challenge, thus conceived, is to buffer the legislature from the pressure of feedback.

One possibility is electoral reform. But limiting the number of congressional terms, the current vogue, makes less sense than expanding the length of terms. The incentive to vote for a responsible budget that's healthful in the long run but painful in the short run depends on whether you face election next year or in three years.

There is another possible solution: leadership. Someone—a President, say—could actually stand up and tell the truth: that various public goods call for widespread sacrifice. But leadership is harder in an age of decentralized media—an age of "demassification," in the Tofflers' term. In the old days a President could give a prime-time talk on all three networks and know that he had everyone's attention. But this sort of forum is disappearing as conservatives watch National Empowerment Television, nature buffs watch the Discovery Channel, sports fans watch ESPN. When Clinton sought to address the nation last December after his party's debacle, the networks, conscious of their competition, were reluctant. But they finally gave him the midsize soapbox they can deliver these days. He used it to promise a tax cut.

This was widely viewed as shameless pandering, not to mention a cheap imitation of Republican pandering. But it wasn't viewed as surprising. Politics *is* pandering in a hyperdemocracy; to lead is to follow. Henry Aaron of the Brookings Institution sees this as one of the great social costs of modern information technology: in a kind of Darwinian process, hyperdemocracy weeds out politicians with the sort of strong internal principles that defy public opinion. "The advantage enjoyed by people willing to trim their views to the tastes of the electorate was smaller back when you couldn't find out what the electorate thought," Aaron says. Today, "few of those with core principles survive." If you don't obey talk-radio or public-opinion polls, you're ushered offstage.

Perversely, though, politicians are also punished if they *do* obey. The classic complaint about President Clinton is that he stands for nothing. Which is to say, he's willing to do just about anything to satisfy voters. Since the 1960s, the number of Americans expressing trust in Washington has dropped from around 70% to near 20%. This is commonly interpreted as a judgment against the growing power of special-interest lobbyists. But it could also be a reaction against the increasingly abject spinelessness of politicians, a byproduct of the very same trend. Indeed, the one clear exception to the number's downward drift are the Reagan years. Aaron says, "Even Democrats like me, who believed Ronald Reagan was a malign force, respected him, because, damn it, there were things he really stood for."

President Clinton, being inside the Beltway, periodically gets accused of being out of touch, of not "getting it." But he has shown that he "gets" the basics: that voters are worried about crime, for example, and that they hate to pay taxes. If there's anything major he doesn't "get," it's that in a hyperdemocracy, "getting it" can be self-defeating. The voters demand slavish obedience, but the more they receive it, the less they respect it. Has this sort of disrespect reached such a level as to be actually auspicious for a politician who leads rather than follows? It is hard to say. Few politicians seem inclined to conduct the experiment.

—With reporting by Wendy Cole/Chicago, John F. Dickerson/Washington, and Edwin M. Reingold/Los Angeles

The Death of Common Sense

Philip K. Howard

In the winter of 1988, nuns of Mother Teresa's Missionaries of Charity were walking through the snow in the South Bronx in their saris and sandals to look for an abandoned building that they might convert into a homeless shelter. They came to two fire-gutted buildings on 148th Street and, finding a Madonna amid the rubble, thought that perhaps Providence itself has ordained the mission. New York City offered the abandoned buildings at $1 each, and the Missionaries of Charity set aside $500,000 for the reconstruction. The only thing unusual about the plan was that the nuns, in addition to their vow of poverty, avoid the routine use of modern conveniences, and there would be no washing machines or other appliances. For New York City, the proposed homeless facility would literally be a godsend.

Although the city owned the buildings, no official had the authority to transfer them except through an extensive bureaucratic process. For 18 months, the nuns were directed from hearing room to hearing room discussing the project with bureaucrats. In September 1989, the city finally approved the plan, and the Missionaries of Charity began repairing the fire damage.

Providence, however, was no match for law. New York's building code, they were told after almost two years, required an elevator. The Missionaries of Charity explained that because of their beliefs they would never use the elevator, which also would add upward of $100,000 to the cost. The nuns were told the law could not be waived even if an elevator didn't make sense.

Mother Teresa gave up. Her representative said: "The Sisters felt they could use the money much more usefully for soup and sandwiches." In a polite, regretful letter to the city, the Missionaries of Charity noted that the episode "served to educate us about the law and its many complexities."

No person decided to spite Mother Teresa. It was the law of government, which controls almost every activity of common interest—fixing potholes, running schools, regulating day-care centers, controlling workplace behavior, cleaning up the environment and deciding whether to give Mother Teresa a building permit. And what it required offends common sense. Law designed to make Americans' lives safer and fairer has now become an enemy of the people.

Government acts like some extraterrestrial power, not an institution that exists to serve us. The bureaucracy almost never deals with real-life problems in a way that reflects an understanding of the situation. We seem to have achieved the worst of both worlds: a system of regulation that goes too far while it also does too little.

This paradox is explained by the absence of the one indispensable ingredient of any successful human endeavor: the use of judgment. In the decades since World War II, we have constructed a system of regulatory law that basically outlaws common sense. Modern law, in an effort to be "self-executing," has shut out our humanity.

The motives to make the law this way had logic. Specific legal mandates would keep government in check and provide crisp guidelines for citizens. Layers of "process"—procedural deliberations—would make sure decisions were responsible. Handing out "rights" would cure injustice. But it doesn't work. Human activity can't be regulated without judgment by humans, adjusting for circumstances and taking responsibility.

The public's fury with government was demonstrated in the November election, and the Republicans who won power now promise to get government off our backs. This rhetoric never turns to reality, though, because the public does not want to cut government essential services. The public is mad at *how* government works—its perpetual ineptitude and staggering waste—not mainly what government aims to do.

Moreover, the GOP's Contract With America proposes to take only small steps in the direction of real reform. One proposal would impose a moratorium on many pending regulations—an idea equivalent to cutting off your leg to lose weight. Another Republican theme is to return government functions to states, which could be a real benefit in certain areas like welfare but disastrous in others like environmental protection. The federalism idea ignores the fact that state governments are typically as ineffective and wasteful as the federal government. To liberate Americans from red tape, real reform must be aimed at simplifying *how* government works. Ending our suffocating legal system should be reformers' goal.

LAW REPLACES HUMANITY

The tension between legal certainty and life's complexities was a primary concern of those who built our legal system. The Constitution is a model of flexible law that can evolve with changing times and unforeseen circumstances. Today, we no

From *U.S. News & World Report,* January 30, 1995, pp. 57-61. Adapted from *The Death of Common Sense* by Philip K. Howard.

longer remember that words can impose rigidity as well as offer clarity. Law had an identity crisis when Oliver Wendell Holmes Jr., then a law professor, suggested in 1881 that law was not certain after all but depended on how the judge and jury saw the facts. This stimulated a wide range of reform movements, especially to codify the common law into statutes. Progressives at the turn of the century, New Dealers in the 1930s and Great Society reformers in the 1960s expanded the role of government in huge ways.

Another form of lawmaking also took hold in the '60s that focused not on government's role but on its techniques. Legal details proliferated. The *Federal Register,* a report of new and proposed regulations, increased from 15,000 pages in the final year of John Kennedy's presidency to over 70,000 pages in the last year of George Bush's.

Precision became the goal. The ideal of lawmaking was to anticipate every situation, every exception and codify it. With obligations set forth precisely, according to this rationale, everyone would know where he stood. But the drive for certainty has destroyed, not enhanced, law's ability to act as a guide. "Regulation has become so elaborate and technical that it is beyond the understanding of all but a handful of mandarins," argued former Stanford Law Dean Bayless Manning. No tax auditor, no building code examiner can possibly know all the rules in thick government volumes. What good is a legal system that cannot be known?

Instead of making law a neutral guidepost protecting against unfairness and abuse, this accretion of law has given bureaucrats almost limitless arbitrary power. A few years ago, the federal Occupational Safety and Health Administration decided workers needed more protection from hazardous chemicals. Bureaucrats decided that everything that could conceivably have a toxic effect should be shipped with a Material Safety Data Sheet describing the possible harmful effects of each item. The list grew and grew until it totaled over 600,000 products. In 1991, OSHA turned its attention to bricks. Bricks can fall on people, of course, but they had never been considered poisonous. The OSHA regional office in Chicago sent a citation to a brick maker for failing to supply an MSDS form with each pallet of bricks. If a brick is sawed, OSHA reasoned, it can release small amounts of the mineral silica. The fact that this doesn't happen much at construction sites was of no consequence. Brick makers thought the government had gone crazy, and they feared a spate of lawsuits. They began sending the form so that workers would know how to identify a brick (a "hard ceramic body with no odor") and giving its boiling point ("above 3,500°" Fahrenheit). In 1994, after three years of litigation, the poison designation was removed by OSHA.

The proliferation of rules may not produce the benefits of certainty and fairness, but it creates endless opportunities for smart lawyers seeking angles and advantages. Law, supposedly the backdrop for society, has been transformed into one of its main enterprises. For some billionaires, cable-TV companies, congressmen and litigators, close scrutiny and manipulation of the rules are a means to an end. The words of law give them lower taxes, a way to circumvent price controls, a secret means of playing favorites and a tool to grind the other side into the ground.

The rest of us feel like law's victims. We divert our energies into defensive measures to avoid tripping over the rules. Knowing for certain that full compliance is impossible, and that the government's reaction may be wholly out of proportion, law has fostered what Prof. Joel Handler has described as a "culture of resistance" where everyone is a potential adversary.

Law that leaves no room for judgment loses its original goal. Safety inspectors wander around without even thinking about safety. The YMCA of New York City, one of the last providers of low-cost, transient housing, gets regular citations for code violations like nonaligning windows and closet doors that do not close tightly. Does the city think that those clean, inexpensive rooms are somehow unworthy of a city that itself provides cots 18 inches apart for those who have no place to sleep? A city inspector recently told the YMCA, after it had virtually completed a renovation, that the fire code had changed and a different kind of fire-alarm system, costing an additional $200,000, would have to be installed. "Don't they realize that the $200,000 can provide yearlong programs for a hundred kids?" asked Paula Gavin, the YMCA's president. In our obsessive effort to perfect a government of laws, not of men, we have invented a government of laws *against* men.

THE NEVER-ENDING PROCESS

In 1962, Rachel Carson shocked the nation by exposing the effects of DDT and other pesticides in her book *Silent Spring.* There was also another side to the issue: Pesticides give us apples without worms and the most productive farms in the world. In 1972, Congress required the newly created Environmental Protection Agency to review all pesticides (about 600 chemical compounds at that time) and decide which should be removed from the market. The deadline was three years. More than 20 years have passed, and yet only 30 pesticides have been judged. Hundreds of others, including some on which there are data suggesting significant risk, continue to be marketed. "At this rate," said Jim Aidala, a onetime congressional pesticide expert, "the review of existing pesticides will be completed in the year 15000 A.D."

Making decisions, it almost seems too obvious to say, is necessary to do anything. Every decision involves a choice and the likelihood that somebody will lose something; otherwise, there would be no need to decide. This is the issue that paralyzes government decision making. "The problem with government," argues economist Charles Schultze of the Brookings Institution, "is that it can't ever be seen to do harm." Bureaucrats find it nearly impossible to say yes. Yet the act of not choosing is not benign: We may eat something bad because the EPA never made a decision.

Sometimes government cannot act even in the face of imminent peril. In the early-morning hours of April 13, 1992, in the heart of Chicago's downtown loop, the Chicago River broke through the masonry of an old railroad tunnel built in the last century. Several hundred million gallons of water from the river were diverted into the basements of downtown office buildings, knocking out boilers, short-circuiting countless electric switches, ruining computers and turning files into wet pulp. Total losses were over $1 billion. Several weeks before the accident, the leak in the tunnel had come to the attention of John LaPlante, then Chicago's transportation commissioner, a public servant with 30 years of exemplary service. He knew that the river was immediately overhead and that a break could be disastrous. He ordered his engineers to shore up the ceiling. As a prudent administrator, he also asked how much it would cost. The initial guess was about $10,000. His subordinates then went to a reputable contractor, who quoted $75,000. Although the amount was paltry, the discrepancy gave LaPlante pause. He put it out for competitive bids. Two weeks later, before the bidding process had even begun, the ceiling collapsed.

Bureaucrats don't even seem capable of looking in the right direction. How things are done has become far more important than what is done. The process has become an end in itself. A weakness of human nature that prompts many to avoid responsibility has become institutionalized in layers of forms and meetings. As a result, government accomplishes virtually nothing of what it sets out to do. It can barely fire an employee who doesn't show up for work.

The actual goals of government are treated like a distant vision, displaced by an almost religious preoccupation with procedural conformity. Public servants who dare take the initiative can be smothered. In the late 1980s, Michael McGuire, a senior research scientist at the University of California at Los Angeles, found himself in trouble. His lab is funded by the Veterans Administration. Its lawn also needs to be cut. When the lawn mower broke, McGuire decided to buy another one. During a subsequent routine audit, the federal auditor asked why the lawn mower was different. McGuire told the truth: He had thrown out a broken federal lawn mower (after saving usable spare parts). That prompted an investigation resulting in several meetings with high-level federal officials. After months, they rendered their findings: They could find no malice, but they determined McGuire to be ignorant of the proper procedures. He received an official reprimand and was admonished to study VA procedure, which he noted was "about the size of an encyclopedia." One other fact: McGuire bought the lab's lawn mower with his own money.

Orthodoxy, not practicality, is the foundation of process. Its credo is for complete fairness: its demons are corruption and favoritism. But concepts like equality and uniformity have no logical stopping point; no place where they say, "The Chicago commissioner shouldn't worry about bidding procedures with the river only a few feet above the leak." No one risks drawing the line. Any potential complaint is answered with one more "review" or "fact finding" procedure.

One destructive message of this is that bureaucrats can't be trusted to exercise their judgment. And the cost of this mistrust is almost inconceivable. The paperwork it generates in the name of "oversight" and "accountability" often costs more than the product it purchases. The Defense Department announced last year that it spent more on procedures for travel reimbursement ($2.2 billion) than on travel ($2 billion).

Setting priorities is difficult in modern government because process has no sense of priorities. Important, often urgent, projects get held up by procedural concerns. Potentially important breakthroughs in medicine wait for years at the Food and Drug Administration. Even obviously necessary safety projects can't break through the thick wall of process. In 1993, during a snowstorm at New York's La Guardia Airport, a Continental Airlines DC-9 had to abort a takeoff and ended up with its nose in Long Island Sound. Another 100 feet and many lives would probably have been lost. Two years earlier, another plane had slid off the runway, killing 27 people. The 7,000-foot runway is about 70 percent as long as those at most commercial airports, and the Port Authority of New York and New Jersey, which runs the airport, had been trying to add 460 feet for six years. But the agency had spent years talking to environmental agencies and community groups whose procedural rights took precedence over making the airport safer.

The irony of our obsession with process is that it has not prevented sharp operators from exploiting the government's contracting system, as the weapons-procurement scandals of the 1980s showed us. Its dense procedural thicket is a perfect hiding place for those who want to cheat. It has also led to a system so inconclusive that fairness is lost: Advocates can bludgeon their adversaries endlessly in public disputes that become too costly to see to a conclusion. And nothing ever gets done.

We must remember why we have process at all. It exists to serve responsibility. Process was not a credit card given out to each citizen for misconduct or delay; nor was it an invisible shield given to each bureaucrat. Responsibility, not process, is what matters.

A NATION OF ENEMIES

Finding a public bathroom in New York City is not easy. To remedy the problem, Joan Davidson, then director of the J. M. Kaplan Fund, a private foundation, proposed in 1991 to finance a test of six sidewalk toilet kiosks in different sections of the city. The coin-operated toilets, which cleaned themselves after every use, were small enough not to disrupt pedestrian traffic and would pay for themselves with the sale of advertising for the side panels. The proposal was greeted with an outpouring of enthusiasm. Then came the problem: Wheelchairs couldn't fit inside them. The director of the mayor's Office for People with Disabilities said the idea was "discrimination in its purest form." The city's antidiscrimination law, she pointed out, made it illegal to deny to the disabled any access to public accommodation. A protracted battle ensued.

The ultimate resolution, while arguably legal, was undeniably silly: Two toilet kiosks would be at each of the three locations, one for the general public and the other, with a fulltime attendant, for wheelchair users only. The test proved how great the demand was. The regular units averaged over 3,000 flushes per month. The wheelchair-friendly units were basically unused; the cost of the attendant was wasted.

Making trade-offs in situations like this is much of what government does. Almost every government act, whether allocating use of public property, creating new programs or granting subsidies, benefits one group more than another, and usually at the expense of everyone else. Most people expected leaders to balance the pros and cons and make decisions in the public interest. The government of New York, however, lacked this power because it had passed an innocuous-sounding law that created "rights" elevating the interests of any disabled person over any other public purpose.

Rights have taken on a new role in America. Whenever there is a perceived injustice, new rights are created to help the victims. Yet these new rights are intended as an often invisible form of subsidy. They are provided at everyone else's expense, but the check is left blank. They give open-ended power to one group, and it comes out of everybody else's hide. The vocabulary of accommodation, the most important language for a democracy, is displaced.

The "rights revolution" did not begin with any of this in mind. It was an effort to give to blacks the freedom the rest of the citizenry enjoyed. The relatively simple changes in law in the Civil Rights Act of 1964 sparked a powerful social change for the good. But that inspired reformers in the 1960s to consider using "rights" as a method to eliminate inequality of all kinds. Reformers zeroed in on the almost nuclear power that "rights" could bring to their causes. People armed with new rights could solve their own problems by going straight to court, bypassing the maddeningly slow process of democracy.

The most influential thinker was Charles Reich, at Yale. In his 1964 article "The New Property," Reich laid out a simple formula to empower citizens: Government decisions should be considered the property of the people affected. Government employees facing termination, professionals licensed by the state and contractors doing government business no longer would be subject to the judgment of government officials. Everyone would have a "right" that government would have no choice but to respect. In a follow-up article, Reich focused on what he thought was the area in which government largess was most important to the individual: welfare. He called for a "bill of rights for the disinherited." His vision heralded a new era of self-determination. Power would be transferred to the wards of the welfare state. Who would draw the line? "Lawyers," he proclaimed, "are desperately needed now."

Reich got his wish. Today, even ordinary encounters—between teachers and students, between supervisors and employees—now involve lawyers. Like termites eating their way through a home, "rights" began weakening the lines of authority of our society. Traditional walls of responsibility—how a teacher manages a classroom or how a social worker makes judgments in the field—began to weaken.

The Supreme Court embraced Professor Reich's concepts in a 1970 decision, *Goldberg v. Kelly,* which held that welfare benefits were "property" and could not be cut off without due process. Congress began handing out rights like land grants. Floodgates opened allowing juveniles, the elderly, the disabled, the mentally ill, immigrants and many others—even animals included under the Endangered Species Act—their days in court.

After 30 years of expanding rights against workplace discrimination, Congress has succeeded in "protecting" over 70 percent of all American workers. But are we witnessing a new age of harmony and understanding in the workplace? Hardly. Even those who are successful are bitter. Ellis Cose, in *The Rage of a Privileged Class,* describes the extraordinary anger of successful blacks—partners in law firms, executives in companies—who feel they are being held back because of race. These feelings, however, mirror those of white professionals who believe blacks are promoted primarily because they are black.

A paranoid silence has settled over the workplace. Only a fool says what he really believes. It is too easy to be misunderstood or to have your words taken out of context. Those hurt most by the clammed-up workplace are minorities and others whom the discrimination laws were intended to help. The dread of living under the cloud of discrimination sensitivity and the lurking fear of potential charges often act as an invisible door blocking any but the most ideal minority applicant.

Beyond the workplace, public schools have been the hardest hit by the rights revolution, especially when it comes to special education. Timothy W. was a profoundly disabled child, born with quadriplegia, cerebral palsy, cortical blindness and virtually no cerebral cortex. His mother thought he should go to school. Experts consulted by the Rochester, N.H., school district concluded he was not "capable of benefiting" from educational services, but a federal judge ruled that the school was obligated to provide a program because under the Individuals with Disabilities Education Act, it didn't matter whether he could benefit. Law books are filled with such cases as local school districts try to stem the hemorrhaging of their budgets. But the districts almost always lose. A right is a right.

Teachers, too, have suffered as the "rights" accorded students have allowed disruptive students to dominate classrooms. Except in the cases of egregious student conduct, most teachers often don't bother to act at all against misbehaving students. The procedures they have to follow are just too onerous. The easiest course is just to do nothing.

Rights are not the language of democracy. Compromise is. Rights are the language of freedom and are absolute because their role is to protect our liberty. By using the absolute power of freedom to accomplish reforms of democracy, we have undermined democracy and diminished our freedom.

THE RETURN TO PRINCIPLES

Like tired debaters, our political parties argue relentlessly over government's goals, as if our only choice is between Big Brother and the laissez-faire state. They miss the problem entirely. Our hatred of government is not caused mainly by what government aims to do. It's how law works that drives us crazy.

Law is hailed as the instrument of freedom because without law there would be anarchy, and we would eventually come under the thumb of whoever gets power. Too much law, we are learning, can have a comparable effect. It is no coincidence that Americans feel disconnected from government: The rigid rules shut out our point of view. By exiling judgment, modern law changed its role from useful tool to brainless tyrant.

Before American law became the world's thickest instruction manual, its goal was to serve general principles. The sunlight of common sense shines high whenever principles control: What is right and reasonable, not the parsing of legal language, dominates the discussion. With the goal always shining before us, the need for lawyers fades. Both regulators and citizens understand what is expected of them and can use their judgment. They can also be held accountable.

We have invented a hybrid government form that achieves nearly perfect inertia. No one is in control. No one makes decisions. This legal experiment hasn't worked out. It crushes our goals and deadens our spirits. Modern law has not protected us from stupidity and caprice but has made stupidity and caprice dominant features of our society. And because the dictates are ironclad, we are prevented from doing anything about it. Our founders would wince; they knew that "the greatest menace to freedom," as the late Chief Justice Earl Warren reminded us in 1972, "is an inert people."

Law cannot save us from ourselves. Waking up every morning, we have to go out and try to accomplish our goals and resolve disagreements by doing what we think is right. Energy and resourcefulness, not millions of legal cubicles, are the things that make America great. Let judgment and personal conviction be important again. There is nothing unusual or frightening about it. It's just common sense.

Reinventing the Corporation

The public gives corporations their right to exist and asks very little in return. It doesn't have to be that way

JONATHAN ROWE

When an act of simple human decency appears heroic, it's time to ask some basic questions about the culture in which that act takes place. That's what happened last December in an old mill town in Massachusetts. AT&T had just announced it was laying off 40,000 workers, even though profits and executive pay were soaring. U.S. corporations had inflicted over three million such layoffs since 1989, and there was a depressing new litany on the evening news: jobs down, stock market up. (More recently, it's been the equally revealing counterpart: jobs up, market down.) The new Republican Congress was giving these corporations the store. Yet the more they got, the less they seemed willing to give back in return.

Amidst this grim backdrop, Aaron Feuerstein's textile mill in Lawrence, Mass. burned down. Without hesitation, he announced that he wasn't going to pull up stakes and move to Mexico. He was going to rebuild the mill right there, in the state conservatives deride as "Taxachusetts." Not only that, he was going to pay his workers a month's wages to get them through the Christmas season.

Soon everyone was talking about Feuerstein. He was an ABC News "Person of the Week." He sat next to Hillary Clinton at the State of the Union address. Yet Feuerstein himself couldn't understand the fuss. "What?" he asked. "For doing the decent thing?" While his modesty may be excessive, his instinct is on the mark. By his example, he raised a pointed question: Why do we expect so little from major businesses these days?

Certainly, that thought is abroad in the land. Not since Ralph Nader's heyday in the early seventies have the words "corporate responsibility" come up so often in political debate. Because the prime messenger this time is Pat Buchanan, much of the mainstream media has dismissed the issue as the benighted economics of Bible-thumping ignorami. But the notion that corporations have responsibilities, just like real people, touches a deep chord; and while the term "corporate responsibility" may strike jaded modern ears as oxymoronic and naive, historically it is exactly right. "The corporation is a creature of the state," the Supreme Court observed back in 1906. "It is presumed to be incorporated for the benefit of the public."

How to get back to that original intent—to traditional moral values in the economic realm—is an urgent question. Conservatives say, correctly, that government should do less and individuals and business more. But if that's so, we have to consider whether the dominant form of business is up to the job. If there is to be less top-down regulation and more voluntary well-doing, then we have to ask whether the Wall Street-oriented corporation of today is capable of such a thing.

The issue here is not the hoary ideological de-

Jonathan Rowe is on the staff of Redefining Progress and is a contributing editor of The Washington Monthly.

bate between the government and the market. Rather, it concerns the kind of entities that will comprise the market. The corporation is an artifice of government, no less than the welfare system or foreign aid. Historically, it has evolved as society has changed. The time has come to ask what the next phase of that evolution should be. In simple terms, how can we reconnect the corporation to the social and community concerns it was originally intended to serve?

The way the corporation drifted from that role is a story that has all the elements of a neoconservative morality play. Corrupt government; self-serving politicians seeking to fill the public coffers and give the voters something for nothing; elite Eastern lawyers riding roughshod over traditional moral values; liberal permissiveness, economic style, and unintended consequences galore—it's all there.

Racing to the Bottom

Today we assume that corporations exist to make money. Ideologists-qua-economists like Professor Milton Friedman of the Hoover Institution assert this as a moral imperative. Yet if we travel back in time five or six hundred years, the European corporations of that era were very different from those of today. They were regulatory bodies, not acquisitive ones, which served to reconcile individual behavior with larger social ends. Gilds, boroughs, monasteries, and the like—today we would call them "mediating institutions," bulwarks of the civil society that has fallen into such disrepair. "Corporations have constituted, for the most part, the framework of society subordinate to that of the state," as John P. Davis put it in his exhaustive two-volume study back in 1905.

When the British Crown was eager to claim the wealth of the New World, it required commercial ventures of enormous scale. But few investors would come forward because they could be held responsible for the enterprise as a whole. The solution was the "joint stock company," the forerunner of today's business corporation. The corporate entity became a legal buffer zone between the enterprise and the actual owners. Ownership and responsibility were severed, so that a larger enterprise could result.

This was a radical step. Individual responsibility is a bedrock principle of the common law tradition. People must stand accountable for their actions and those taken on their behalf. To compromise this principle, something had to be given in return; specifically, the enterprise that gained this exemption had to serve the public in concrete ways.

Accordingly, in the early days, corporate charters were not granted to all comers the way they are today. They were granted selectively, one by one, for ventures that seemed worthy of public promotion and support. The trading companies that served as commercial agents of British foreign policy were prime examples; in today's terms they were much like Amtrak or the Tennessee Valley Authority.

This basically was the form of corporation that existed in Adam Smith's day. When Smith called England a "nation of shopkeepers," he was speaking literally. His notion of a divine market mechanism guiding individual ambition towards the betterment of all was premised on a world of individual business people, rooted in locality and place and subject to social mores and conscience. In one of the less prescient passages in the *Wealth of Nations*, Smith contended that corporations would never amount to much in the international marketplace. They were too cumbersome and bureaucratic, he said. Individual business people, with their superior "dexterity and judgment," would run rings around them.

In other words, the notion of the invisible hand is premised on a pre-corporate world that no longer exists. So too were the founding premises of the American republic. The colonists were extremely suspicious of corporations, which were seen as oppressive agents of the Crown and potential usurpers of the public will. At the time of the Constitutional Convention, only some 40 business corporations had been chartered in all the colonies. Most of these were for bridges, toll roads, and similar public-works endeavors. So it's not surprising the Founding Fathers omitted the corporation from the scheme of checks and balances by which they hoped to keep institutional power under restraint.

"The corporation is a creature of the state," the Supreme Court observed in 1906. "It is presumed to be incorporated for the benefit of the public."

Even as business corporations became more common, they stayed grounded in the premise that they were agents of a larger public good. Charters typically spelled out that the corporation in question was created to serve "the public interest and necessity." Some required shareholders to be local residents, and some even vested part ownership in the public. Before 1842, for example, the State of Maryland chose one third of the directors in the Baltimore and Ohio Railroad. There were also mandates to serve the public in specific ways: A bank charter in New Jersey, for example, required the company to help local fisheries.

As decades passed, the nation's surging commerce pushed against these restraints. Legislatures were besieged by supplicants seeking the privilege of operating as a corporation. In addition, the corporate charter process had gotten a taint of special privilege. The result was the general incorporation laws, which made the corporate form available to everyone.

This didn't mean that the suspicion of agglomerated power had died, nor the conviction that the corporate privilege was connected to a public purpose. Until 1837, for example, every state still required that corporations be chartered only for a particular kind of business. It took almost half a century for states to permit blank-check incorporation "for any lawful purpose." Restrictions on size were common too. New York, which was not unfriendly to business, limited corporations to $2 million in capital until 1881; and to $5 million until 1890.

Similarly, as late as 1903, almost half the states limited the duration of corporate charters to 20 to 50 years. Legislatures would actually revoke charters when the corporation wasn't fulfilling its responsibilities. With the rise of corporate megatrusts and the robber barons, the role of the corporation in American life became a topic of almost obsessive concern. The public wanted more accountability. They ended up with less.

The reason was an outbreak of corporate charter-mongering among the states that eventually dragged them all down to the lowest common denominator. The downward spiral actually began when John D. Rockefeller's lawyer concocted a way to evade the state charter laws. (The story is laid out in one of Ralph Nader's less-noted studies, called "Constitutionalizing the Corporation," which he wrote 20 years ago with Mark Green and Joel Seligman.) The device was the infamous secret trust agreement which enabled the Standard Oil empire to grow far beyond the size the state laws permitted. Rockefeller's conniving set off a wave of trusts—whiskey, sugar, lead, and others—which came to control much of the commerce in their industries.

Eventually these agreements came to light, and the state courts struck them down as exceeding the powers granted in the charters of the individual corporations that comprised them. The charter laws had done their job. Therefore, the charter laws would have to go, and the first to fall was New Jersey.

As Nader's study recounts, in 1890 a young New York lawyer named James B. Dill made an offer that the governor of New Jersey couldn't refuse. Enact the most liberal and permissive law in the land, Dill said. Let corporate managements do whatever they want, shareholders and public be damned. Corporations will flock to your state for new charters; revenues will pour into the treasury. Plus the clincher: Dill would form a company to handle the paperwork for the incorporating process, and the governor would get a cut.

Soon, Standard Oil, U.S. Steel, and other major companies were lining up for New Jersey charters. Prompted by the permissive new laws, there was an orgy of mergers and combinations, which hastened America's transition from a nation of entrepreneurs to one of corporate employees.

But politically, the New Jersey regime reaped its reward. By 1905 the state was running a surplus of almost $3 million. "Of the entire income of the government, not a penny was contributed directly by the people," the governor boasted. These revenues enabled him to push a rash of new social programs and public works projects, all without burdening ordinary taxpayers. In other words, tax-and-spend liberalism was boosted by the movement that set corporations free of every vestige of social accountability and restraint.

Other states were helpless to counter New Jersey's dirty deal. So, why not get a piece of the action? Those revenues were pretty attractive, as were the other benefits that flowed to politicians more directly. West Virginia was among the first; the "Snug Harbor for roaming and piratical corporations," a contemporary legal treatise called it. Maine, Delaware, Maryland and Kentucky followed in this new race to set the lowest standards and collect the most booty. At one point, the New York legislature enacted a special charter for the General Electric Company, based on the lax New Jersey standards, to prevent the company from absconding across the Hudson River. The Commission on Uniform Incorporation Law declared that the evolving system ensured "the maximum protection of fraud" and "the minimum of protection and cover ... for honest dealing."

There was a brief flurry of rectitude in New Jersey when Woodrow Wilson became governor in 1910. But others were only too ready to fill the temporary gap at the bottom—Delaware most of all. By the time of the Great Depression, Delaware had become home to more than one third of the industrial corporations on the New York Stock Exchange; 12,000 corporations claimed legal resi-

What "Pitchfork Pat" (and You) Can Do For Workers

Rare is the Brooklyn pensioner who can say she changed the lives of thousands of workers, but Marie Walsh did just that. She didn't run for office or write articles. She merely inquired about where her pension money was invested.

In 1984, the trustees of the $70 billion New York City Comptroller pension funds were meeting to discuss the Sullivan principles, which guided corporate investment in South Africa. Walsh went to the meeting. Why, she asked the trustees, weren't they also looking at companies doing work in Northern Ireland, where Catholic workers were being heavily discriminated against? The trustees considered her question, and then investigated her charge. Soon after, they drew up the MacBride Principles, which insisted that American companies in Ireland halt discriminatory employment practices. They convinced other funds, state legislatures, and hundreds of companies to adopt the principles. Ultimately, the principles led to an anti-discrimination employment law in Britain.

What Walsh did—raising a question at a trustee meeting—is within the rights of every American with money invested in a pension fund. Yet few workers or pensioners exercise those rights; most Americans have no idea which companies they own stock in.

So American workers watch the nightly news and mutter about the layoffs from corporations; sometimes they themselves are laid off. The irony is that together, they *own* corporate America. More than 50 million Americans own stock, enough that if they exercised their rights and responsibilities as company owners, the results could be revolutionary.

The real players are the "institutional investors," such as public and private pension funds, and most workers have a stake in them. Pension funds alone owned 27 percent of all outstanding U.S. equities as of the end of 1994, according to *The Brancato Report on Institutional Investment*. Other institutional investors with ready-made constituencies among their shareholders include religious institutions such as the Catholic archdioceses; large foundations; and even some mutual funds set up specifically for stockholders in search of socially responsible investing. Most institutional investors are invested for the long haul—10 or 20 years—and their holdings are so large they can't do the "Wall Street walk" and quickly sell out. Their only hope of influence is to encourage or force a company to reform. They are long-term owners, not short-term traders.

It was the corporate raiders of the 1980s who laid the foundations for shareholder influence. Until then, institutional investors were sleeping giants—they put up money, then shut up. But suddenly, shareholders saw management using *their* money to entrench itself against hostile takeovers. Up sprang a shareholders' movement in which the owners, and not just the managers, of capital began calling the shots.

Led by the California Public Employees Retirement System (CalPERS)—which currently has $96.9 billion in its portfolio—public pension funds began targeting companies who were underperforming in comparison to their industry average. The investors would first try quiet diplomacy, then pull out the big guns: *Wall Street Journal* ads and other bad publicity, shareholder resolutions, proxy votes. They have lobbied for more responsive directors, for disclosure of executive compensation—and sometimes for new executives. As such targeting has become routine, companies from Sears to Kodak have felt investors' wrath.

By performance measures, shareholder activists have been extraordinarily successful. Companies pushed to reform by members of the Council of Institutional Investors subsequently outperformed the Standard & Poor 500 by 14 percent, according to a study by the Council. Institutional investors—particularly pension funds—essentially stopped Kirk Kerkorian's takeover attempt at Chrysler because they opposed his plan to deplete the company's cash reserves.

But the activism, to date, has been almost exclusively about shareholders' rights—specifically, the right to maximize investment return by influencing corporate governance. For those who despair at corporate callousness, the real potential of shareholder activism lies in getting shareholders to recognize their *responsibilities* as company owners.

To date, this type of activism has been limited in scope, but effective where it has weighed in. The most visible campaigns have involved human rights in countries where American companies do business, such as South Africa or Nigeria. But there have also been shareholder campaigns on everything from glass ceilings to environmental protection.

One recent, and inspiring, example came from the New-York based Jesse Smith Noyes Foundation. The foundation was heavily invested in the computer chip maker, Intel. At the same time, it was helping the Southwest Organizing Project, a community-based organization in New Mexico, battle Intel's usurpation of water and land at its New Mexico plant. The foundation decided to take an obvious step: to use its status as an Intel shareholder—meeting with management, bringing shareholder resolutions—to force Intel to negotiate with Southwest. Intel eventually agreed to the foundation's demands.

Another unexpected flare of shareholder activism came in 1992, when Sire/Time Warner released rapper Ice-T's "Body Count" album, with the song "Cop Killer." There was a national outcry, but the most intense pressure came from Time Warner shareholders—particularly police pension funds. Some divested; others protested at the annual meeting. Distribution of "Cop Killer" was stopped; Ice-T and Time Warner later parted ways.

Labor unions, meanwhile, have brought the spirit of labor organizing to their status as major investors. The carpenters' union pension funds, for example, pressed Dow Chemical to improve workplace safety, and even visited plants to check on conditions. Several years ago, the AFL-CIO adopted a set of proxy guidelines that are a model for shareholder responsibility. They advocate the consideration of employee security and compensation in proxy voting; and they encourage funds to prioritize domestic over foreign investment.

In their concern for American workers and communities, though, labor unions too often are the exception. Socially responsible institutional investors are reluctant to confront corporate management on bread-and-butter issues relating to American workers. Shareholders can't—and shouldn't—micromanage the companies they invest in. But they could push for a set of guiding principles, such as avoiding cataclysmic layoffs or investing in retraining, that make clear their commitment to balancing the interest of other corporate stakeholders, from employees to the local community, with profitability.

Larger numbers of institutional investors aren't more active in part because of political dynamics in the investment community. Private pension funds are the largest institutional investors out there (they own 16 percent of all outstanding equity), and they are supposed to be independent of company management. Clubby corporate culture ensures that's rarely the case. Few private pension funds ever attempt to influence the corporate governance of the companies they hold stock in. And at public pension funds, the state or local politicians who sit on the board—or control who does—tend to dissuade activism that might alienate corporations. When one Wisconsin state pension fund targeted T. Boone Pickens, it got an angry call from Gov. Tommy Thompson saying "lay off"—he was trying to lure Pickens to do business in the state.

Other barriers to shareholder activism are regulatory. In 1992, for example, under pressure from the Business Roundtable, the Securities and Exchange Commission ruled that *all* issues pertaining to the general workforce (any worker below top management, in other words) were "ordinary business," and therefore beyond shareholder domain under SEC regulations; that means everything from whether a company discriminates in hiring to issues of fair compensation and safe working conditions. The ruling seemed designed to muzzle investors who *should* monitor a company's employment principles—because as owners they are responsible for the welfare of employees and communities.

Under President Clinton, the SEC has not reversed the 1992 ruling. "The SEC [is] bending over to encourage companies to dodge the scrutiny of their own shareholders on polices and responsibilities the administration is encouraging," says Diane Bratcher of the Interfaith Center on Corporate Responsibility, which represents 275 religious institutional investors. She's referring to Robert Reich's support for tax-code carrots and sticks to make corporations behave responsibly. As Bratcher suggests, doesn't it make more economic and political sense to encourage shareholders to wield the sticks themselves?

Wary money managers and trustees tend to portray such activism as a threat to profits; it's either social responsibility or maximum profitability, they think, and a fund manager's duty is to the latter. That dichotomy, though, is a false one: In the long run, how a company treats its workforce has direct impact on the value of its stock. A company that's managed well should never have to resort to the draconian layoffs that AT&T enacted. And insecure, underpaid, or undertrained employees aren't going to be the most productive.

Edward Durkin of the United Brotherhood of Carpenters and Joiners of America notes that when his funds look at a company's health and safety activities, they are acting as a representative of labor, but also as investors: They don't want to absorb liability for regulatory violations or workplace accidents. (The same concern, incidentally, may shake up the tobacco industry; worried about a massing stormfront of anti-tobacco litigation, some institutional investors are beginning to rethink their investments, or press conglomerates to spin off tobacco units or eliminate marketing practices that seem to target underage smokers.) Corporate accounting procedures, unfortunately, are geared to short-term, quantifiable measures—investment in labor is a cost, not an asset. Reforming those methods could help; so could AFL-CIO efforts to survey companies on issues like employee training.

Individual investors, of course, have far fewer shares, and therefore less clout. But they can still make a difference; any shareholder can bring a resolution or attend an annual meeting. Indeed, as the spring season of annual meetings approaches, who better than "Pitchfork Pat" Buchanan—who owns stock in AT&T and other companies whose labor policies he's deplored—to lead stockholding peasants in storming corporate castles?

—*Amy Waldman*

dence in a single office in downtown Wilmington.

When other states made their own runs for the bottom, Delaware dropped standards even further. In the 1960s, it simply turned over the drafting of a new law to a bevy of corporate lawyers. The legislature rubber stamped the results. By the mid-1970s, half the nation's largest 500 corporations were chartered in tiny Delaware. Now only directors, rather than shareholders, could propose amendments to the corporate charter. On top of that, corporate officers and directors could be indemnified for court costs and settlements of criminal and civil cases without shareholder approval.

In other words, the concept of individual responsibility for corporate management was entirely out the window. This trend had troubled the upholders of traditional morality from the very beginning. "The pernicious movement has decreased the personal responsibility on which the integrity of democratic institutions depends," Professor Davis observed seven decades earlier. William Carey, former chairman of the Securities and Exchange Commission and author of a leading textbook on corporate law, declared that the only public policy left in Delaware's corporation law was "raising revenue."

Such developments did not go unnoticed politically. Theodore Roosevelt, a Republican, actually established a Federal Bureau of Corporations to monitor the impact of these new and disruptive entities. Presidents Roosevelt, Taft, and Wilson all proposed federal chartering for large corporations, in order to stop the state charter-mongering and set minimum standards for national businesses. (Most corporations chartered in Delaware had little presence there besides a file in a lawyer's office.) These corporations "are in fact federal," Taft said, "because they are as wide as the country and are entirely unlimited in their business by state lines."

But Congress chose instead the routes of antitrust and regulation, trying to restrain what the permissive state charter laws had set loose. The first big growth of federal government came from new agencies, such as the Interstate Commerce Commission and the Federal Trade Commission, that were supposed to keep the burgeoning corporations in check.

In the New Deal-era, people like David Lillienthal, the first chairman of the TVA, tried to tilt the balance back toward individual entrepreneurs and local enterprise. But this group lost out, first to the megaplanners, and then to the new Keynesian technocrats, who reduced the economic problem to the manipulation of the valves and levers of taxation, expenditure, the money supply, and the maintenance of "consumer demand." Those who raised questions about corporate governance and the scale of enterprise were dismissed as descendants of the bumpkins and small-town nostalgics whom Richard Hofstadter ridiculed in *The American Political Tradition.*

Chartering A New Course

That is pretty much where things stand today. The Keynesian policy nostrums no longer hold, but the fixation on scientistic "macro" policy still dominates the national debate. Just get taxes right, cut federal spending, and the Red Sea will open wide. When a Robert Reich or a Pat Buchanan suggest that something more is involved—that the economic entities that do so well in America perhaps owe America something in return—they are dismissed as demagogues and know-nothings.

But Buchanan has let the genie out of the political bottle. Whether you agree with his remedies or not, he has tapped a genuine feeling of betrayal. Americans think the high and mighty as well as welfare mothers have responsibilities—and that corporate America has been obscenely derelict in this regard. *Business Week* put it well when it observed, "U.S. corporations may have to strike a new balance between the need to cut costs to be more globally competitive, and the need to be more responsible corporate citizens."

The problem, of course, is that corporations today aren't constituted to be responsible. The large corporation whose stock is traded publicly on stock exchanges has become an extension of the Wall Street mind. A CEO who did what Aaron Feuerstein did—that is, who forsook a measure of profit for acts of decency to employees and the community—could have furious portfolio managers to contend with. Shareholders might have his or her scalp. The publicly-traded corporation does to the economic realm what the political action committee does to politics—it reduces people to the lowest common denominator of self-seeking, and subordinates their best instincts to an institutional mandate to maximize pecuniary gain.

This might have been tolerable for a period in our history. But in a global economy, with the sense of community coming apart, institutional self-interest has flown out of orbit, sweeping up even smaller companies in its centrifugal pull. Yvon Chouinard, founder of Patagonia clothing, has put the problem eloquently. The goal in the entrepreneurial world today is to "grow (your business) as quickly as you can until you cash out, and retire to the golf courses of Leisure World," he wrote. "When the company becomes the fatted calf, it's sold for a profit and its resources and holdings are often ravaged and broken apart, disrupting family ties and jeopardizing the long-term health of local communities."

Chouinard, a mountain-climber who happened into the gear-and-clothing business, has gone the opposite route. Instead of cashing out, he decided to keep the company at a scale at which it can still embody the values he seeks to live.

You can't legislate that kind of decency. But it is possible to encourage the kind of enterprise that gives it room to operate. Individual and family owners, for example, at least have the ability to temper their profit-seeking with civic and other concerns. (The Cleveland Browns' owner, Art Modell, is a good reminder that not all of them will.) Family-owned newspapers have done this for decades, which is one reason that the corporatizing of the media is a tragedy.

Local ownership also can have a salutary effect, even from a business standpoint. Consider the Green Bay Packers football team, which is owned by about 2,000 individual shareholders, most of them residents of Green Bay. The Packers have a stability that is rare in business today, let alone pro sports. Packer fans don't worry that a greedy owner will skip town—because they are the owner. The Packers won the first two Superbowls and made it to the conference championship this year. Yet NFL rules now bar franchises from Packers-style local ownership.

If community-centered ownership works in pro sports, which have become the ultimate business, then why not in other businesses? Inner cities, for example, have trouble luring supermarkets and other essential services. There's lots of money to be made, but it takes a level of patience and hands-on commitment that most major corporations aren't willing to expend when there's such easy pickings in the suburbs. Local and community-based ownership can fill in the gap.

Employee ownership works much the same way. It's not a panacea, but an employee-owned company is less likely to move jobs abroad or lay off 40,000 workers when business is booming. It is more likely to take seriously the impact of business decisions on the community at large. *The New York Times* recently highlighted this balance at United Airlines, where employee owners resisted taking over USAir because it would have caused layoffs there. But United has also found that employee owners are also willing to make sacrifices for the company, such as accepting pay freezes, in return for job security.

Unfortunately, the most common form of employee ownership is the Employee Stock Ownership Plan (ESOP), which is a passive investment scheme that often denies workers a real say in policy. Management frequently uses ESOPS as a financial ploy to fend off takeover bids. That's not good enough. Ownership and control need to go together, as they do at United.

The nation can encourage socially-cohesive forms of ownership—family, local, and employee—in any number of ways. Taxes are an obvious example. Currently, estate tax laws push heirs to sell a family business, such as a newspaper, to generate the cash to pay the tax. That's insane; it should be possible to keep family businesses—up to a certain size—in the family, as long as there's active management and ownership. Similarly, the tax laws currently encourage mobility instead of stability. If there's a deduction for moving expenses, for example, shouldn't there be one for staying expenses, as when a firm stays in the inner city when it would be less expensive to operate elsewhere? Deductions for moving expenses should not be permitted at all when a profitable company is enticed away by public subsidies offered elsewhere.

By the 1960s, Delaware simply turned over the drafting of corporate law to corporate lawyers; the legislature rubber stamped the results.

With the largest corporations, we must address the problem directly, and revisit the corporate charter laws themselves. Presidents Theodore Roosevelt, Taft, and Wilson were right. The largest corporations should be chartered at the federal level. Decentralization is great for some things, but it just doesn't work for dealing with the largest economic institutions on the planet. A Delaware charter has become the business equivalent of a Liberian flag of convenience or a Haitian divorce. At the very least there should be a minimal federal standard, as with estate taxes, so the Delawares can't drag everyone down.

That standard should include individual responsibility for corporate officials, of the kind that existed before Delaware's lax and permissive regime. Charters should specify particular kinds of business, the way they used to. And charters should expire after a given period of years, for review under fair standards that ensure renewal except for egregious bad behavior. Nothing would do more to insure a minimum level of decent conduct—without a multitude of new regulations —than the knowledge that sooner or later, the corporate charter would come up for review.

Right-wing ideologues will fume about government "tinkering." But the corporation itself is a form of tinkering; and if the government is going to establish something, shouldn't there be some built-in accountability to the people? Let's not forget: the period of American history that is most associated with rugged individualism and the frontier enterprise spirit—that is, the era of President Andrew Jackson—was one in which corporate charter restrictions were still strong. Individualism thrived when institutional economic power was held in check.

Finally, there's a need to bring back a healthy dose of good old-fashioned shame, the kind that used to operate in small-town business settings. The way to do this, for sprawling corporations, is through public information regarding their behavior. Currently, the SEC collects elaborate data on corporate finances. Now we need to add information on community involvement, treatment of workers, investment in the U.S. and the rest. One of the environmental success stories of recent years has been the toxics inventory. Corporations have made significant steps towards cleaning up their operations, simply because they had to be good neighbors and make known the chemicals they used and were emitting. This approach works without cumbersome regulations and bureaucracy, and it can work in the broader arena of social responsibility as well.

An economy can't thrive for long if the underlying social structure is falling apart. After two decades of reinventing the corporation to be more efficient, we have to ask whether the result is merely a more efficient machine for corroding the nation's social glue. If Americans were asked which they thought the nation needed more right now, more corporate profits or more social cohesion and trust, there's not much question which they would choose. Aaron Feuerstein, explaining why he chose to keep his mill in Lawrence, gave his version of the ancient Jewish teaching: "In a place where there's moral depravity and no feeling of moral responsibility, do your damnedest to be a man." Wouldn't it be something if the chairman of AT&T would say that too?

THE REVOLUTION

The Revolution in the Workplace: What's Happening to Our Jobs?

Millions of workers are being displaced in today's information revolution, causing them enormous stress and pain. We may, like the Luddites before us, rebel—or we may join the revolution and reinvent ourselves.

David Pearce Snyder

Recent books with titles like *The Jobless Future* and *The End of Work* present strikingly unattractive portraits of the future. They claim that there is already a worldwide surplus of 850 million workers, and this number will rise in the future. The authors claim that, as we get better and better at using information technology to make us all more efficient, productivity will keep on rising and we will need fewer and fewer workers. This, in turn, means that we will have permanent mass underemployment and unemployment.

This vision of the future leads to talk of things like a guaranteed income, shorter workweeks, vouchers for community service, consumption taxes, or a "negative" income tax to redistribute work and income equitably.

The principal concern of many Americans is not simply that there will be fewer and fewer jobs, because, in actual fact, the number of jobs in the United States has been growing robustly for the past 15 to 20 years, and U.S. unemployment rates are much lower than those of most other mature industrial economies. Rather, the real concern is that there will be fewer and fewer "good" jobs than there used to be.

"Good" jobs may be loosely defined as jobs that pay enough for a reasonably comfortable lifestyle, and in the United States that has become a problem, because average U.S. wages have fallen more than 20% in the last 22 years. Today, nearly one-fifth of America's 85 million full-time, year-round workers earn *less* than a poverty wage. For young Americans, it's even worse: 47% of the 18- to 24-year-old U.S. workers hold full-time jobs that *pay less than a poverty income!*

This helps to explain why young Americans are not leaving home as fast as they used to—a serious problem for their parents, who begin to realize in their mid-40s that gracious living really begins when the last kid has left home and the dog has finally died. In 1980, half of America's 18- to 24-year-olds lived with their parents; today, it's two-thirds. In 1980, only 8% of our 25- to 34-year-olds lived with their parents; today, it's nearly 25%. What's more, when young people *do* leave home, more than half return within 30 months, often with spouses and offspring in tow!

The decline in average U.S. wages since 1973 has been the principal force behind both the kids returning home *and* the explosive growth of two-income households. In order to maintain middle-class lifestyles without middle-income jobs, nearly three-quarters of all U.S. married couples aged 20 to 55 now *both* work full time, up from one-third just 30 years ago. This socioeconomic adap-

This article originally appeared in *The Futurist*, March/April 1996, pp. 8-13. © 1996 by the World Future Society, Bethesda, MD. Reprinted by permission.

tation has sustained average household income in the United States, but it has also led to an increase in unattended children in society, which has been linked to rises in teen pregnancies, crime, and drug use and to declining academic performance in public schools.

Alarmed at the apparent intractability of unsatisfactory economic conditions—unemployment, underemployment, and falling wages—policy makers from 41 industrial nations gathered in Jackson Hole, Wyoming, in 1994 to determine how best to restore the vitality of their economies. They met for a week and adjourned without a solution. They concluded that the future appeared to confront the world's mature industrial economies with two alternative "end-states," neither of which is satisfactory. We can end up like Europe, with high wages and benefits but also high unemployment, they said, or like the United States, with low unemployment but with many workers who earn poverty wages with few benefits.

Meanwhile, planned corporate staffing reductions and projected government cutbacks strongly suggest that downsizing and wage deflation will continue in North America and last for at least another five years in Europe. Japan appears to be headed down the same path. Furthermore, these developments in the industrialized countries will likely reduce consumer demand for the exports of the developing world, slowing the latter's economic growth and progress as well.

The diminished prosperity of the industrial economies can reasonably be expected to have volatile sociopolitical consequences in rich and poor nations alike. In the United States, in particular, the devolution of the labor-intensive industrial economy has been accompanied by a widening income disparity between the top 20% of all households, who now earn 55% of all the money, and the bottom 80% of workers, who split the remaining 45% of all the money.

Class Warfare

The rhetoric of class warfare, regarded as politically incorrect in the United States since the onset of the Great Depression, has already raised its ugly head, along with appeals to economic nationalism and cultural chauvinism. In this context, it is worth noting a speech that presidential candidate Patrick Buchanan made in 1995 before the Republican National Committee in Philadelphia. Buchanan described the plight of many workers in these terms:

> These are fellow Americans, backed into low-wage jobs while owners of corporate stock make huge profits. These are people who work with their hands and tools; they don't work with word processors, and their jobs are threatened by machines and immigrant labor.

There is a striking parallel between Buchanan's statement and one of the first major political speeches ever made about technological innovation and job displacement: In July 1812, the poet Lord Byron addressed the British Parliament, seeking clemency for the leaders of the Luddite uprisings, who had been sentenced to hang for incitement to riot. Speaking of the new cloth-making machines, Byron declared:

> These machines were an advantage to the proprietors, inasmuch as they obviated the necessity of employing a number of workers, who were left in consequence to starve. You now call these men a mob. But this is the mob that labor in your fields and mills, that serve in your houses, that man your navy and your army, and whom neglect and calamity have driven to despair.

Byron's plea fell on deaf ears; the Luddites were hanged....

Byron and Buchanan were describing the effects of similar historic events: genuine techno-economic revolutions. We have long accepted the nineteenth-century Industrial Revolution as a historic event with fundamental long-term impacts on Western civilization and the world at large. Buchanan's language should serve notice that *we today* are currently passing through an equally revolutionary moment in history.

To fully appreciate the implications of such a revolution, we need to remind ourselves that, during the original Industrial Revolution in Britain, the central government in London had to nationalize local militias for seven straight years to keep the peace. We also know that there arose poorhouses and workhouses that were so terrible that the mortality rates among children reached 90%! To fully comprehend the future that confronts us today, it is important for us to remember that, although the Industrial Revolution of the past ultimately led to the mass-creation of high-value jobs and made most of us prosperous, it took more than a century for it all to work out. Moreover, there were terrible social costs before things got better.

Today, corporate reengineering is producing dramatic changes in the workplace. Michael Hammer, the current guru of management change, has said, "Reengineering takes 40% of the labor out of most processes. For middle managers, it is even worse; 80% of them either have their jobs eliminated or cannot adjust to a team-based organization that requires them to be more of a coach than a taskmaster." Hammer has also described the reengineering process as an organizational "civil war" waged by two allied groups—the rank and file plus the executives—against the middle managers.

To realize the economic benefits of the original Industrial Revolution, the largely self-disciplined primary producers of the existing agrarian economy—farmers—had to subject themselves to the rudimentary but rigorous *external* discipline of the factory system. Rank-and-file anguish over this perceived loss of personal freedom led most rural societies to characterize the coming of industrialization as a degrading and inescapable force of the future, a force that agricultural communities should forestall as long as possible.

The prophets of the Information Age have asserted that, to realize the economic benefits of the computer revolution, it will be necessary for essentially all citizens to subject themselves to even greater external discipline, in the form of rigorous academic skills, technical competencies, and flexible, collaborative workstyles. While millions are energized by the prospects of a collegial, informated workplace and the notion of sequential careers, other millions are hostile to what they perceive as a further loss of personal freedom and the subjugation of their individuality and independence to standards and expectations set by others. These people initially dismissed the early visions of informated, team-based operations as utopian and idealistic. Now that this revolutionary future has actually arrived, their denial is turning to fear and anger.

"While millions are energized by the prospects of a collegial, informated workplace, . . . other millions are hostile to what they perceive as a further loss of personal freedom."

Futurists Foresaw Our Troubled Times

While it may be of little comfort to the growing millions of displaced, underpaid workers of the industrial world, a number of futurists anticipated this historic moment 25 to 30 years ago, and they accurately described the social and economic consequences of the Information Revolution in considerable detail.

Many of those forecasts appeared in the pages of THE FUTURIST. For example, the April 1967 issue includes social psychologist Donald N. Michael's forecast that middle-level professional personnel in government and industry will increasingly have to be fired as rapidly shifting requirements make their jobs obsolete. He suggested that industry leaders would need to recruit managers for their hard-nosed ability to fire when firing is needed. In the same issue, Sir Leon Bagrit, a British cyberneticist, was quoted as saying, "The pace of technical change is so fast now that we must be prepared for a man to change not only his job, but his entire skills, three or four times in a lifetime."

Warren Bennis and Phillip Slater, in their wonderful little book *The Temporary Society* (1968), described how we were going to move from stable organizations to more adaptive, flexible organizations. These futurists talked about how uncomfortable people were going to be in that setting and how difficult it would be for people to adjust to the new organizational culture. Another scholar, Donald Schön, in his excellent book *Technology and Change*, noted that technology does not avalanche overnight into the mainstream of daily life; it takes time, and there are all kinds of fights and rear-guard actions as individuals and organizations try to avoid the pain and discomfort of change.

While *process* futurists like Schön and Michael were describing the dynamics of our present transformation, *content* futurists did an impressive job of depicting the evolving features of work and life in America during the final quarter of the twentieth century. Ian Wilson's characterization of the future for General Electric, *Our Future Business Environment* (1969), served as one of the crucial inputs to that company's successful transformation over the past 25 years. Most remarkable of all, almost certainly, was sociologist Daniel Bell's *The Coming of Post-Industrial Society* (1973), which foresaw a society increasingly divided between a small, prosperous technocratic elite and a huge population of lower-middle-income service drones. From the very month it came out, the number of middle-class jobs as a share of *all* the jobs in America fell, just as Bell had projected; it was almost as if he had programmed the economy.

So, 25 years ago, futurists accurately described many of the components and characteristics of our current techno-economic transition from labor-intensive to information-intensive production and management, complete with institutional realignments, displacement of workers, major social disruptions, and political turbulence, etc. But no one is particularly eager to stand up and take credit for having predicted all of that!

In any event, the 1960s' future has become the 1990s' present for most Americans. People now are enveloped in the revolution that was forecast in the 1960s. It's all around us!

Goodbye to the Safe Career

Back in the good old days of stable economic expansion—the 1950s and 1960s—a person could choose to do something new, exciting, and innovative in life but could also choose to say, "That's not for me: I am going to play it safe in life. I am going to stay in my home town and have a nice comfortable career in a salaried job." *That second choice no longer exists for the vast majority of Americans.* All of us are going to be innovators and pioneers over the next 10 years whether we like it or not, and many of us don't like it.

Just look at what the attitude surveys tell us. In the United States, three-quarters of the adults surveyed by the Harris poll and two-thirds of all high-school seniors surveyed by *Scholastic* magazine say they believe that the United States will be a worse place 10 years from now than it is today. No wonder young people are disaffected. No wonder they are not motivated to learn. They think the world in which they are going to spend their lives won't be a very satisfactory place.

Young men, in particular, are not happy with their prospects for the future. When pollsters ask U.S. female high-school students what they are going to do when they graduate, they list all kinds of roles they want to fill, like doctors, lawyers, engineers, accountants, civil servants, police and fire personnel, and fighter pilots. In short, they want to do all the things that men have always done. Moreover, fewer than 10% of female high-school seniors expect to spend their adult lives solely as mothers and domestic managers, while nearly 90% are committed to having *both* a career *and* an egalitarian marriage.

By comparison, nearly half of male high-school students express their preference for a traditional, male-headed, one wage-earner, nuclear family, where the wife stays home as mother and homemaker. And when male high-school students are asked what kinds of careers they would like to have, the only two job fields that consistently receive large numbers of responses in open-ended surveys are "professional athlete" and "media personality." A large proportion of America's young men—one-third or more—simply say they don't know what they're going to do as adults.

If these people do not acquire some constructive vision of purpose for themselves, they are likely to be very destructive, counterprogressive forces in society throughout their lives. We already see that. One recent estimate is that one-sixth of all 16- to 24-year-olds in America—mostly males—are currently "disaffected and disconnected." They are not associated with any formal role in society, nor are they in any formal relationship with another person. These are the folks who are joining the gangs in center cities and swelling the ranks of the rural militias. They see no roles for themselves in an informated society, and they are angry about their empty future.

So this is a very pregnant moment, not only for the future of America, but for all of the mature industrial economies and, ultimately, for the world at large. It is an uncertain moment, a scary moment. It is the kind of moment in history when, to paraphrase Alfred North Whitehead, familiar patterns fade, familiar solutions fail, and familiar options foreclose. Of course, the books and periodicals that are warning society about "dejobbing," "the end of work," and wage deflation only serve to increase public anxiety—a slow-motion variation of shouting "Fire!" in a crowded theater.

These alarming forecasts are largely naive extrapolations of the past two or three decades of workplace trends. However, in the absence of plausible alternative explanations for the gloomy economic news of the past 15–20 years and the gloomier prospects implicit in the extrapolation of those trends, industrial societies—fearful for the future—might very well take retrograde steps that will principally serve the interests of the economically dominant groups who want to protect their assets and resources from the forces of change. Nations that take such steps will falter: Social and economic progress will grind to a halt, and more and more jobs will be eliminated by the downside of this transformation. The anger and frustration displayed by people who do not understand what is happening to them will be a terrible and dangerous force in all the major industrial economies.

What Futurists Should Do Now

At this catalytic moment, the futurists of the world need to refocus their attention—and their expertise—from the distant horizon to the immediate. To people who are struggling just to get through today, a bunch of people who only talk about tomorrow are likely to appear irrelevant. But futurists *do* have useful knowledge for individuals, organizations, communities, and whole societies being confronted with multiple imperatives and opportunities for innovation and change. We can help voters, teachers, consumers, middle managers, etc., to better understand the interplay of large systems and multiple long-term trends in which we are all caught up.

We also ought to help the politicians to understand what's going on. Political leadership today involves continuous networking with all stakeholders and constituencies, through meetings, committees, task forces, and memberships in associations and ad hoc groups. This means that most politicians must rely on their staffs for substantive knowledge and subject-matter expertise to determine what they should actually say or do to be leaderly. On those occasions when a politician actually tries to explain the "big picture" to the electorate or talks to people *in detail* about what must be done for the nation or the community to cope with this dramatic and revolutionary moment, the entire apparatus of the political system seeks to reject the effort. Opponents, from both left and right, quickly condemn any discussion of "how the system works" as representing "discredited socialist central planning" or "fascist state capitalism" or "industrial policy" promoted by ivory tower academics and the Trilateral Commission.

"The anger and frustration displayed by people who do not understand what is happening to them will be a terrible and dangerous force."

Most striking of all, however, is the response of the professional political service providers to those politicians who seek to educate the public about the dynamics of economy, technology, and society, or to campaign on the basis of a specific program of long-term investments and commitments. The entire array of political handlers—the pollsters, the fundraisers, the PR people, the "spin doctors," media experts, and campaign strategists—tell every political candidate: "Don't try to talk about reality! Reality is too complicated. The average person can't understand what is really going on and will get confused." (Indeed, there is always the danger that the *candidate* might become confused.) "You can talk about principles and values, or about integrity and character. But whatever you do, don't talk about taking specific actions to produce specific long-term economic or social benefits. It will only get people upset with you."

Yet, without leadership from politicians, who will shape the U.S. public's understanding of the long-term, historic patterns of development so that they can gain a view of the future that will give meaning and purpose to this moment—*our* moment in history? We are experiencing a genuine techno-economic revolution: the sort of event that historians write entire chapters about.

Economist Joseph Schumpeter once described these recurring historic events as "waves of creative destruction." And in 1990, Paul David and his staff at the Center for Economic Policy Research at Stanford University detailed the dimensions and dynamics of such a technologic revolution in their enlightening report, *Computer and Dynamo: The Modern Productivity Paradox in a Not Too Distant Mirror.* [See note at end of article.]

The Stanford study describes the pace and scale of the U.S. economy's transformation from steam to electromechanical technology. In doing so, the study provides us with three powerful insights about *our* moment in time.

The first insight is that it takes one to two generations—50 to 70 years—to fully mature and assimilate the productive potential of a fundamental new technology, such as electricity or computers. It takes several decades for a new technology to mature to its full potential, for institutions to identify and establish the technology's most-productive applications, and for the work force to master those applications.

The second insight is that, during the first half of this transformational period, general levels of economic performance and prosperity plunge before they go up. Since the early 1970s, U.S. productivity has improved very slowly and real wages have been falling as the nation's employers have struggled to get undertrained workers to generate improved performance with immature technologies. But, by 2010 to 2015, the United States will become a mature information-intensive economy and surpass the levels of general prosperity and upward mobility experienced during the 1950s and 1960s.

Until then, however, we will be passing through a painful period of diminished national prosperity, including a reduction in public services and the social safety net. But there *is* light at the end of the tunnel! We are beginning to invent and mass-create high-value (and high-paying) jobs: As more rank-and-file operations are "informated," employers are increasingly required to pay the 15% to 20% labor market wage premium typically commanded by computer-competent workers.

The third insight from the Stanford study is that the really beneficial impacts of a new technology don't arise until two-thirds of the way through the transition. For America, that moment is just ahead, in the next five to 10 years. The new integrated information technology, highly refined and matured, is about to avalanche into all of our homes and workplaces, enriching and complicating daily life for everybody. It will be wonderful, but it is not quite here yet, and most people need some hope. Yet most of the new leaders in the mature industrial economies are short on talking about hope. The new leaders talk about bottom lines and biting bullets, about cutting back and getting back to basic values. But we cannot really go "back to the future."

These are revolutionary times. To survive and prosper in a revolution, you must plan for a revolution. Nothing less will do! This is what every U.S. organization and community should be doing right now. To assist in these efforts, practitioners in the futures-research fields need to become more engaged in our own communities—with local schools, public-utility reform, land-use planning, and economic development schemes, etc.—to help the participants in all of these activities understand that the difficulties that confront us all today are *not* the product of failed policies or faltering leadership. What's happening is part of a recurring historic cycle. We are living through a great technological revolution.

We in America are reinventing our corporations, reinventing government, reinventing labor relations, reinventing health care and public education. We are reinventing all of our great institutions, and when we are all done, we will have reinvented America. The other industrial nations are beginning to reinvent themselves as well. Eventually, the whole world will be reinvented.

Of course, it will always be important to talk about long-term future concerns. But right now, a large percentage of the world's population needs assistance in getting through this immediate revolutionary transformation. To address this need, we should all begin to "learn locally and share globally," as my partner, Gregg Edwards, likes to say. That is what the motto of all futurists should be for the next five years. Futurists should be at the heart of a collegial worldwide learning-sharing network, so that the lessons learned by the innovators and risk-takers around the world can quickly be learned by everyone else, accelerating the pace of progressive change. In this way, we can help the world rediscover a future that is hopeful, creative, and constructive—a long-term future that we can all begin to make happen today.

David Pearce Snyder is the Lifestyles editor of THE FUTURIST and is principal partner of Snyder Family Enterprise, 8628 Garfield Street, Bethesda, Maryland 20817. Telephone 301/530-1028. He will speak and give a course at the World Future Society's Eighth General Assembly, July 14–18, 1996.

For a copy of the report *Computer and Dynamo: The Modern Productivity Paradox in a Not Too Distant Mirror,* contact the Center for Economic Policy Research at Stanford University, Stanford, California 94305. Telephone: 415/725-1874. This report describes how long it took the United States to shift from steam to electro-mechanical technology. It contains 30 pages of text and 30 pages of graphs, tables, and charts. The cost of the report is $3 plus postage.

Seeking Abortion's Middle Ground

Why My Pro-Life Allies Should Revise Their Self-Defeating Rhetoric

Frederica Mathewes-Green

Frederica Mathewes-Green is a columnist for Religion News Service and does commentary on National Public Radio. She is the author of "Real Choices," a book on alternatives to abortion (Questar, 1994). She lives in Baltimore with her husband and three teenaged children.

I WAS pro-choice at one point in my life, but I came over to a pro-life position years ago. I've been there ever since. Perhaps because of my background, I think there's a logic to the pro-choice position that deserves respect, even as we engage it critically. It is possible to disagree with somebody without calling them baby-killers, without believing that they are monsters or fiends. It is possible to disagree in an agreeable way.

The abortion argument is essentially an argument among women. It's been a bitter and ugly debate, and I find that embarrassing. For me, that gives a special urgency to this conference.

To reach agreement in any kind of conflict, you need to be able to back up and see far enough into the distance to locate a point you can actually agree on. What the two sides have in common is this: Each of us would like to see a world where women no longer want abortions. I don't believe that even among the most fervent pro-choice people there is anybody who rejoices over abortion. I think we both wish that there were better solutions that could make abortion unnecessary, or prevent pregnancies in the first place. We'd like to see the demand for the procedure reduced, by resolving women's problems and alleviating the pressure for abortion. We can go along this road together as far as we can, and there will come a time when pro-choicers are satisfied, and pro-lifers want to keep going, but that doesn't mean we can't go together for now.

A few years ago, quite by accident, I discovered an important piece of common ground. Something I wrote in a conservative think-tank journal was picked up and quoted widely. I had written: "There is a tremendous sadness and loneliness in the cry 'A woman's right to choose.' No one wants an abortion as she wants an ice-cream cone or a Porsche. She wants an abortion as an animal, caught in a trap, wants to gnaw off its own leg."

What surprised me was where it appeared: I started getting clips in the mail from friends, showing the quote featured in pro-choice publications. I realized I had stumbled across one of those points of agreement: We all know that no one leaves the abortion clinic skipping. This made me think that there was common ground, that instead of marching against each other, maybe we could envision a world without abortion, a world we could reach by marching together.

The problem thus far, and I believe the pro-life movement has been especially complicit in this, is that we have focused only on abortion, and not on women's needs. We in the pro-life movement have perpetuated a dichotomy where it's the baby against the woman, and we're on the baby's side. You can look over 25 years of pro-life rhetoric and basically boil it down to three words: "It's a baby." We have our little-feet lapel pins, our "Abortion stops a beating heart" bumper stickers, and we've pounded on that message.

In the process we have contributed to what I think is a false concept—an unnatural and even bizarre concept—that women and their unborn children are mortal enemies. We have contributed to the idea that they've got to duke it out, it's going to be a fight to the finish. Either the woman is going to lose control of her life, or the child is going to lose its life.

It occurred to me that there's something wrong with this picture. When we presume this degree of conflict between women and their own children, we're locating the conflict in the wrong place. Women and their own children are not naturally mortal enemies, and the problem is not located inside women's bodies, it's within society. Social expectations make unwanted pregnancy more likely to occur and harder for women to bear. Unwed mothers are supposed to have abortions, to save the rest of us from all the costs of bringing an "unwanted" child into the world.

There are three drawbacks to emphasizing "It's a baby" as the sole message. One is that it contributes to the present deadlock in this debate. We say "It's a baby," and our friends on the pro-choice side say, "No, it's her right," and the arguments don't even engage each other. It's an endless, interminable argument

The following article is adapted from a talk the author gave May 31 in Madison, Wis., at the first national conference of the Common Ground Network for Life and Choice, an organization of antiabortion activists and abortion-rights supporters who are seeking new ways to discuss their differences.

In news coverage of the abortion controversy, The Washington Post has adopted the terms "anti-abortion" and "abortion rights" as more neutral descriptions of the opposing points of view. Frederica Mathewes-Green's article, however, is drawn from her talk at the network's recent conference and therefore is similar to a quotation; it would change the meaning and context of her remarks to use Post style, so the terms "pro-life" and "pro-choice" appear throughout.
—The Editors

that can go on for another 25 years if we don't find a way to break through.

Second, the "It's a baby" message alienates the woman distressed by a difficult pregnancy. There's a pro-life message that I sometimes hear which makes me cringe: "Women only want abortions for convenience. They do this for frivolous reasons. She wants to fit into her prom dress. She wants to go on a cruise." But this alienates the very person to whom we need to show compassion. If we're going to begin finding ways to live without abortion, we need to understand her problems better.

Of course, there has been a wing of the pro-life movement that has been addressing itself to pregnant women's needs for a long time, and that is the crisis pregnancy center movement. Centers like these have been giving women maternity clothes, shelter, medical care, job training and other help for 30 years. But you wouldn't know that from the things the movement says. I once saw a breakdown of the money and time spent on various sorts of pro-life activities, and over half the movement's energy was going into direct aid to pregnant women. Yet you don't hear this in the rhetoric.

The third problem with this rhetoric is that it enables the people in the great mushy middle, the ones who are neither strongly pro-life or strongly pro-choice, to go on a shrugging off the problem. While both sides know that women don't actually want abortions in any positive sense, the middle is convinced they do. And that's because both sides are telling it they do. Pro-lifers say, "She wants an abortion because she's selfish"; pro-choicers say, "She wants an abortion because it will set her free." No wonder the middle believes us; it's one of the few things we appear to agree on.

But both sides know that abortion is usually a very unhappy choice. If women are lining up by the thousands every day to do something they do not want to do, it's not liberation we've won. But our rhetoric in the pro-life movement, our insistence that "It's a baby and she's just selfish," keeps the middle thinking that abortion really is what women want, so there's no need for change and nothing to fix. I want to recognize my side's complicity in contributing to this deadlock and confusion.

I can understand why my pro-life allies put the emphasis on "It's a baby." It's a powerful and essential message. Visualizing the violence against the unborn was the conversion point for me and many others. But it cannot be the sole message. Polls on American attitudes toward abortion show that between 70 and 80 percent already agree that it's a baby—especially since the advent of sonograms. So when we say, "It's a baby," we're answering a question nobody's asking any more. I believe there is a question they are asking about abortion, and the question is, "How could we live without abortion?"

The abortion rate in this country is about a million and a half a year, a rate that has held fairly stable for about 15 years. Divide that figure by 365 and that equals about 4,100 abortions every day.

Now imagine for a moment that in the middle of the country there is a big abortion store, and outside it 4,100 women got in a long line, one behind the other—and that's just today. It's a sobering image. And the short-sighted pro-life response has been, "Put a padlock on the abortion store." But that's not going to solve the problem. You cannot reduce the demand by shutting off the supply. If 4,100 women were lining up every day to get breast implants, we'd be saying, "What's causing this demand? What's going on here?"

How can we solve the problems that contribute to the demand for abortion? If this were easy, we would have done it by now. It's not easy. There are two obvious components: preventing the unwanted pregnancy in the first place, and assisting women who slip through the cracks and become pregnant anyway.

The obvious tool for pregnancy prevention is contraception, but the pro-life movement has been very reluctant to support the contraceptive option. I come from a religious tradition that permits some forms of contraception, so it's not been a theological problem for me. So when I started considering this, I thought, "This is great! I'll get a helicopter, fill it with condoms, get a snow shovel, and just fly over the country tossing 'em out. We'll close all of the abortion clinics tomorrow!"

But then I began to analyze it a little deeper. While I believe the pro-life movement needs to make a strong stand in favor of preventing these unplanned pregnancies, I became skeptical of the contraceptive solution. For example, there's the recent study showing about two-thirds of births to teenage moms in California involved a dad who was an adult, and another one that found teen mothers had been forced into sex at a young age and that the men who molested them had an average age of 27. Closer to home, a friend of mine was brought to an abortion clinic by her older brother, who molested her when she was 12; they gave her a bag of condoms and told her to be more careful. You're not going to solve problems like these by tossing a handful of condoms at [them].

But leaving aside the question of sexual abuse, I think we need to look hard at the consequences of the sexual revolution that began in the 1960s. When I entered college in the early 1970s, the revolution was in full bloom. It seemed at the time a pretty care-free enterprise. Condoms, pills and diaphrams were readily available and abortion had just been legalized by the Supreme Court. But I gradually began to think that it was a con game being played on women. We were "expected to behave according to men's notions of sexuality," to use author Adrienne Rich's phrase. Instead of gaining respect and security in our bodies, we were expected to be more physically available, more vulnerable than before, with little offered in return.

What women found out is that we have hearts in here along with all our other physical equipment, and you can't put a condom on your heart. So in answering the question, "How do we live without abortion?," I'd say we need to look at restoring respect and righting the balance of power in male-female sexual relationships.

What can we do to help women who get pregnant and would rather not be? For a book I was writing, I went around the country talking to women who have had an abortion and to women who provide care for pregnant women. I had presumed that most abortions are prompted by problems that are financial or practical in nature.

But to my surprise, I found something very different. What I heard most frequently in my interviews was that the reason for the abortion was not financial or practical. The core reason I heard was, "I had the abortion because someone I love told me to." It was either the father of the child, or else her own mother, who was pressuring the woman to have the abortion.

Again and again, I learned that women had abortions because they felt abandoned, they felt isolated and afraid. As one woman said, "I felt like everyone would support me if I had the abortion, but if I had the baby I'd be alone." When I asked, "Is there anything anyone could have done? What would you have needed in order to have had that child?" I heard the same answer over and over: "I needed a friend. I felt so alone. I felt like I didn't have a choice. If only one person had stood by me, even a stranger, I would have had that baby."

We also must stop thinking about abortion in terms of pregnancy. We harp on pregnancy and forget all about what comes next. Getting through the pregnancy isn't nearly the dilemma that raising a child for 18 years is. In most families, marriage lightens the load, but for some people that isn't the best solution. A neglected option is adoption, which can free the woman to resume her life, while giving the child a loving home.

The numbers on this, however, are shocking. Only 2 percent of unwed pregnant women choose to place their babies for adoptions. Among clients at crisis pregnancy centers, it's 1 to 2 percent. Adoption is a difficult sell to make for a number of complex reasons, but the bottom line is that 80 to 90 percent of the clients who go through pregnancy care centers and have their babies end by setting up single-parent homes. This is very serious. Pregnancy care centers know this, but aren't sure what to do about it. I've been strongly encouraging that there be more emphasis on presenting adoption to clients, and equipping center volunteers so they feel comfortable with the topic and enabled to discuss it. Adoption is not a one-size-fits-all solution, but it's got to fit more than 1 or 2 percent. More women should try it on for size.

Let me finish with these thoughts. I want to encourage us to view the pregnant woman and child as a naturally-linked pair that we strive to keep together and support. Nature puts the mother and the child together, it doesn't make them enemies, it doesn't set one against the other in a battle to the death. If our rhetoric is tearing them apart, we're the ones who are out of step. The pro-life movement should be answering the question "How can we live without abortion?" by keeping mother and child together, looking into pregnant women's needs and examining how to meet them, and encouraging responsible sexual behavior that will prevent those pregnancies in the first place.

Health Unlimited

Willard Gaylin

WILLARD GAYLIN, M.D., *is professor of psychiatry at Columbia University Medical School and cofounder and president of the Hastings Center, an institution devoted to bioethical research. His most recent book, with Bruce Jennings, is* The Perversion of Autonomy, *published by the Free Press.*

The debate over the current crisis in health care often seems to swirl like a dust storm, generating little but further obfuscation as it drearily goes around and around. And no wonder. Attempts to explain how we got into this mess—and it is a mess—seem invariably to begin in precisely the wrong place. Most experts have been focusing on the failures and deficiencies of modern medicine. The litany is familiar: greedy physicians, unnecessary procedures, expensive technologies, and so on. Each of these certainly adds its pennyweight to the scales. But even were we to make angels out of doctors and philanthropists out of insurance company executives, we would not stem the rise of healthcare costs. That is because this increase, far from being a symptom of modern medicine's failure, is a product of its success.

Good medicine keeps sick people alive. It increases the percentage of people in the population with illnesses. The fact that there are proportionally more people with arteriosclerotic heart disease, diabetes, essential hypertension, and other chronic—and expensive—diseases in the United States than there are in Iraq, Nigeria, or Colombia paradoxically signals the triumph of the American healthcare system.

There is another and perhaps even more important way in which modern medicine keeps costs rising: by altering our very definition of sickness and vastly expanding the boundaries of what is considered the domain of health care. This process is not entirely new. Consider this example. As I am writing now, I am using reading glasses, prescribed on the basis of an ophthalmologist's diagnosis of presbyopia, a loss of acuity in close range vision. Before the invention of the glass lens, there was no such disease as presbyopia. It simply was expected that old people wouldn't be able to read without difficulty, if indeed they could read at all. Declining eyesight, like diminished hearing, potency, and fertility, was regarded as an inevitable part of growing older. But once impairments are no longer perceived as inevitable, they become curable impediments to healthy functioning—illnesses in need of treatment.

To understand how the domain of health care has expanded, one must go back to the late 19th century, when modern medicine was born in the laboratories of Europe—mainly those of France and Germany. Through the genius of researchers such as Wilhelm Wundt, Rudolph Virchow, Robert Koch, and Louis Pasteur, a basic understanding of human physiology was established, the foundations of pathology were laid, and the first true understanding of the nature of disease—the germ theory—was developed. Researchers and physicians now had a much better understanding of what was going on in the human body, but there was still little they could do about it. As late as 1950, a distinguished physiologist could tell an incoming class of medical students that, until then, medical intervention had taken more lives than it had saved.

Even as this truth was being articulated, however, a second revolution in medicine was under way. It was only after breakthroughs in the late 1930s and during World War II that the age of therapeutic medicine began to emerge. With the discovery of the sulfonamides, and then of penicillin and a series of major antibiotics, medicine finally became what the laity in its ignorance had always assumed it to be: a lifesaving enterprise. We

From *The Wilson Quarterly,* Summer 1996, pp. 38-42. © 1996 by Willard Gaylin. Reprinted by permission.

in the medical profession became very effective at treating sick people and saving lives—so effective, in fact, that until the advent of AIDS (acquired immune deficiency syndrome), we arrogantly assumed that we had conquered infectious diseases.

The control of infection and the development of new anesthetics permitted extraordinary medical interventions that previously had been inconceivable. As a result, the traditional quantitative methods of evaluating alternative procedures became outmoded. "Survival days," for example, was traditionally the one central measurement by which various treatments for a cancer were weighed. If one treatment averaged 100 survival days and another averaged 50 survival days, then the first treatment was considered, if not twice as good, at least superior. But today, the new antibiotics permit surgical procedures so extravagant and extreme that the old standard no longer makes sense. An oncologist once made this point using an example that remains indelibly imprinted on my mind: 100 days of survival without a face, he observed, may not be superior to 50 days of survival with a face.

Introducing considerations of the nature or quality of survival adds a whole new dimension to the definitions of sickness and health. Increasingly, to be "healthy," one must not only be free of disease but enjoy a good "quality of life." Happiness, self-fulfillment, and enrichment have been added to the criteria for medical treatment. This has set the stage for a profound expansion of the concept of health and a changed perception of the ends of medicine.

I can illustrate how this process works by casting stones at my own glass house, psychiatry, even though it is not the most extreme example. The patients I deal with in my daily practice would not have been considered mentally ill in the 19th century. The concept of mental illness then described a clear and limited set of conditions. The leading causes of mental illness were tertiary syphilis and schizophrenia. Those who were mentally ill were confined to asylums. They were insane; they were different from you and me.

Let me offer a brief (and necessarily crude) history of psychiatry since then. At the turn of the century, psychiatry's first true genius, Sigmund Freud, decided that craziness was not necessarily confined to those who are completely out of touch with reality, that a normal person, like himself or people he knew, could be partly crazy. These "normal" people had in their psyches isolated areas of irrationality, with symptoms that demonstrated the same "crazy" distortions that one saw in the insane. Freud invented a new category of mental diseases that we now call the "neuroses," thereby vastly increasing the population of the mentally ill. The neuroses were characterized by such symptoms as phobias, compulsions, anxiety attacks, and hysterical conversions.

In the 1930s, Wilhelm Reich went further. He decided that one does not even have to exhibit a neurosis to be mentally ill, that one can suffer from "character disorders." An individual could be totally without symptoms of any illness, yet the nature of his character might so limit his productivity or his pleasure in life that we might justifiably (or not) label him "neurotic."

Still later, in the 1940s and '50s, medicine "discovered" the psychosomatic disorders. There are people who have no evidence of mental illness or impairment but have physical conditions with psychic roots, such as peptic ulcers, ulcerative colitis, migraine headache, and allergy. They, too, were now classifiable as mentally ill. By such imaginative expansions, we eventually managed to get some 60 to 70 percent of the population (as one study of the residents of Manhattan's Upper East Side did) into the realm of the mentally ill.

But we still were short about 30 percent. The mental hygiene movement and preventive medicine solved that problem. When one takes a preventive approach, encompassing both the mentally ill and the potentially mentally ill, the universe expands to include the entire population.

Thus, by progressively expanding the definition of mental illness, we took in more and more of the populace. The same sort of growth has happened with health in general, as can be readily demonstrated in surgery, orthopedics, gynecology, and virtually all other fields of medicine. Until recently, for example, infertility was not considered a disease. It was a God-given condition. With the advances in modern medicine—in vitro fertilization, artificial insemination, and surrogate mothering—a whole new array of cures was discovered for "illnesses" that had to be invented. And this, of course, meant new demands for dollars to be spent on health care.

One might question the necessity of some of these expenditures. Many knee operations, for instance, are performed so that the individual can continue to play golf or to ski, and many elbow operations are done for tennis buffs. Are these things for which anyone other than the amateur athlete himself should pay? If a person is free of pain except when playing tennis, should not the only insurable prescription be—much as the old joke has it—to stop playing tennis? How much "quality of life" is an American entitled to have?

New technologies also exert strong pressure to expand the domain of health. Consider the seemingly rather undramatic development of the electronic fetal monitor. It used to be that when a pregnant woman in labor came to a hospital—if she came at all—she was "observed" by a nurse, who at frequent intervals checked the fetal heartbeat with a stethoscope. If it became more rapid, suggesting fetal distress, a Caesarean section was considered. But once the electronic fetal monitor came into common use in the 1970s, continuous monitoring by the device became standard. As a result, there was a huge increase in the number of Caesareans performed in major teaching hospitals across the country, to the point that 30 to 32 percent of the pregnant women in those hospitals were giving birth through surgery. It is ridiculous to suggest that one out of three pregnancies requires surgical intervention. Yet technology, or rather the seductiveness of technology, has caused that to happen.

Linked to the national enthusiasm for high technology is the archetypically American reluctance to acknowledge that there are limits, not just limits to health care but limits to anything. The American character is different. Why this is so was suggested some years ago by historian William Leuchtenberg in a lecture on the meaning of the frontier. To Europeans, he explained, the frontier *meant* limits. You sowed seed up to the border and then you had to stop; you cut timber up to the border and then you had to stop; you journeyed across your country to the border and then you had to stop. In America, the frontier had exactly the opposite connotation: it was where things began. If you ran out of timber, you went to the frontier, where there was more; if you ran out of land, again, you went to the frontier for more. Whatever it was that you ran out of, you would find more if you kept pushing forward. That is our historical experience, and it is a key to the American character. We simply refuse to accept limits. Why should the provision of health care be an exception?

To see that it isn't, all one need do is consider Americans' infatuation with such notions as "death with dignity," which translates into death without dying, and "growing old gracefully," which on close inspection turns out to mean living a long time without aging. The only "death with dignity" that most American men seem willing to accept is to die in one's sleep at the age of 92 after winning three sets of tennis from one's 40-year-old grandson in the afternoon and making passionate love to one's wife twice in the evening. This does indeed sound like a wonderful way to go—but it may not be entirely realistic to think that that is what lies in store for most of us.

During the past 25 years, health-care costs in the United States have risen from six percent of the gross national product to about 14 percent. If spending continues on its current trajectory, it will bankrupt the country. To my knowledge, there is no way to alter that trajectory except by limiting access to health care and by limiting the incessant expansion of the concept of health. There is absolutely no evidence that the costs of health-care services can be brought under control through improved management techniques alone. So-called managed care saves money, for the most part, by offering less—by covert allocation. Expensive, unprofitable operations such as burn centers, neonatal intensive care units, and emergency rooms are curtailed or eliminated (with the comforting, if perhaps unrealistic, thought that municipal and university hospitals will make up the difference).

Rationing, when done, should not be hidden; nor should it be left to the discretion of a relative handful of health-care managers. It requires open discussion and wide participation. When that which we are rationing is life itself, the decisions as to how, what, and when must be made by a consensus of the public at large through its elected and other representatives, in open debate.

What factors ought to be considered in weighing claims on scarce and expensive services? An obvious one is age. This suggestion is often met with violent abuse and accusations of "age-ism," or worse. But age *is* a factor. Surely, most of us would agree that, *all other things being equal*, a 75-year-old man (never mind a 92-year-old man) has less claim on certain scarce resources, such as an organ transplant, than a 32-year-old mother or a 16-year-old boy. But, of course, other things often are not equal. Suppose the 75-year-old man is president of the United States and the 32-year-old mother is a drug addict, or the 16-year-old boy is a high school dropout. We need, in as dispassionate and disinterested a way as possible, to consider what other factors besides age should be taken into account. Should political position count? Character? General health? Marital status? Number of dependents?

Rationing is already being done through market mechanisms, with access to kidney or liver transplants and other scarce and expensive procedures determined by such factors as how much money one has or how close one lives to a major health-care center. Power and celebrity can also play a role—which explains why politicians and professional athletes suddenly turn up at the top of waiting lists for donated organs. A fairer system is needed.

The painful but necessary decisions involved in explicit rationing are, obviously, not just medical matters—and they must not be left to physicians or health-care managers. Nor should they be left to philosophers designated as "bioethicists," though these may be helpful. The population at large will have to reach a consensus, through the messy—but noble—devices of democratic government. This will require legislation, as well as litigation and case law.

In the late 1980s, the state of Oregon began to face up to the necessity of rationing. The state legislature decided to extend Medicaid coverage to more poor people but to pay for the change by curbing Medicaid costs by explicitly rationing benefits. (Eventually, rationing was to be extended to virtually all Oregonians, but that part of the plan later ran afoul of federal regulations.) After hundreds of public hearings, a priority list of services was drawn up to guide the allocation of funds. As a result, dozens of services became difficult (but not impossible) for the poor to obtain through Medicaid. These range from psychotherapy for sexual dysfunctions and severe conduct disorder to medical therapy for chronic bronchitis and splints for TMJ Disorder, a painful jaw condition. Although the idea of explicit rationing created a furor at first, most Oregonians came to accept it. Most other Americans will have to do the same.

Our nation has a health-care crisis, and rationing is the only solution. There is no honorable way that we Americans can duck this responsibility. Despite our historical reluctance to accept limits, we must finally acknowledge that they exist, in health care, as in life itself.

A new vision for city schools

DIANE RAVITCH & JOSEPH VITERITTI

Diane Ravitch is a senior research scholar at the School of Education of New York University. Joseph Viteritti is a research professor at NYU's Robert F. Wagner Graduate School of Public Service.

YES, there is hope for urban education. A wave of reform is spreading from city to city and state to state. Rather than aiming to alter isolated practices or to fix one piece of a jerry-built system, these changes are meant to transform the basic character of public schooling. When taken together, the ambitious range of initiatives currently under way can be structured into an integrated program for reforming urban education—one that shifts from a bureaucratic system that prizes compliance to a deregulated system that focuses on student performance.

A century ago, progressive reformers reshaped big-city schools according to the era's widely shared vision of efficient administration. To get schools "out of politics," they created tightly controlled bureaucracies. At the apex of authority were "professional experts," who managed a top-down system designed to impose uniform rules on teachers and students alike. The model for this system was the factory, which, at that time in history, was considered the acme of scientific management. The raw materials for these educational factories were the children of immigrants, who were pouring into American cities in unprecedented numbers, in need of instruction in literacy, hygiene, and basic Americanization. The workers in these factories were teachers, whose views about what or how to teach were not solicited. Nor did the experts see any need to consult parents about anything regarding their children, since many of them were barely literate.

These turn-of-the-century efforts to create what historian David Tyack called "the one best system" were remarkably effective for at least the first half of the century. Big-city schools offered unparalleled educational opportunity to millions of children and helped to generate a vast middle class. At mid-century, the nation's urban schools were considered to be a great success. But no more. The system that transformed an earlier generation of impoverished children into prosperous adults has become sclerotic; the bureaucratic organization created to impose efficiency and order has grown tired and inefficient, tangled helplessly in rules and regulations devised by the courts, state governments, federal government, union contracts, and its own minions.

In city after city, reports of corruption, disorder, neglect, and low educational achievement are legion. Urban education is in deep trouble, in part because of inept big-city bureaucracies inherited from the past, but also because the public's expectations for the schools are higher now than they were earlier in the century. Fifty years ago, the public was neither surprised nor alarmed by the large numbers of young people who did not graduate from high school; they believed that the numbers would continually improve over time. Today, the public expects a large majority of students to complete high school, especially since the jobs available for high-school dropouts are diminishing.

Convinced that the structure of public education contributes to its ineffectiveness in educating a larger proportion of students, imaginative leaders in cities and states across the country are implementing systemic changes. The new reforms proceed on the conviction that the century-old bureaucratic structures of urban education cannot succeed in today's society. The century-old system of schools cannot work today because it was designed to function in a very different society, with different social mores and different problems, where supervisors instructed teachers, teachers instructed students, and parents expected their children to mind what they were told.

The factory is no longer a useful model for urban education; teachers and children are not interchangeable parts to be moved around to fit the requirements of administrators. The reforms that are now being enacted in many cities incorporate such principles as diversity, quality, choice, and accountability. Instead of a system that regulates identical schools, reformers seek a system in which academic standards are the same for all but where schools vary widely. In the new reform vision, the schools are as diverse as teachers' imagination and will; students and their families choose the school that best meets their needs and interests; and central authorities perform a monitoring and auditing function to assure educational quality and fiscal integrity. In such a reconfigured system, the role of the local superintendent shifts from regulating behavior to auditing results. The bottom line is not whether everyone has complied with the same rules and procedures but whether children are learning.

Charter schools

Not since the beginning of the twentieth century has there been such a burst of bold experimentation in the organization and governance of schools as there has been in just the last half decade. Among the most notable initiatives on the current scene are charter schools, the contracting of instructional services, and a variety of school-choice programs. These innovations are driven by demands from parents and elected officials for higher levels of educational success. However, few of these innovations are based on hard evidence that they will succeed. But that is the nature of innovation: one purpose of these experiments is to identify what will work and what will not. Most of these initiatives, however, are based on well-documented evidence that the current institutional arrangement does not work very well for large numbers of children.

One of the most promising ideas to appear on the national horizon is charter schools. Charter schools are semi-autonomous, public entities that are freed from most bureaucratic rules and regulations by state and local authorities in return for a commitment to meet explicit performance goals. They are established under a contract between a group that manages a school and a sponsoring authority that oversees it. The contractor might consist of parents, teachers, a labor union, a college, a museum, or other nonprofit or for-profit entities. The sponsor might be a school board, a state education department, a state university campus, or a government agency. In Arizona, the legislature has created a special governing body authorized to grant or deny a request for a charter; thus such power is not limited to the state or local school boards, which may have a stake in restricting the number of these institutions.

Charter schools may be either new schools or existing ones. Their development will contribute to both the number and variety of quality institutions. In Detroit, the Drug Enforcement Administration is creating a residential school for 200 at-risk students; a school called Metro Deaf serves the hearing impaired in St. Paul. In Wilmington, Delaware, five corporations and a medical center have cooperated in a joint venture to run a new high school for math and science. Boston University runs a school for homeless children in Massachusetts; and the Denver Youth Academy was created for at-risk, middle-school children and their families in Colorado. The possibilities seem endless.

A charter serves as a negotiated, legal agreement that sets standards and expectations for the school. School professionals are authorized to manage their own budget and to choose their own staff, but the degree of autonomy varies from state to state. The Education Commission of the States, in conjunction with the Humphrey Institute at the University of Minnesota, recently completed a survey of 110 charter schools in seven states. It found that educators at these schools are quite willing to be held more accountable for improved student performance, so long as they are permitted to enjoy more autonomy. A majority of these schools focus their attention on at-risk populations.

Presently 19 states have charter-school laws. Minnesota passed the first one as recently as 1991, with eight schools participating; there are now 40 participating. California approved the establishment of 100 charter schools in 1992. Michigan has 30. Among the states that grant the most autonomy to charter schools are Arizona, California, Colorado, Delaware, Massachusetts, Michigan, Minnesota, New Hampshire, and Texas. (The other states with charter laws are Alaska, Arkansas, Georgia, Hawaii, Kansas, Louisiana, New Mexico, Rhode Island, Wisconsin, and Wyoming.)

Charter schools are public schools. They are accountable to a public authority. In fact, the charter, which defines academic expectations and other legal responsibilities, often serves as a more powerful instrument for accountability than anything that exists for most ordinary public schools. If a charter school fails educationally or misuses its funds, the charter can be revoked, as was the case with one Los Angeles school last year. Most charter schools must accept any student who applies, or they select students by lottery if there are more applicants than places. All are bound by the usual state laws and regulations requiring schools to be nondiscriminatory and protective of civil rights.

Contracts for performance

Unlike charter-school laws, which begin as state initiatives, contracting-out arrangements usually originate with a local school board. It is not unusual for school boards to contract with private vendors for the performance of non-instructional functions—e.g., transportation, food, supplies, facilities, and custodial and administrative services. What is novel about recent developments is for school boards to arrange to have instructional programs provided by outsiders. This approach has given rise to new entrepreneurial organizations on the education scene. Educational Alternatives, Inc. (EAI), for example, is under contract to run nine public schools in Baltimore, as well as a single school in Duluth. It will also overhaul six schools in Hartford (eventually all 32) and assume general responsibility for the management of that district. The Edison Project has contracted to operate individual schools in Wichita, Kansas, Mt. Clemens, Michigan, Sherman, Texas, and Boston. Washington, D.C., has recently contracted with Sylvan Learning Systems to offer remedial reading for a limited number of students; and Sabis, an International group, runs a school in Springfield, Massachusetts.

These arrangements are similar to charter schools in that they are brought into being by a performance agreement between a school organization and a public authority. Some contracts allow more autonomy than others. EAI, for instance, ran into great difficulty implementing changes in Baltimore after it agreed to hire all the existing teachers in what were supposed to be reconstituted schools; moreover, the teachers' union was antagonistic to the project from the beginning. And some argue that EAI committed a major strategic error when it took on general responsibility for running the entire Hartford school district. The approach adopted by the Edison Project, involving the development of new schools, one at a time, with a staff that it has hired and trained, seems to hold more promise.

As with most innovations in public education, the more profound changes exacted through contracting tend to generate the strongest opposition. Wilkinsburg, Pennsylvania, a working-class suburb of Pittsburgh, where 78 percent of the children qualify economically for a free-lunch program, is a case in point. It became the scene of an intense political and legal battle when a newly elected, reform-minded school board announced its intention to contract with Alternative Public School Strategies to operate one of its three elementary schools. The local teachers' union and the National Education Association fiercely opposed its reform proposals: an extended school year, new after-school programs, merit pay for teachers.

Some observers have confused the contracting approach with privatization. Contract schools are public institutions, supported with public funds, accountable to a public authority—usually a local school board. As is the case with charter schools, they are expected to meet specific standards of academic performance defined in a legal agreement. If they do not perform adequately, they can be put out of business. When Baltimore Mayor Kurt Schmoke became dissatisfied with student performance at EAI-operated schools in Baltimore, he said he would rethink the contract. It is a rare occurrence for a city public school to be shut down for poor performance, regardless of its record over time. Contracting arrangements, whether they result from state charter laws or local initiatives, mark a new threshold of aspiration and accountability for public education.

Dimensions of choice

Choice programs are designed to enhance the options made available to parents in selecting a school for their children. The most common form of choice program allows parents to choose a public school that lies outside the ordinary range of geographical options. The objective is to improve the chances for students to be placed in settings that suit their needs. It is also assumed that giving parents choice will induce competition among schools. Minnesota adopted the first statewide inter-district choice program in 1985. By 1991, 10 states had approved some form of open-enrollment program; and now, more than two-thirds of the states have enacted public school-choice programs. The first city-wide choice program was developed in 1981 in Cambridge, Massachusetts; perhaps the most celebrated success story at the local level is found in District 4 of East Harlem.

The basic shortcoming of public school-choice programs is that those jurisdictions that have the greatest need for expanding opportunities usually offer the fewest number of satisfactory options. For example, in Massachusetts, where voluntary inter-district choice has existed since 1991, only 25 percent of the districts participate, and none of the 29 on the suburban rim of Boston is included in this group. Supposedly, New York City has had a city-wide choice program since 1992; but, in reality, choice is permitted in only six of 32 districts, and the availability of space is extremely limited. Without measures designed to increase the total number of quality institutions, public school choice promotes competition among parents and children, not educators. It raises expectations but often leads to disappointment.

In 1990, Wisconsin passed innovative legislation that would expand parental choice among low-income parents in Milwaukee. Families who met income criteria ($18,000 or less) were given a state voucher for $2,987, which they might use in either a public school or a participating private school. By the end of the 1994–1995 school year, there were 1,500 children participating in the program that involved 12 nonpublic schools. Last spring, the legislation was amended to increase the value of the voucher to $3,600 and to permit schools with religious affiliations to participate. By next year, 15,000 low-income students are expected to take advantage of this unusual opportunity.

Similarly, last spring, the Ohio legislature enacted a law that will permit up to 2,000 low-income students in Cleveland to use a $2,500 state voucher in a school of choice—public, private, or sectarian. As in Wisconsin, the law was passed at the urging of minority parents dissatisfied with the quality of education in inner-city public schools. Like parents in Milwaukee, they had been frustrated with court-imposed integration plans that led to longer bus rides, rather than better schools. For the first time, many poor children whose life chances would have been determined by assignment to a failing public school were given the opportunity for real choices that gave them access to quality institutions (choices that were formerly available only to the middle class). In the meantime, the Milwaukee program is being challenged in state court, and a legal contest is expected in Cleveland.

Opponents of these programs claim that they violate requirements for the Constitutional separation of church and state. But there is nothing in the First Amendment of the Constitution that prohibits parents who want to send their children to religious schools from receiving public support. Since 1983, rulings by the Supreme Court have held that such support is legally permissible provided that aid goes directly to the parents (not the school), that the choice of school is freely made by parents, and that the system of funding is neutral. Cognizant of these rulings, opponents of choice have resorted to legal arguments based on provisions found in state law, many of which may be incompatible with federal Constitutional standards.

Some critics of choice fear that providing families with private-school options will spell the doom of public education. They predict a mass exodus of children and dollars from public schools. This is highly unlikely, indeed impossible, since the number of children permitted to participate has been limited. Let us keep in mind that every choice program that has gotten serious consideration by policy makers thus far—including those in Wisconsin and Ohio—has targeted a limited portion of the school population, those on the lowest rung of the economic ladder. Most public-school children were not even eligible according to these criteria.

Traditionally, school reformers have asked how we can improve the existing system; today, many ask, instead, what we can do that is in the best interest of students who are at risk of failing. Since the ground-breaking work by the late sociologist James S. Coleman and his colleagues at the University of Chicago in 1982,[1] there is evidence that private and parochial schools are more educationally effective than public schools. Some scholars have attributed the difference to the selectivity of private schools. However, more recent research on Catholic schools by Bryk, Lee, and Holland[2] indicates that the differences are more the result of characteristics identified within the schools themselves, e.g., high standards, a strong academic curriculum, autonomy, an orderly environment, and a sense of community. Other studies by Coleman, Greeley, and Hoffer[3] have demonstrated rather persuasively that Catholic schools have been particularly effective in educating at-risk, inner-city students who have performed poorly in public schools.

An agenda for change

In light of the wide range of reforms currently under way in cities across the country, we propose a six-point agenda to improve educational opportunities for all children. Some of these proposals will require strong legislative action at the state level. Implementation will require an "hourglass strategy," allowing schools to escape one by one from the bureaucratic system. The best schools would function as charter schools and the worst schools would be replaced by institutions with performance contracts. Over time, more and more schools will seek the autonomy and performance agreements that charter schools have, and educational authorities will incrementally replace ineffective schools with new schools that have committed themselves to meet performance goals. The net effect of this approach would be to increase the number of desirable schools that children can attend.

1. Setting Standards. What matters most is whether children are learning, and this can only be assured by having real accountability at the school level. Each school district should establish clear performance standards and administer regular assessments to determine whether students are learning what they should at each grade level. By standards, we mean objective outcome measures that prescribe what should be expected from every school at regular intervals. For example, we would focus on such items as test scores in reading and math, attendance rates, and dropout rates. We would be especially interested in measuring "value added" or "gain scores"—the progress made over a given academic year—rather than unfairly comparing schools with children from vastly different social circumstances. We would not involve district-level administrators in defining basic inputs like instructional approaches or building specifications beyond code requirements designed to protect safety.

2. School Closings. The public school that once served as a gateway of opportunity for immigrant populations now serves as a custodial institution for disadvantaged children. Even as national achievement-test scores creep slowly upward, the gap between black and white students' scores remains shamefully wide, and those of Hispanic children are actually declining. As a matter of public policy, no child should be forced to go to a failing school. We cannot ask parents of children who are trapped in floundering institutions to be patient while we work things out. Educators and political leaders should not expect poor parents to accept educational standards for their children that the middle class would not tolerate for their own.

Any school that shows a consistent record of failure over several years should be a candidate for closure. The first step must be to define objective standards for placing a school on probation. Most school systems already have the basic data needed to develop appropriate criteria. A combination of attendance rates, test scores, dropout rates, improvement ratings, and similar markers will identify those schools that must be placed on probation or eventually closed.

3. School Autonomy. Schools that are working well should receive a performance contract and control over their budget and personnel. Subject to due process, principals should be able to select and remove staff. They should be allowed to purchase supplies and support services of their choosing, though central authorities will audit these purchases.

All schools, whether autonomous or not, should be liberated from unnecessary and cumbersome mandates. In New York, the state Education Department recently identified more than 120 regulations that are not related to health, safety, or civil rights. California, Michigan, and Florida have taken the lead on regulatory reform; their governors have called for sunsetting the entire education code and starting over, enacting only those regulations that are essential. Illinois is one of 10 states that has set up a procedure where local school districts can request waivers from outdated requirements. School superintendents should conduct a top-to-bottom review of local regulations to eliminate those that are not necessary. The goal should be to minimize the burden on teachers and principals. and to grant schools greater independence in providing education.

Many school systems in the United States are experimenting with site-based management that moves decision making out of the administrative structure of the bureaucracy down to the school level. When implemented seriously, site-based management can improve both flexibility and accountability. However, a key question is whether central school authorities will actually concede the power to which they have become accustomed. A recent evaluation of the Los Angeles autonomy program by McKinsey & Company identified significant delays at the district level in implementing the reforms. Several years ago, New York City launched a modest experiment in school-based management without reducing the power of the central bureaucracy. It resulted in neither real autonomy for participating schools nor increased accountability for student performance.

Worth consideration is the city of London's opt-out program, in which a majority of parents in a school can vote to remove the school from the supervision of the local school district. Independence means that the school gets control over its own budget and a portion of the administrative overhead, so long as it continues to meet well-defined performance standards based on the national curriculum and national tests. All big-city school districts should have a similar plan, enabling schools to "opt out" of the present bureaucratic structure. We would propose, however, that approval require a majority vote of both parents and teachers. Teachers are important members of the school community, and their support is essential for success.

Not all schools are ready for such autonomy. But, if the parents and professionals at a school apply for approval as a charter school, then such a request should be evaluated by the local superintendent, the state education commissioner, or by an independent chartering agency on the basis of established criteria. In exchange for autonomy, the school administration would be required to sign a contract defining educational and financial standards to which the school would be held accountable. Greater autonomy would permit the school to hire staff, choose its teaching materials, set its fiscal priorities, and decide where it wants to purchase supplies and support services. Decisions about what to buy and where to buy it would be made by educators at the school level, not bureaucrats at central headquarters.

4. New Schools. If we intend to close failing schools, we need to provide alternatives to students who attend them. School superintendents should be given the power to solicit proposals for new institutions, either to replace failing schools or to grant contracts for increasing the total number of quality institutions. Proposals would be received from groups of teachers or parents, universities, libraries or museums, nonprofit organizations, or private entrepreneurs who demonstrate a professional capacity to administer a school. They may be progressive schools, family-style schools, Outward Bound schools, single-sex schools, back-to-basics schools, classical academies: The range is as vast as the imagination of creative educators.

These should, of course, be schools of choice. If they attract enough students, they will succeed; if they don't, they won't. These new schools would be established under the same terms as the charter schools described above. They would be granted autonomy in exchange for signing a compact outlining the educational and financial standards to which they would be held accountable. But, if the marketplace prevails, parents will become the ultimate judges of success or failure at the school level.

5. Central Administration. In a recent article in the *Wall Street Journal*, Peter Drucker predicts that in 10 to 15 years, most organizations will be "outsourcing" all of their support activities to specialist groups, thus allowing executives to avoid distractions and focus on functions that are directly related to their central mission. We share a similar vision for public education, where the role of central authorities will be transformed. Over time, the central administrative institution will significantly reduce its role as a provider of support services to schools. It would not be in charge of supplies, leasing, meals, building repairs, transportation, personnel, and other functions that can be performed better by others. Depending on the outcome of the competitive market, support services will either be provided by a private vendor or administered by an appropriate municipal agency. Marriot, for example, already provides food services to schools in Baltimore and Salt Lake City; and most municipal governments are well equipped to assume responsibility for such functions as personnel administration, procurement, transportation, or building maintenance.

According to our plan, the central school administration will become a monitoring agency with clearly focused and limited responsibilities. It will be responsible for educational standards, city-wide assessments, fiscal accountability, capital improvements, the authorization of new schools, and negotiation of union contracts that are specific enough to protect members' rights but flexible enough to permit school-by-school variations. The school system's chief executive should concentrate on setting standards, monitoring performance, and identifying those schools that either should be put on probation or closed. The chief executive would also be responsible for financial monitoring to protect against corruption and malfeasance. In a system where every school has its own budget, this is a formidable task, and it will probably require some form of administrative decentralization.

6. Real Choice for the Poor. Parents whose children attend the worst schools—those targeted for closure—should be given scholarships on a means-tested basis to use in any accredited school, be it public, private, or religious. Middle-class parents exercise such options for their children all the time; poor children should have the same. Priority for financial aid should go first to children in failing institutions whose families are on public assistance. The amount of a scholarship should not exceed the per capita cost of sending a child to public school. Schools receiving scholarship students should accept the award as a full fee for tuition. Students who get public scholarships should be regularly tested to assess their progress, in order to assure accountability and to exclude inadequate schools from participation.

Schools for the twenty-first century

We believe that these proposals, taken together, will strengthen and energize public education—freeing professionals and students from counterproductive regulations, shifting resources from the district level to the schools, providing alternative means for the delivery of vital support services, assuring choice for the students who now receive the least educational opportunity, rewarding success, encouraging creativity, requiring accountability for results, phasing out schools that are not conducive environments for teaching or learning, and placing institutions on the line by putting students first.

The public education system as currently structured is archaic. It cannot reform itself, nor can it be reformed by even the most talented chief executive. Trying to do so would be like trying to convert an old-fashioned linotype machine into a word-processor: It can't be done. They perform the same function, but their methods and technologies are so different that one cannot be turned into the other. Instead of a school system that attempts to impose uniform rules and regulations, we need a system of schools that is dynamic, diverse, performance based, and accountable. The school system that we now have may have been right for the age in which it was created; it is not right for the twenty-first century.

Notes

1. James S. Coleman, Thomas Hoffer, and Sally Kilgore, *High School Achievement: Public, Catholic, and Private Schools* (New York: Basic Books, 1982).
2. Anthony B. Bryk, Valerie E. Lee, and Peter B. Holland, *Catholic Schools and the Common Good* (Cambridge: Harvard University Press, 1993).
3. Andrew M. Greeley, *Catholic High Schools and Minority Students* (Rutgers: Transaction, 1982); Thomas Hoffer, Andrew M. Greeley, and James S. Coleman. "Achievement Growth in Public and Catholic Schools," *Sociology of Education* 58 (1985): 74–97.

CAN CHURCHES SAVE AMERICA?

Some politicians say faith-based programs cure social ills. But the needs are huge and a backlash grows

In the beginning came the governor's challenge: If each of Mississippi's churches would help just one poor family back on its feet, welfare could end. "God, not government, will be the savior of welfare recipients," Gov. Kirk Fordice likes to say. That's how Frances White, a jobless mother facing eviction from a crumbling house, suddenly found herself adopted by a suburban church. There were good deeds (church members paid White's bills), communion (they took her kids shopping) and finally redemption (through one congregant, White found a job as a hospital records clerk and now is partly off welfare).

Not long ago, no governor or politician could get away with asking churches to do the antipoverty work that is normally the responsibility of government. And it would have been the height of hokeyness to suggest that churches, synagogues and mosques could cure the drug addicted, feed the hungry, house the homeless, rehabilitate the criminal and lift the poor better than government. Yet with President Clinton and Congress agreeing to end welfare as we know it, there is talk of a second welfare revolution: Let churches and charities, not government, provide more of the social safety net.

Republican presidential nominee Bob Dole is a convert. Dole – who relied on private hometown charity when he returned from war severely injured – has endorsed a plan to shift tens of billions of social welfare dollars to direct tax credits. The money would reimburse taxpayers who donate up to $500 to poverty-fighting charities. And while it is largely conservatives and Republicans who want faith programs to do more, there are Democratic and liberal believers, too, like Joseph Califano, an architect of Lyndon Johnson's Great Society and cabinet secretary to Jimmy Carter. Califano, now head of Columbia University's Center on Addiction and Substance Abuse, says he was surprised when, on a tour of center programs, nearly every ex-drug addict he met cited religious belief as a key to rehabilitation. "I don't see anything wrong with public funding for a drug-treatment program that provides for spiritual needs," says Califano, "if that's what an individual needs to shake cocaine, to shake alcohol, to shake heroin."

Supporters of turning over social programs to churches and charities say their plan will reawaken American compassion. They claim neighbors no longer help neighbors, as they did before the rise of big government, because they assume their taxes pay for assistance programs. Further, advocates say charitable programs are more effective and efficient than government. Religion is often crucial to turning lives around, they contend.

Detractors see a dangerous trend that will distribute money unfairly, further cutting scarce resources to the poor. Worse, they say, counting on charities ignores history: The government social safety net grew because churches and volunteers could no longer deal with the entrenched poverty, the demands of a mobile society and the runaway health care costs of the late 20th century. Further, charities are already overburdened trying to respond to existing cuts in government spending on the poor, disabled and needy. Finally, nonprofit institutions are not a substitute for government, critics maintain, because government already provides 37 percent of the funding for charities, which is used to run programs from housing for the elderly to job training for the homeless and unemployed.

Here are the biggest questions shaping the debate:

CAN CHURCHES REPLACE THE SOCIAL SAFETY NET?

Frances White's journey off welfare is the kind of tale that supporters of Mississippi's Faith and Families highlight. But like most such odysseys, it included false starts and setbacks that show both the promise and the limitations of reliance on church-based programs.

It was Christmas 1994 when White, a divorced mother of three, got laid off from her third temporary job. ("Oh, God, please, my mother is out of a job again. Don't let this be happening," wrote White's 9-year-old daughter in her diary.) Governor Fordice had proposed Faith and Families just months before, and critics charged it was political gimmickry.

But for White, the program was a lifeline. In a state where about 25 percent of residents get some public assistance, Fordice (who in 1992 controversially declared America a "Christian nation") challenged churches to assist some of the 141,000 people on Aid to Families With Dependent Children. "Government screwed it up," says Fordice of welfare. "People, one on one, can fix it." White

From *U.S. News & World Report*, September 9, 1996, pp. 46-53.

volunteered to be "adopted" and was taken in by members of the Crossgates Baptist Church in Brandon, a middle-class suburb east of Jackson. "They paid my phone bill, found me a refrigerator, got my car running and some tires for it, caught up my rent, bought food. If it weren't for them, I would've been evicted," says the 40-year-old White.

Yet for White there was something more meaningful than the church's financial help: "The personal contact was better than anything. When you apply for welfare, you are a case number. But the church officials would call me, and we'd just talk."

The suburban church provided a ready-made job network, too. Through a church member, she landed work as a records clerk at a Jackson hospital. Making more than minimum wage, White left her substandard home. Although her rent jumped from $175 a month to $465, she doesn't have to worry, in her new home, about a leaky roof or cold winds whipping through the walls.

Still, White has climbed only a few tenuous rungs up the ladder from poverty. When she signed up with Faith and Families, she was receiving $144 per month in welfare and $360 per month in food stamps. She still gets $247 in food stamps because her salary is low and her new job does not provide health benefits. The Mississippi program promises to keep providing Medicaid a full year after a participant finds a job, but White and others found themselves bumped, because other state health officials were unaware of the pledge. For White, the results were disastrous. She lost her Medicaid just as her daughter was about to enter the hospital for a tonsillectomy. Unable to pay the $4,000 bill, White filed for bankruptcy in June.

At Crossgates Baptist Church, White counts as a success. But it's not one parishioners will try to replicate soon. The church spent $1,400 on her, far more than anticipated, and some members were miffed about aiding someone outside the congregation. "We're holding off on helping others right now," says Ken Box, chairman of the congregation's Benevolence Committee. "We have so many of our own church members who need help. We should take care of our own first."

The biggest disappointment with Faith and Families, Fordice concedes, is that there have been too few Frances Whites. In a year and a half, only 21 participants got off welfare. One problem is that just 98 welfare families—the most motivated—volunteered for the program. Explaining such reluctance, state coordinator Margaret Luckett notes, "The No. 1 question is always, 'Do we have to go to church?' " But White says that although she was often invited to Crossgates, she never felt compelled to attend. Congregations, too, are hesitant. Only 267 of the state's 5,500 signed up, and most of those never followed through on their commitment. Today, only about 15 churches are matched with families. Critics like the Rev. Rims Barber, who dismisses the program as "Faith and Foofoo," say that despite breakthroughs with mothers like White, it stigmatizes welfare seekers. "It's based on myths," says Barber, a Jackson Presbyterian minister. "The main one is that the problem with poor people is that they don't love Jesus."

As much as churches can add to the safety net, it is doubtful they can ever replace more than a fraction of government assistance. In White's case, for example, perhaps the most long-lasting aid she received was state-paid tuition to a local college to seek a bachelor's degree in clinical psychology.

Making matters harder, government is slashing funding for social services. Overall, each of the nation's 258,000 churches, synagogues and mosques would have to donate at least $225,000 a year to make up for congressionally proposed cuts, estimates Father Fred Kammer of Catholic Charities USA. But the average total budget of a congregation today, Kammer notes, is only around $100,000.

Paradoxically, at the exact moment churches are being asked to do more to save the poor, they face a backlash. This November, for example, voters in Colorado could make theirs the first state to tax the property of churches and many charities. In Hartford, Conn., the City Council has imposed a moratorium on the expansion of any new charity-run social services, like soup kitchens and drug clinics. A badly needed health center in one of the most distressed parts of town has been blocked. And although the moratorium and consequent zoning changes seem to hurt the poor the most, it was the city's Puerto Rican deputy mayor and a black state representative, herself a former welfare mother, who pushed hardest for the moratorium. They say the impoverished city's many social services drive private businesses out of downtown.

DOES FAITH WORK BEST?

What's the surest guarantee that an African-American urban youth will not fall to drugs or crime? Regular church attendance turns out to be a better predictor than family structure or income, according to a study by Harvard University economist Richard Freeman. Call it the "faith factor." The link between religious participation and avoidance of drug abuse, alcoholism, crime and other social pathologies is grist for some intriguing new research. Says Brookings Institution political scientist John DiIulio, "It's remarkable how much good empirical evidence there is that religious belief can make a positive difference." Policy makers are loath to promote faith because of their intellectual bias, he argues. But in most inner cities, where government, schools and other institutions fail the poor, says DiIulio, it is church programs that are "leveraging 10 times their own weight and solving social problems for us." And they offer personal salvation. A survey by John Gartner of Loyola College of Maryland and David Larson of Duke University Medical Center found over 30 studies that show a correlation between religious participation and avoidance of crime and substance abuse.

New research goes further, exploring how faith can turn around already troubled lives. Federal prisoners who got leadership training from Prison Fellowship, a prison ministry started by Watergate conspirator Charles Colson, were 11 percent less likely to be rearrested after 14 years, according to one survey. A more rigorous follow-up found similar drops in recidivism among inmates who regularly attended prison Bible classes. Similarly, Alcoholics Anonymous invokes a "higher power" to overcome addiction. "Religion is the forgotten factor," says criminologist Byron Johnson of Lamar University in Texas. He studied Prison Fellowship ministries and concluded faith is essential to preventing recidivism. Family support, education, job training and other factors matter, too, he notes. But religion is the one piece of the puzzle that most troubles government. One counselor at a Kentucky prison, on state orders, left the room when inmates in an alcohol-treatment group ended their daily meetings by standing up, holding hands and reciting the Serenity prayer. Complains Johnson: "We use pet therapy, horticulture therapy, acupuncture in prisons, but if you mention God, there's a problem."

Religion provides a set of values and moral beliefs. That's obvious. And churches provide a supportive community. That can be crucial. But belief seems to work at a more redeeming level. "To overcome addictions, you have to have phenomenal motivation," notes Patrick Fagan, who has reviewed studies on the impact of religion for the Heritage Foundation. Most addicts have so often disappointed loved ones that no one trusts them. So there is powerful blessing, says Fagan, in "a knowledge of God's acceptance, that he accepts the person and their sinfulness, that even as they fail they will not face rejection from God."

That seems clear at a singular drug rehab program in San Antonio. It is midday on the city's blighted West Side and the streets are ominously empty while some nearby buildings house cocaine and heroin dealers. But on one block, buzzing voices fill the street. More than 100 men and women sit in a parched yard, praying and discussing the Bible. At a tiny outdoor chapel, men crowd together in gospel song, prayer and witness. These are addicts, killers, thieves and gang members. But there are no guards and no security. "There's no fence. I could leave, but I still don't leave," says Joe, a founder of the Mexican Mafia, a brutal Texas prison gang. A tattoo of a snarling panther crawls up his muscular forearm, the better to hide the needle tracks of 32 years of drug addiction. "I can't explain it," says Joe, "but whatever it is, is working." To Freddie García, the ex-heroin addict who started the Victory Fellowship with his wife, Ninfa, the answer is simple: "The human spirit needs God."

It's hard to say what makes the program work for a man like Joe. Belief? "Now, I know a little bit about Jesus and he's nobody to fool around with," he offers reluctantly. Or the fact that Joe finds García's approach more culturally relevant than his too-many-to-count stays at professionalized drug clinics? Unlike most government programs, for example, an addicted mother can come to Victory Fellowship without placing her kids with relatives or in foster care (even if it means the mother sleeps on the floor). On this day, García has reunited Joe with his skeptical wife, a reunion that has left the hulking ex-con fighting tears.

If faith is the forgotten factor that can turn around lives, García knows government is too often needlessly meddlesome. In the past year, Texas state officials objected to his referring to his program as "rehabilitation," since it did not hire licensed counselors. Another San Antonio faith-based treatment program, Teen Challenge of South Texas, faced a more serious threat last year when Texas revoked its license. Director James Heurich says state officials insisted it use more traditional therapeutic approaches and could not understand the group's view of drug addiction as sin, not disease. After members of the successful program protested at the Alamo, the state backed off.

That raises the tough constitutional question: Should public funding go to proselytizing groups? The separation of church and state, a valued constitutional principle, has allowed religious diversity to flourish. Still, some legal scholars, like Yale University law professor Stephen Carter, author of *The Culture of Disbelief,* say courts interpret this separation too strictly, thereby depriving churches of their proper "involvement in the public square." Indeed, silly bureaucracy can stifle the efficacy of faith, like New York officials who threatened to sue the City Mission in Schenectady because it refused to let the homeless and addicted men who lived there keep pornography.

Yet balancing religion and rights can be very tricky. In Fort Worth, the sheriff created a special living unit in the county jail for religious inmates to hold daily Bible studies and nightly chapel. But only fundamental Christianity is accepted and other mainstream religions, including Unitarianism, the Church of Jesus Christ of Latter-day Saints and Christian Science, are taught to be "cults." Other inmates sued, and a Texas court soon will consider whether the so-called God Pod creates a preferential living setup.

CAN GOVERNMENT HELP CHARITIES FLOURISH?

Betty Christian got job training from a state program—and she got some from an evangelical ministry. "Government training programs are so impersonal," she says. But at the Joy of Jesus, there were daily Bible lessons to talk about diligence, responsibility and other values that make good workers. And there was prayerful discussion of the loneliness and financial woes in her life. "It renewed my faith; it made me feel everything was not lost," she says.

The Detroit ministry was so successful at finding jobs for unemployed workers like Christian that two years ago Michigan officials asked it to accept state funding. There was one catch: The prayers and Bible studies had to go. The result, says Kevin Feldman of the ministry, was disastrous. The program, which placed 60 percent of its students in jobs while it relied on a faith orientation, saw the rate drop to near zero. Classroom sessions were disrupted by noisy, uninterested enrollees. Last December, Joy of Jesus closed its work program and returned the state money. It has been unable to reopen because few donors—other than government—provide money for adult job training.

One proposed, but controversial, solution: vouchers. The new welfare reform law would allow states to give vouchers to unemployed workers, who, in turn, could choose a faith-based work program like Joy of Jesus. But letting individuals, not government, direct money to religious groups may still violate the separation between church and state. Ohio last week became the first state to give public money to students to pay for religious schools. But the Wisconsin Supreme Court last year stopped a similar plan in that state. Bob Dole champions school vouchers; President Clinton opposes them.

An idea also in vogue is using tax credits. Sen. Dan Coats would give taxpayers a credit equal to their donations to antipoverty charities – up to $500 for individuals and $1,000 for couples. The Indiana Republican estimates his tax credit would cost the Treasury $25 billion a year, and those funds would come largely from cuts in social programs.

Some 20 state experiments with tax credits show they sometimes infuse charities with needed funds. But tax credits can be costly to government, create unwanted paperwork for charities and, ultimately, distribute money unevenly. In Michigan, groups that aid the homeless get the credit, but only if that is their sole mission. So while other food pantries qualify, the American Red Cross Regional Food Distribution Center in Lansing is excluded. Further, food pantry officials say, it is unclear whether the credit stimulates substantial new giving or simply takes state revenues to reward those who give anyway.

Coats says tax credits let taxpayers, instead of government, decide which charities get money, "creating a free market of charitable giving." Citing a Beacon Hill Institute study, he argues that 67 percent of federal social service dollars go to overhead and public employees, instead of those who need services.

For many of the nation's most successful religious charities, however, whether to take public or private donations is not an either-or question. For them, government is a crucial partner. When blocks of central Newark, N.J., burned down during 1967 riots, William Linder was a parish priest. Today, Monsignor Linder presides over one of the nation's most astonishing urban rebirths. Linder started the New Community Corp., a community development corporation that used public and private money to revitalize the city's economy by building housing, nursing homes, day-care centers, restaurants – even a shopping mall with a supermarket so residents no longer have to leave the neighborhood to shop.

Faith fuels a uniquely successful type of compassion. That truth is also on display at Hope House, a residential facility for 51 abandoned and severely abused children in Nampa, Idaho. The 12 full-time staffers work for free, allowing the ranch to operate at a fraction of the cost of state facilities. "We believe in our heart we are called to serve," says founder Donnalee Velvick. Of the nine residents who will graduate from the facility's Christian academy in June, two are headed for military service and five for college. The remaining two, with mental retardation, will live in the program's adult house. Thousands of such faith-based charities across the country fill holes in the social safety net. The issue is whether they can fill even more.

BY JOSEPH P. SHAPIRO WITH
ANDREA R. WRIGHT IN MISSISSIPPI

Social Change and the Future

- Population, Environment, and Society (Articles 39 and 40)
- Technology and Society (Articles 41 and 42)
- The Future (Articles 43 and 44)

Fascination with the future is an enduring theme in literature, art, poetry, and religion. Human beings are anxious to know if tomorrow will be different from today and in what ways it might differ. Coping with change has become a top priority in the lives of many. One result of change is stress. When the future is uncertain and the individual appears to have little control over what happens, stress can be a serious problem. On the other hand, stress can have positive effects on people's lives if they can perceive changes as challenges and opportunities.

Any discussion of the future must begin with a look at the interface of population and the environment. Sandra Postel claims, "As a result of our population size, consumption patterns, and technology choices, we have surpassed the planet's carrying capacity." Natural capital is depleting, and the environment is degrading. We need to change consumption patterns, refocus technology, and change the incentive structure governing the behavior of individuals and businesses. Another demographic issue that worries many Americans is the influx of immigrants who compete for scarce jobs with citizens and increase costs for public services. Luis Gutierrez provides a careful examination of both the costs and benefits of immigrants and shows that they contribute more than they cost.

The next section addresses the linkage between technological change and society. One big story in the technology field concerns the new information technologies. Andrew Kupfer explores the many resulting social changes that can be expected in our work, home life, and residential location. The potentials for both good and evil are tremendous. Next, Bruce Nelan assesses the horrifying potential of nuclear, chemical, and biological technologies for terrorism. It is now clear that extremist groups are willing to murder millions of people, so just when the fear of nuclear war fades, fear of terrorist mass destruction skyrockets.

The final section looks at the future largely in political and economic terms. Robert Kaplan observes the social disorganization that is occurring in African countries where population growth is crashing into environmental limits and causing economic hardships and political instability. He predicts that this pattern will spread to many other less-developed societies, so that anarchy will dramatically increase throughout the world. Finally, Lester Thurow concludes in an article from his book about the future of capitalism that American capitalism is in deep trouble. The growing inequality and the decline of support for the welfare state, which helps those who are inadequately served by capitalism, will eventually undermine capitalism's legitimacy. Unfortunately, the problems with capitalism evolve slowly, and appropriate actions are not taken because American democracy acts decisively only in crises. Television teaches us to "consume," when sacrificing to build the future is needed. Therefore, Thurow says, capitalism faces stagnation unless it changes dramatically.

Looking Ahead: Challenge Questions

What are the advantages of slowing world population growth? How can it be done?

Why are people concerned about current immigration patterns? Do you think their fears are largely imaginary? Explain your answer.

What are some of the major problems that technology is creating?

How well is America preparing for the future? What are its major failures in this regard?

What are the significant factors bringing about social change at present?

UNIT 6

GEORGE BRAZIL
Plumbing·Heating
Air Conditioning
TIP
TRANSPORT INTERNATIONAL POOL
DODGE

Carrying Capacity: Earth's Bottom Line

As a society, we have failed to discriminate between technologies that meet our needs in a sustainable way and those that harm the earth.

SANDRA POSTEL

Sandra Postel is Vice President for Research at Worldwatch Institute and a co-author of State of the World 1994, *from which this article is adapted.*

It takes no stretch of the imagination to see that the human species is now an agent of change of geologic proportions. We literally move mountains to mine the earth's minerals, redirect rivers to build cities in the desert, torch forests to make way for crops and cattle, and alter the chemistry of the atmosphere in disposing of our wastes. At humanity's hand, the earth is undergoing a profound transformation—one with consequences we cannot fully grasp.

It may be the ultimate irony that, in our efforts to make the earth yield more for ourselves, we are diminishing its ability to sustain life of all kinds—humans included. Signs of environmental constraints are now pervasive. Cropland is scarcely expanding any more, and a good portion of existing agricultural land is losing fertility. Grasslands have been overgrazed and fisheries overharvested, limiting the amount of additional food from these sources. Water bodies have suffered extensive depletion and pollution, severely restricting future food production and urban expansion. And natural forests—which help stabilize the climate, moderate water supplies, and harbor a majority of the planet's terrestrial biodiversity—continue to recede.

These trends are not new. Human societies have been altering the earth since they began. But the pace and scale of degradation that started about mid-century—and continues today—is historically new. The central conundrum of sustainable development is now all too apparent: Population and economies grow exponentially, but the natural resources that support them do not.

Biologists often apply the concept of "carrying capacity" to questions of population pressures on an environment. Carrying capacity is the largest number of any given species that a habitat can support indefinitely. When that maximum sustainable population level is surpassed, the resource base begins to decline; sometime thereafter, so does the population.

The earth's capacity to support humans is determined not just by our most basic food requirements but also by our levels of consumption of a whole range of resources, by the amount of waste we generate, by the technologies we choose for our varied activities, and by our success at mobilizing to deal with major threats. In recent years, the global problems of ozone depletion and greenhouse warming have underscored the danger of overstepping the earth's ability to absorb our waste products. Less well recognized, however, are the consequences of exceeding the sustainable supply of essential resources, and how far along that course we may already be.

As a result of our population size, consumption patterns, and technology choices, we have surpassed the planet's carrying capacity. This is plainly evident by the extent to which we are damaging and depleting natural capital. The earth's environmental assets are now insufficient to sustain both our present patterns of economic activity and the life-support systems we depend on. If current trends in resource use continue, and if world population grows as projected, by 2010 per capita availability of rangeland will drop by 22 percent and the fish catch by 10 percent. Together, these provide much of the world's animal protein. The per capita area of irrigated land, which now yields about one-third of the global food harvest, will drop by 12 percent. And cropland area and forestland per person will shrink by 21 and 30 percent, respectively.

The days of the frontier economy, in which abundant resources were available to propel economic growth and living standards, are over. We have entered an era in which global prosperity increasingly depends on using resources more efficiently, on distributing them more equitably, and on reducing consumption levels overall. Unless we accelerate this transition, powerful social tensions are likely to arise from increased competition for the scarce resources that remain. There likely will be, for example, a surge in hunger, cross-border migration, and conflict—trends already painfully evident in parts of the world.

The roots of environmental damage run deep. Unless they are unearthed soon, we risk exceeding the planet's carrying capacity to such a degree that a future of economic and social decline will be impossible to avoid.

Driving forces

Since mid-century, three trends have contributed most directly to the excessive pressures now being placed on the earth's natural systems—the doubling of world population, the quintupling of global economic output, and the widening gap in the distribution of income. The environmental impact of the world's population (now numbering 5.5 billion) has been vastly multiplied by economic and social systems that strongly favor growth and ever-rising consumption over equity and poverty alleviation; that fail to give women equal rights, education, and economic opportunity, and thereby perpetuate the conditions under which poverty and rapid population growth persist; and that do not discriminate between means of production that are environmentally sound and those that are not.

Table 1 **Global Income Distribution, 1960-89**

Year	Share of Global Income Going to: Richest 20% (percent)	Poorest 20% (percent)	Ratio of Richest to Poorest
1960	70.2	2.3	30 to 1
1970	73.9	2.3	32 to 1
1980	76.3	1.7	45 to 1
1989	82.7	1.4	59 to 1

Source: United Nations Development Programme, *Human Development Report 1992*, Oxford University Press, 1992.

• *Growing Inequality in Income*: Of the three principal driving forces, the growing inequality in income between rich and poor stands out in sharpest relief. In 1960, the richest 20 percent of the world's people absorbed 70 percent of global income; by 1989 (the latest year for which comparable figures are available), the wealthy people's share had climbed to nearly 83 percent. The poorest 20 percent, meanwhile, saw their share of global income drop from an already meager 2.3 percent to just 1.4 percent. The ratio of the richest fifth's share to the poorest's thus grew from 30 to 1 in 1960 to 59 to 1 in 1989 (see *Table 1*).

This chasm of inequity is a major cause of environmental decline. It fosters overconsumption at the top of the income ladder and persistent poverty at the bottom. People at either end of the income spectrum are often more likely than those in the middle to damage the earth's ecological health—the rich because of their high consumption of energy, raw materials, and manufactured goods, and the poor because they must often cut trees, grow crops, or graze cattle in ways harmful to the earth, merely to survive from one day to the next.

Families in the western United States, for instance, often use as much as 3,000 liters of water a day—enough to fill a bathtub 20 times. Overdevelopment of water there has contributed to the depletion of rivers and aquifers, has destroyed wetlands and fisheries, and, by creating an illusion of abundance, has led to excessive consumption. Meanwhile, nearly one out of every three people in the developing world (some 1.2 billion people in all) lack access to a safe supply of drinking water.

Disparities in food consumption are revealing as well (see *Table 2*). As many as 700 million people do not eat enough to live and work at their full potential. The average African, for instance, consumes only 87 percent of the calories needed for a healthy and pro-

Table 2 **Grain Consumption Per Person in Selected Countries, 1990**

Country	Grain Consumption Per Person (kilograms)
Canada	974
United States	860
Soviet Union	843
Australia	503
France	465
Turkey	419
Mexico	309
Japan	297
China	292
Brazil	277
India	186
Bangladesh	176
Kenya	145
Tanzania	145
Haiti	100
World Average	323

Sources: Worldwatch Institute estimate, based on U.S. Department of Agriculture, *World Grain Database* (unpublished printout), 1992; Population Reference Bureau, *1990 World Population Data Sheet*, 1990.

ductive life. Meanwhile, diets in many rich countries are so laden with animal fat as to cause increased rates of heart disease and cancer. Moreover, the meat-intensive diets of the wealthy usurp a disproportionately large share of the earth's agricultural carrying capacity, since producing one kilogram of meat takes several kilograms of grain. If all people in the world required as much grain for their diet as the average American does, the global harvest would need to be 2.6 times greater than it is today—a highly improbable scenario.

• *Economic Growth*: The second driving force—economic growth—has been fueled in part by the introduction of oil onto the energy scene. Since mid-century, the global economy has expanded fivefold. As much was produced in two-and-a-half months of 1990 as in the entire year of 1950. World trade, moreover, grew even faster. Exports of primary commodities and manufactured products rose elevenfold.

Unfortunately, economic growth has most often been of the damaging variety—powered by the extraction and consumption of fossil fuels, water, timber, minerals, and other resources. Between 1950 and 1990, the industrial roundwood harvest doubled, water use tripled, and oil production rose nearly sixfold. Environmental damage increased proportionately.

• *Population Growth*: Compounding the rises in both poverty and resource consumption in relation to the worsening of inequality and rapid economic expansion, population growth has added greatly to pressures on the earth's carrying capacity. The doubling of world population since 1950 has meant more or less steady increases in the number of people added to the planet each year. Whereas births exceeded deaths by 37 million in 1950, the net population gain in 1993 was 87 million—roughly equal to the population of Mexico.

The U.N. median population projection now shows world population reaching 8.9 billion by 2030, and leveling off at 11.5 billion around 2150.

The resource base

The outer limit of the planet's carrying capacity is determined by the total amount of solar energy converted into biochemical energy through plant photosynthesis minus the energy those plants use for their own life processes. This is called the earth's net primary productivity (NPP), and it is the basic food source for all life.

Prior to human impacts, the earth's forests, grasslands, and other terrestrial ecosystems had the potential to produce a net total of some 150 billion tons of organic matter per year. Stanford University biologist Peter Vitousek and his colleagues estimate, however, that humans have destroyed outright about 12 percent of the terrestrial NPP and now directly use or co-opt an additional 27 percent. Thus, one species—*Homo sapiens*—has appropriated nearly 40 percent of the terrestrial food supply, leaving only 60 percent for the millions of other land-based plants and animals.

It may be tempting to infer that, at 40 percent of NPP, we are still comfortably below the ultimate limit. But this is not the case. We have appropriated the 40 percent that was easiest to acquire. It may be impossible to double our share, yet theoretically that would happen in just 60 years if our share rose in tandem with population growth. And if average resource consumption per person continues to increase, that doubling would occur much sooner.

Perhaps more important, human survival hinges on a host of environmental services provided by natural systems; for example, forests regulate the hydrological cycle and wetlands filter pollutants. As we destroy, alter, or appropriate more of these natural systems for ourselves, these environmental services are compromised. At some point, the likely result is a chain reaction of environmental decline—widespread flooding and erosion brought on by deforestation, for example, or worsened drought and crop losses from desertification, or pervasive aquatic pollution and fisheries losses from wetlands destruction. The simultaneous unfolding of several such scenarios could cause unprecedented human hardship, famine, and disease. Precisely when vital thresholds will be crossed, no one can say. But as Vitousek and his colleagues note, those "who believe that limits to growth are so distant as to be of no consequence for today's decisionmakers appear unaware of these biological realities."

How have we come to usurp so much of the earth's productive capacity? In our efforts to feed, clothe, and house ourselves, and otherwise satisfy our evergrowing material desires, we have steadily converted diverse and complex biological systems to more uniform and simple ones that are managed for human benefit. Timber companies cleared primary forests and replaced them with monoculture pine plantations to make pulp and paper. Migrant peasants torched tropical forests in order to plant crops merely to survive. And farmers plowed the prairie grasslands of the United States' Midwest to plant corn, thereby creating one of the most productive agricultural regions in the world. Although these transformations have allowed more humans to be supported at a higher standard of living, they have come at the expense of natural systems, other plant and animal species, and ecological stability.

Continuing along this course is risky. But the flip side of the problem is equally sobering. What do we do when we have claimed virtually all that we can, yet our population and demands are still growing?

• *Cropland*: Cropland area worldwide expanded by just 2 percent between 1980 and 1990. That means that gains in the global food harvest came almost entirely from raising yields on existing cropland. Most of the remaining area that could be used to grow crops is in Africa and Latin America; very little is in Asia. The most sizable near-term additions to the cropland base are likely to be a portion of the 76 million hectares of savanna grasslands in South America that are already accessible and potentially cultivable, as well as some portion of African rangeland and forest. These conversions, of course, may come at a high environmental price, and will push our 40-percent share of NPP even higher.

Moreover, a portion of any cropland gains that do occur will be offset by losses. As economies of developing countries diversify and as cities expand to accommodate population growth and migration, land is rapidly being lost to industrial development, housing, road construction, and the like. Canadian geographer Vaclav Smil estimates, for instance, that between 1957 and 1990, China's arable land diminished by at least 35 million hectares—an area equal to all the cropland in France, Germany, Denmark, and the Netherlands combined. At China's 1990 average grain yield and consumption levels, that amount of cropland could have supported some 450 million people, about 40 percent of its population.

In addition, much of the land we continue to farm is losing its inherent productivity because of unsound agricultural practices and overuse. The "Global Assessment of Soil Degradation," a three-year study involving some 250 scientists, found that more than 550 million hectares are losing topsoil or undergoing other forms of degradation as a direct result of poor agricultural methods (see *Table 3*).

On balance, unless crop prices rise, it appears unlikely that the net cropland area will expand much more quickly over the next two decades than it did between 1980 and 1990. Assuming a net expansion of 5 percent (which may be optimistic), total cropland area would climb to just over 1.5 billion hectares. Given the projected 33-percent increase in world population by 2010, the amount of crop-

Table 3 **Human-Induced Land Degradation Worldwide, 1945 to Present**

Region	Overgrazing	Deforestation	Agricultural Mismanagement	Other[1]	Total	Degraded Area as Share of Total Vegetated Land (percent)
	(million hectares)					
Asia	197	298	204	47	746	20
Africa	243	67	121	63	494	22
South America	68	100	64	12	244	14
Europe	50	84	64	22	220	23
North & Central America	38	18	91	11	158	8
Oceania	83	12	8	0	103	13
World	679	579	552	155	1,965	17

1 Includes exploitation of vegetation for domestic use (133 million hectares) and bioindustrial activities, such as pollution (22 million hectares).
Sources: Worldwatch Institute, based on "The Extent of Human-Induced Soil Degradation," Annex 5 in L.R. Oldeman et al., *World Map of the Status of Human-Induced Soil Degradation* (Wageningen, Netherlands: United Nations Environment Programme and International Soil Reference and Information Centre, 1991).

land per person would decline by 21 percent (see *Table 4*).

• *Pasture and Rangeland*: They cover some 3.4 billion hectares of land, more than twice the area in crops. The cattle, sheep, goats, buffalo, and camels that graze them convert grass (which humans cannot digest) into meat and milk (which they can). The global ruminant livestock herd, which numbers about 3.3 billion, thus adds a source of food for people that does not subtract from the grain supply, in contrast to the production of pigs, chickens, and cattle raised in feedlots.

Much of the world's rangeland is already heavily overgrazed and cannot continue to support the livestock herds and management practices that exist today. According to the "Global Assessment of Soil Degradation," overgrazing has degraded some 680 million hectares since mid-century. This suggests that 20 percent of the world's pasture and range is losing productivity and will continue to do so unless herd sizes are reduced or more sustainable livestock practices are put in place.

During the 1980s, the total range area increased slightly, in part because land deforested or taken out of crops often reverted to some form of grass. If similar trends persist over the next two decades, by 2010 the total area of rangeland and pasture will have increased 4 percent, but it will have dropped 22 percent in per capita terms. In Africa and Asia, which together contain nearly half the world's rangelands and where many traditional cultures depend heavily on livestock, even larger per capita declines could significantly weaken food economies.

• *Fisheries*: Another natural biological system that humans depend on to add calories, protein, and diversity to human diets is our fisheries. The annual catch from all sources (including aquaculture) totaled 97 million tons in 1990—about 5 percent of the protein humans consume. Fish account for a good portion of the calories consumed overall in many coastal regions and island nations.

The world fish catch has climbed rapidly in recent decades, expanding nearly fivefold since 1950. But it peaked at just above 100 million tons in 1989. Although catches from both inland fisheries and aquaculture (fish farming) have been rising steadily, they have not offset the decline in the much larger wild marine catch, which fell from a historic peak of 82 million tons in 1989 to 77 million in 1991, a drop of 6 percent.

With the advent of mechanized hauling gear, bigger nets, electronic fish detection aids, and other technologies, almost all marine fisheries have suffered from extensive overexploitation. Under current practices, considerable additional growth in the global fish catch overall looks highly unlikely. Indeed, the U.N. Food and Agriculture Organization (FAO) now estimates that all seventeen of the world's major fishing areas have either reached or exceeded their natural limits, and that nine are in serious decline.

Table 4 **Population Size and Availability of Renewable Resources, Circa 1990, With Projections for 2010**

	Circa 1990 (million)	2010 (million)	Total Change (percent)	Per Capita Change (percent)
Population	5,290	7,030	+33	
Fish Catch (tons)[1]	85	102	+20	-10
Irrigated Land (hectares)	237	277	+17	-12
Cropland (hectares)	1,444	1,516	+ 5	-21
Rangeland and Pasture (hectares)	3,402	3,540	+4	-22
Forests (hectares)[2]	3,413	3,165	-7	-30

1 Wild catch from fresh and marine waters, excludes aquaculture.

2 Includes plantations; excludes woodlands and shrublands.

Sources: Population figures from U.S. Bureau of the Census, Department of Commerce, *International Data Base*, unpublished printout, November 2, 1993; 1990 irrigated land, cropland, and rangeland from U.N. Food and Agriculture Organization (FAO), *Production Yearbook 1991*; fish catch from M. Perotti, chief, Statistics Branch, Fisheries Department, FAO, private communication, November 3, 1993; forests from FAO, *Forest Resources Assessment 1990*, 1992 and 1993. For detailed methodology, see *State of the World* 1994, among other sources.

FAO scientists believe that better fisheries management might allow the wild marine catch to increase by some 20 percent. If this could be achieved, and if the freshwater catch increased proportionately, the total wild catch would rise to 102 million tons; by 2010, this would nonetheless represent a 10-percent drop in per capita terms.

• *Fresh Water*: It may be even more essential than cropland, rangeland, and fisheries; without water, after all, nothing can live. Signs of water scarcity are now pervasive. Today, twenty-six countries have insufficient renewable water supplies within their own territories to meet the needs of a moderately developed society at their current population size. And populations are growing fastest in some of the most water-short countries, including many in Africa and the Middle East. Rivers, lakes, and underground aquifers show widespread signs of degradation and depletion, even as human demands rise inexorably.

Water constraints already appear to be slowing food production, and those restrictions will only become more severe. Agricultural lands that receive irrigation water play a disproportionate role in meeting the world's food needs. The 237 million hectares of irrigated land account for only 16 percent of total cropland but more than one-third of the global harvest. For most of human history, irrigated area expanded faster than population did, which helped food production per person to increase steadily. In 1978, however, per-capita irrigated land peaked, and it has fallen nearly 6 percent since then.

• *Forests and Woodlands*: They are the last key component of the biological resource base. They contribute a host of important commodities to the global economy—logs and lumber for constructing homes and furniture, fiber for making paper, fruits and nuts for direct consumption, and, in poor countries, fuelwood for heating and cooking. More important even than these benefits, however, are the ecological services forests perform—from conserving soils and moderating water cycles to storing carbon, protecting air quality, and harboring millions of plant and animal species.

Today, forests cover 24 percent less area than in 1700—3.4 billion hectares compared with an estimated 4.5 billion about 300 years ago. Most of that area was cleared for crop cultivation, but cattle ranching, timber and fuelwood harvesting, and the growth of cities, suburbs, and highways all claimed a share as well. Recent assessments suggest that the world's forests declined by about 130 million hectares between 1980 and 1990, an area larger than Peru.

Redirecting technology

Advances in technology—which is used broadly here to mean the application of knowledge to an activity—offer at least a partial way out of our predicament. In most cases, "appropriate" technologies will no longer be engineering schemes, techniques, or methods that enable us to claim more of nature's resources but, instead, systems that allow us to benefit more from the resources we already have. As long as the resulting gains are directed toward bettering the environment and the lives of the less fortunate instead of toward increased consumption by the rich, such efforts will reduce human impacts on the earth.

The power of technology to help meet human needs was a critical missing piece in the world-view of Thomas Malthus, the English curate whose famous 1798 essay postulated that the growth of human population would outstrip the earth's food-producing capabilities. His prediction was a dire one—massive famine, disease, and death. But a stream of agricultural advances combined with the productivity leaps of the Industrial Revolution made the Malthusian nightmare fade for much of the world.

Without question, technological advances have steadily enhanced our capacity to raise living standards. They not only helped to boost food production—the main concern of mothers—they also increased our access to sources of water, energy, timber, and minerals.

As a society, however, we have failed to discriminate between technologies that meet our needs in a sustainable way and those that harm the earth. We have largely let the market dictate which technologies move forward, without adjusting for its failure to take proper account of environmental damages. Now that we have exceeded the planet's carrying capacity and are rapidly running down its natural capital, such a correction is urgently needed.

In the area of food supply, it remains an open question whether technological advances will continue to raise crop yields fast enough to meet rising demand, and whether such gains will be sustainable. Given the extent of cropland and rangeland degradation and the slowdown in irrigation expansion, it may be difficult to sustain the past pace of yield increases. Indeed, per capita grain production in 1992 was 7 percent lower than the historic peak in 1984. Whether this is a short-term phenomenon or the onset of a longer-term trend will depend on what new crop varieties and technologies reach farmers' fields and whether they can overcome the yield-suppressing effects of environmental degradation. Another factor is whether agricultural policies and prices will encourage farmers to invest in raising land productivity further.

In many agricultural regions—including northern China, parts of India, Mexico, the western United States, and much of the Middle East—water may be more of a constraint to future food production than land, crop yield potential, or most other factors. Developing and distributing technologies and practices that improve water management is critical to sustaining the food production capability we now have, much less to increasing it for the future.

Water-short Israel is a front-runner in making its agricultural economy more water-efficient. Its current agricultural output could probably not have been achieved without steady advances in water management—including highly efficient drip irrigation, automated systems that apply water only when crops need it, and the setting of water allocations based on predetermined optimal water applications for each crop. The nation's success is notable: Between 1951 and 1990, Israeli farmers reduced the amount of water applied to each hectare of cropland by 36 percent. This allowed the irrigated area to more than triple with only a doubling of irrigation-water use.

Matching the need for sustainable gains in land and water productivity is the need for improvements in the efficiency of wood use and reductions in wood and paper waste, in order to reduce pressures on forests and woodlands. A beneficial timber technology is no longer one that improves logging efficiency—the number of trees cut per hour—but rather one that makes each log harvested go further. Raising the efficiency of forest product manufacturing in the United States, the world's largest wood consumer, roughly to Japanese levels would reduce timber needs by about one-fourth, for instance. Together, available methods of reducing waste, increasing manufacturing efficiency, and recycling more paper could cut U.S. wood consumption in half; a serious effort to produce new wood-saving techniques would reduce it even more.

With the world's paper demand projected to double by the year 2010, there may be good reason to shift production toward "treeless paper"—that made from nonwood pulp. Hemp, bamboo, jute, and kenaf are among the alternative sources of pulp. The fast-growing kenaf plant, for example, produces two to four times more pulp per hectare than southern pine, and the pulp has all of the main qualities needed for making most grades of paper. In China, more than 80 percent of all paper pulp is made from nonwood sources. Treeless paper was manufactured in forty-five countries in 1992, and accounted for 9 percent of the world's paper supply. With proper economic incentives and support for technology and market development, the use of treeless paper could expand greatly.

The role of trade

Consider two countries, each with a population of about 125 million. Country A has a population density of 331 people per square kilometer, has just 372 square meters of cropland per inhabitant (one-seventh the world average), and imports almost three-fourths of its grain and nearly two-thirds of its wood. Country B, on the other hand, has a population density less than half that of Country A and nearly five times as much cropland per person. It imports only one-tenth of its grain and no wood. Which country has most exceeded its carrying capacity?

Certainly it would be Country A—which, as it turns out, is Japan—a nation boasting a real gross domestic product (GDP) of some $18,000 per capita. Country B, which from these few indicators seems closer to living within its means, is Pakistan—with a real GDP per capita of only $1,900. By any economic measure, Japan is far and away the more successful of the two. So how can questions of carrying capacity be all that relevant?

The answer, of course, lies in large part with trade. Japan sells cars and computers, and uses some of the earnings to buy food, timber, oil, and other raw materials. And that is what trade is supposed to be about—selling what one can make better or more efficiently, and buying what others have a comparative advantage in producing. Through trade, countries with scarce resources can import what they need from countries with a greater abundance.

Imports of biologically based commodities like food and timber are, indirectly, imports of land, water, nutrients, and the other components of ecological capital needed to produce them. Many countries would not be able to support anything like their current population and consumption levels were it not for trade. To meet its food and timber demands alone, the Netherlands, for instance, appropriates the production capabilities of 24 million hectares of land—10 times its own area of cropland, pasture, and forest.

In principle, there is nothing inherently unsustainable about one nation relying on another's ecological surplus. The problem, however, is the widespread perception that all countries can exceed their carrying capacities and grow economically by expanding manufactured and industrial goods at the expense of natural capital—paving over agricultural land to build factories, for example, or clear-cutting forest to build new homes. But all countries cannot continue to do this indefinitely. Globally, the ecological books must balance.

Many economists see no cause for worry. They believe that the market will take care of any needed adjustments. As cropland, forests, and water grow scarce, all that is necessary, they say, is for prices to rise; the added incentives to conserve, use resources more productively, alter consumption patterns, and develop new technologies will keep output rising with demand. But once paved over for a highway or housing complex, cropland is unlikely to be brought back into production—no matter how severe food shortages may become. Moreover, no mechanism exists for assuring that an adequate resource base is maintained to meet needs that the marketplace ignores or heavily discounts—including those of vital ecosystems, other species, the poor, or the next generation.

Trade in forest products illuminates some of these trends. East Asia, where the much-touted economic miracles of Japan and the newly industrializing countries have taken place, has steadily and rapidly appropriated increasing amounts of other nations' forest resources. In Japan, where economic activity boomed after World War II, net imports of forest products rose eightfold between 1961 and 1991. The nation is now the world's largest net importer of forest products by far. Starting from a smaller base, South Korea's net imports have more than quadrupled since 1971, and Taiwan's have risen more than sevenfold.

Like technology, trade is inherently neither good nor bad. One of its strengths is its ability to spread the benefits of more efficient and sustainable technologies and products, whether they be advanced drip irrigation systems, nontimber products from tropical forests, or the latest paper recycling techniques. Trade can also generate more wealth in developing countries, which conceivably could permit greater investments in environmental protection and help alleviate poverty. So far, however, the potential gains from trade have been overwhelmed by its more negative facets—in particular, by its tendency to foster ecological deficit-financing and unsustainable consumption.

In light of this, it is disturbing, to say the least, that negotiators involved in the seven-year-long Uruguay Round of the General Agreement on Tariffs and Trade (GATT) seemed barely interested in the role trade plays in promoting environmental destruction. While the reduction of government subsidies and other barriers to free trade—the main concern of the GATT round—could make international markets more efficient and increase the foreign exchange earnings of developing countries, that offers no guarantee that trade will be more environmentally sound or socially equitable.

As part of the newly created World Trade Organization, a committee will probably be formed to address the trade/environment nexus more directly, although probably not as broadly as is needed. With short-term considerations such as slow economic growth and high unemployment taking precedence over long-term concerns, a coordinated effort to make trade more sustainable through cost-internalizing measures is not high on the agenda. If action is delayed too long, however, the future will arrive in a state of ecological impoverishment that no amount of free trade will be able to overcome.

Lightening the load

Ship captains pay careful attention to a marking on their vessels called the Plimsoll line. If the water level rises above the Plimsoll line, the boat is too heavy and is in danger of sinking. When that happens, rearranging items on the ship will not help much. The problem is the total weight, which has surpassed the carrying capacity of the ship.

Economist Herman Daly sometimes uses this analogy to underscore that the scale of human activity can reach a level that the earth's natural systems can no longer support. The ecological equivalent of the Plimsoll line may be the maximum share of the earth's biological resource base that humans can appropriate before a rapid and cascading deterioration in the planet's life-support systems is set in motion. Given the degree of resource destruction already evident, we may be close to this critical mark. The challenge, then, is to lighten our burden on the planet before "the ship" sinks.

More than 1,600 scientists, including 102 Nobel Laureates, underscored this point by collectively signing a "Warning to Humanity" in late 1992. It states: "No more than one or a few decades remain before the chance to avert the threats we now confront will be lost and the prospects for humanity immeasurably diminished. . . . A new ethic is required—a new attitude towards discharging our responsibility for caring for ourselves and for the earth. . . . This ethic must motivate a great movement, convincing reluctant leaders and reluctant governments and reluctant peoples themselves to effect the needed changes."

A successful global effort to lighten humanity's load on the earth would directly address the three major driving forces of environmental decline—the grossly inequitable distribution of income, resource-consumptive economic growth, and rapid population growth—and would redirect technology and trade to buy time for this great movement. Although there is far too much to say about each of these challenges to be comprehensive here, some key points bear noting.

Wealth inequality may be the most intractable problem, since it has existed for millennia. The difference today, however, is that the future of rich and poor alike hinges on reducing poverty and thereby eliminating this driving force of global environmental decline. In this way, self-interest joins ethics as a motive for redistributing wealth, and raises the chances that it might be done.

Important actions to narrow the income gap include greatly reducing Third World debt, much talked about in the 1980s but still not accomplished, and focusing foreign aid, trade, and international lending policies more directly on improving the living standards of the poor. If decisionmakers consistently asked themselves whether a choice they were about to make would help the poorest of the poor—that 20 percent of the world's people who share only 1.4 percent of the world's income—and acted only if the answer were yes, more people might break out of the poverty trap and have the opportunity to live sustainably.

A key prescription for reducing the kinds of economic growth that harm the environment is the same as that for making technology and trade more sustainable—internalizing environmental costs. If this is done through the adoption of environmental taxes, governments can avoid imposing heavier taxes overall by lowering income taxes accordingly. In addition, establishing better measures of economic accounting is critical. Since the calculations used to produce the gross national product do not account for the destruction or depletion of natural resources, this popular economic measure is extremely misleading. It tells us we are making progress even as our ecological foundations are crumbling. A better beacon to guide us toward a sustainable path is essential. The United Nations and several individual governments have been working to develop better accounting methods, but progress with implementation has been slow.

In September 1994, government officials will gather in Cairo for the "International Conference on Population and Development," the third such gathering on population. This is a timely opportunity to draw attention to the connections between poverty, population growth, and environmental decline; and to devise strategies that simultaneously address the root causes. Much greater efforts are needed, for instance, to raise women's social and economic status and to give women equal rights and access to resources. Only if gender biases are rooted out will women be able to escape the poverty trap and choose to have fewer children.

The challenge of living sustainably on the earth will never be met, however, if population and environment conferences are the only forums in which it is addressed. Success hinges on the creativity and energy of a wide range of people in many walks of life. The scientists' "Warning to Humanity" ends with a call to the world's scientists, business and industry leaders, the religious community, and people everywhere to join in the urgent mission of halting the earth's environmental decline.

FOR FURTHER READING

HERMAN DALY, *Steady-State Economics,* Island Press, 1991.

ALAN THEIN DURNING, *How Much Is Enough? The Consumer Society and the Future of the Earth,* W. W. Norton, 1992.

LLOYD T. EVANS, *Crop Evolution, Adaptation and Yield,* Cambridge University Press, 1993.

DONELLA H. MEADOWS, DENNIS L. MEADOWS, AND JORGEN RANDERS, *Beyond the Limits,* Chelsea Green, 1992.

SANDRA POSTEL, *Last Oasis: Facing Water Scarcity,* W. W. Norton, 1992.

FRANCIS URBAN AND RAY NIGHTINGALE, *World Population by Country and Region, 1950–1990 and Projections to 2050,* USDA Economic Research Service, 1993.

The New Assault on Immigrants

Luis Gutierrez

Luis Gutierrez is US Representative from the Illinois 4th Congressional District.

In *Boss,* Mike Royko's account of the life of Mayor Richard J. Daley and the city that he ruled, he describes the neighborhoods of Chicago vividly and accurately.

> Chicago, until as late as the 1950s, was a place where people stayed put for a while, creating tightly knit neighborhoods, as small-townish as any village in the wheat fields. The neighborhood-towns were part of larger ethnic states. To the north of the Loop was Germany. To the northwest Poland. To the west were Italy and Israel. To the southwest were Bohemia and Lithuania. And to the south was Ireland. It wasn't perfectly defined because the borders shifted as newcomers moved in on the old settlers, sending them fleeing in terror and disgust. . . . But you could always tell, even with your eyes closed, which state you were in by the odors of the food stores and the open kitchen windows, the sound of the foreign or familiar language, and by whether a stranger hit you in the head with a rock.

Chicago, like the rest of urban America, has changed. It has experienced several waves of new settlers since Royko wrote those words in 1971. In my city, Germany and Poland have been replaced by Puerto Rico and Central America. Bohemia, Lithuania and Ireland have moved out and Mexico has moved in. Settling around the edges of the old neighborhoods are countries as diverse as Korea, Cuba and Vietnam. Like the Germans and Poles, and the Irish and Lithuanians before them, these new settlers have brought with them distinct traditions of culture and language. And like the settlers before them, they are often the target of disgust and terror, of blame and fear.

Anti-immigrant backlash, too, seems to flow in waves, and in recent years, we've seen public discussion about immigration grow increasingly ugly and more divisive. Listen closely to the rhetoric coming out of state capitals from Sacramento to Tallahassee. Note the tone of radio talk shows. Study the legislation being proposed in Washington and across the country. From welfare reform to health care, immigrants—these new immigrants—are the target of the legislative and rhetorical equivalent of rocks aimed at their heads.

Today, the debate about immigration is becoming more personal and pointed. It is threatening our national identity and crying out for leaders who are willing to stand up and defend the rights, contributions and importance of immigrants to our nation. If the voices of reason are not heard now, the voices of intolerance and division are ready to drown them out for good.

As a veteran of the Chicago City Council, I came to Washington believing I was pretty accustomed to belligerent arguments, scream-

From *Social Policy,* Summer 1995, pp. 56-63.

ing debates and divisive legislation. I didn't take into account that I was entering Congress during the year of the immigrant as scapegoat.

The list of anti-immigrant legislation proposed in Congress this year is far too long to be catalogued here. That list might be headed by a current Republican welfare reform proposal that denies the benefits of 61 federal programs to non-citizen, permanent residents of the US. The bill would affect people who are in the US perfectly legally. Many of them have lived and worked here, and have contributed to the US economy for 10, 20 or 30 years.

Pick a day at random from the past session of Congress and you will probably find a bill or amendment introduced with the sole purpose of prohibiting federal aid to immigrants. You might find a proposal to deny eligibility for postsecondary-education student aid, to deny eligibility for the Job Training Partnership Act, to deny eligibility for Aid to Families with Dependent Children or to deny eligibility for food stamps. A low was hit last June with an amendment to deny Federal Emergency Management Assistance to "illegal aliens," an amendment that would have the practical effect of denying food, clothes and shelter to immigrants who are the victims of a natural disaster.

And while proponents of most federal legislation argue that their restrictions are aimed at undocumented immigrants, not those that are here through legal channels, most of their proposals are widely drawn and open to interpretation. Anti-immigrant legislation makes any person who doesn't look "American" an immediate suspect, whether that person is applying for a student loan or Social Security benefits. Almost all of these proposals would turn every local agency, from schools to employment offices, into regulated outposts of the Immigration and Naturalization Service, forcing them to constantly check every applicant's immigration status.

Finding Scapegoats

At the state level, anti-immigrant backlash has been even more severe. California and Florida are suing the federal government for reimbursement for funds spent on immigrants. While the question of who should pay the costs associated with immigration is legitimate, what their claims have instead focused on is whether immigrants deserve any services at all. In fact, the centerpiece of Pete Wilson's campaign for re-election, and his resurgence in popularity, has been his blame of immigrants for everything from crime to budget deficits.

The cumulative effect of this assault has been to make blaming immigrants for our nation's economic and social woes an accepted part of our political and cultural debate.

There is certainly plenty of blame to go around. Economic woes in California. Overcrowding in Florida. Crime in Texas. Budget problems from coast to coast.

The circumstances that lead to this situation are not hard to understand. The economic problems our nation is facing are tremendous. Immigrants are easy targets for the blame. More that a few elected officials are ready to reap the easy dividends that come from saying we can solve problems from crime to the deficit if we just didn't let so many immigrants across our borders.

The substance of the legislation—while in many cases exceedingly offensive to anyone who still subscribes to the notion that immigrants contribute to our national well-being—is often not as bad as the rhetoric that accompanies it. But the rhetoric is critically important: it is divisive, intolerant, unresponsive to the realities of immigration today, and ignorant of the history of immigration to our country. Creating a climate in which it is socially acceptable for public leaders to talk this way sets the stage for escalation of fears and distortions.

If you think I am exaggerating, let me relate a couple of statements I have come across recently.

Take this statement, made by a representative in the US Congress:

> ... the fact is that many people are coming into our society and they are participating in services, they are consuming money that was supposed to go for the benefits of our own people, and when we struggle to come up with the funds for different programs here, and we know that we are not even

> providing all the funds we need for our own people, and then we find out that someone who has come out illegally from another society is consuming those resources, it is not right.

Or this one:

> ... we are sending an unmistakable invitation to needy people everywhere in the world and on every continent of the world. We are inviting them to ignore our immigration laws, and to come to our country and receive a host of benefits provided by the taxpayers. The taxpayer cannot continue to provide such largess.

I can tell you what people my colleague believes are coming here illegally from other societies and consuming resources meant for "our people." They are not the Germans and Poles and Irish that Royko wrote about. I think an awful lot of people he is talking about have names ending in "ez" and look and talk a lot like I do.

Lawbreakers and Derelicts

Another statement I came across recently put it this way: "It is hopeless to think of civilizing these new immigrants, or keeping them in order, except by the arm of the law." This, a carbon copy in tone if not in substance to the words spoken on the floor of the US Congress today, was written in the *New York Times.* In 1875. The lawbreakers and derelicts they were writing about were Italians.

The lesson of these statements is simple—and it is *not* that it has taken 120 years for the long arm of the law to civilize Italians.

The lesson is that the current wave of intolerance sweeping our legislators in Washington—and Springfield, and Sacramento, and Austin and Tallahassee—is not a new invention. We've heard the charges for years: immigrants are draining our resources. They force Americans into unemployment. They are tearing the fabric of our society apart by demanding to speak their own language and have their own culture.

Before I heard this line of argument on the floor of the US House of Representatives, I heard it growing up in the Lincoln Park neighborhood of Chicago as a young Puerto Rican in an ethnically evolving community.

Public debate about immigration is growing increasingly ugly and more divisive. It is threatening our national identity and crying out for leaders who are willing to stand up and defend the rights, contributions and importance of immigrants to our nation.

And had I been in California in 1910, I would have heard it about the Chinese. Had I been in New York in 1890, I would have heard it about Italians and Poles. Had I been in Pennsylvania in 1795, I would have heard it about the Germans who were taking over William Penn's commonwealth from the only real Americans—those who had come directly from England.

The economic blame, the unemployment worries, the corrupting of culture and language—this fingerpointing is as old as the Mayflower. The intensity of the debate regarding the perceived problems of immigration blinds far too many of us to learning from history. Keeping in mind a historical perspective would serve well to remind us of the foolishness of targeting a single group of immigrants or a particular wave of immigration as "different." How seriously can we take the opponents of immigration today when we realize that the same urgent crisis they believe Mexicans are bringing across our border today their forbears vehemently argued were brought by Poles and Italians many decades ago.

The intolerance we see today is grounded in real problems—eroding tax bases, budget deficits, increasing crime and the difficulty of controlling our borders—that need real solutions. These problems have made the voices of intolerance grow louder and more shrill; perhaps worse, they have given them more widespread credibility.

But something else, too, may contribute to the increased backlash. The ethnic rivalries Royko described were primarily among Europeans. Unfortunately, today many of the voices of intolerance have a pronounced racial edge less evident in the battles between the Germans and English in Pennsylvania, or the Poles, Lithuanians and Bohemians of Chicago's past. That immigrant-bashing has become meaner and more personal seems to me clearly related to the fact that so many of our new immigrants are traveling here not from Europe but from Asia and Latin America.

The stakes have been raised, and that is why any of us who cares about the future of immigration—and particularly Latino leaders—must meet the challenge that the voices of intolerance and division are raising.

In the past, the rights of those who came to America to take part in the chance for a better future were protected by leaders who had the courage to stand up and say that newcomers always have and always will be a benefit—not a detriment—to America.

Immigrants to this country are not fighting for a handout; they are not here merely for a student loan or food stamps or access to a free clinic. What they are interested in is something much simpler, much more basic. They are fighting to have all of the opportunities that they are denied in their native countries: a chance to work at a decent job and make a decent living, the opportunity to educate their children, the opportunity to have decent shelter and clothing. They want to contribute to a better America.

The facts speak clearly. The stereotype of immigrants, particularly Latino immigrants, as uneducated, unwilling to work and unable to contribute to our economy and our society simply does not hold true. By any objective standard, the immigrant population in the US today is as productive and beneficial to the national economy and culture as it has ever been in our nation's history.

Far too many of the studies that opponents of immigration rely on look *exclusively* at the costs of the immigrants' population to the government, without considering immigrants *contributions* to the treasury. Those studies are misleading at best.

A major study recently released by the Urban Institute belies the fear that immigration is a threatening drain on public resources. In fact, the report shows, immigration since 1970 has resulted in a government surplus of some $27 billion, estimating some $43 billion in costs to government and $70 billion in revenues collected in federal, state, and local taxes, including income tax, property tax, Social Security payments, and other taxes. Even looking just at undocumented immigrants—a grouping that includes many people who will eventually gain or may already have legal status—it is clear that the numbers have been vastly exaggerated. The Urban Institute study estimates that undocumented immigrants cost the government $9 billion, and pay taxes (including Social Security taxes, which are collected even if under a false Social Security number) of $7 billion. States may have some basis for complaint, since the federal government collects most of the money but the states pay for the majority of the services, but the fact is, all in all, immigrants contribute more than they cost.

Do immigrants take jobs from native-born Americans and act as a drag on the economy? Last September, *The New York Times* quoted Princeton economist David Card, who studied the results of the arrival of 125,000 Cubans during the Mariel boat lift in 1980, as saying that "the Mariel influx appears to have had virtually no effect on the wages or unemployment rates of less skilled workers, whether white or black." And a 1990 American Immigration Institute survey of economists, including ex-presidents of the American Economic Association and then-members of the Council of Economic Advisers, found that four out of five estimated that immigrants had a favorable impact on economic growth. None said that immigrants had an adverse impact on economic growth. You can choose your study. With virtual unanimity, objective studies show that while immigration has many costs, it has benefits that outweigh those costs.

According to *Business Week*, one-quarter of immigrant workers are college graduates—slightly higher than the rate for native-born Americans. The reality, according to the *Wall Street Journal*, is that immigrant children frequently outperform their schoolmates. On average, immigrant children in the eighth and ninth grades spend two to three hours per day on homework. Native children average less than one hour.

Immigrants, who make up 22 percent of California's population, are only 12 percent of AFDC recipients. Only 2.3 percent of immigrants entering from nonrefugee-sending countries during the 1980s were receiving public assistance, lower than the 3.3. percent welfare-participation rate of natives. And immigrants are revitalizing our cities. In the past decade, population in the nation's ten largest cities grew by 4.7 percent, but would have shrunk by 6.7 percent if it were not for new immigrants.

The voices of intolerance ignore the Cuban businessman who has worked so hard to rebuild Miami, the Indian doctors working in our nations' best clinics, the Russian engineers working in successful industries, the Mexican families that are revitalizing so many of our neighborhoods.

In my career as a public official, I have been fortunate to witness again and again the improvements that immigrants can bring to a community. I have met and worked with social service agencies that helped victims of political oppression come to America from Cuba and El Salvador and Nicaragua. I was fortunate enough to be at the Ukrainian Cultural Center in Chicago when almost 100 orphans from Ukraine were met by their new adoptive parents. I have sponsored citizenship workshops that have been attended by Poles, Koreans, Latinos—all of whom were thrilled to finally reach their goal of becoming US citizens.

I have also witnessed the tremendous pride and celebration that Mexican Independence Day brings to the city of Chicago. It binds together a community and helps people share in its culture and accomplishments. It helps the rest of our city understand their neighbors' native country a little better. Is this celebration of Mexico somehow anti-American, a threat to our country? No more than when German Americans gather to celebrate Oktoberfest or Irish Americans celebrate St. Patrick's Day.

Anti-immigrant legislation makes any person who doesn't look "American" an immediate suspect, whether that person is applying for a student loan or Social Security benefits.

Talking About Leadership

When I leave Chicago, I love to tell people about a street in my congressional district: one of the longest uninterrupted commercial strips in the Midwest. Commercial vacancies are virtually nonexistent. It boomed and prospered after the Poles and Lithuanians left it for dead. That street is not located in Chicago's Gold Coast. It is 26th Street, and it has been rebuilt almost exclusively by Mexican immigrants. It is a success story that is now the heart of an economically vibrant immigrant community. If America and Chicago had followed the shortsighted and small-minded path of closing its doors to immigration, it would be nothing more than a three-mile stretch of urban decay.

Chicago has no shortage of immigrant success stories. Immigrants from southeast Asia have transformed economically struggling sections of the Uptown neighborhood, Koreans have prospered in struggling sections of Ravenswood and Cubans and Puerto Ricans have helped to keep Logan Square alive.

Every city in America is filled with exactly the same kind of success stories, but we don't hear nearly enough about them. Our country is moving toward bad immigration policy because not enough leaders—particularly political leaders—are willing to stand up for immigrants.

In the past, America has struggled and fought and eventually overcome the voices of intolerance to implement immigration policies that do not break our budget and allow immigrants to grow and thrive in our country. Chicago has a history of ethnic leaders who have fought for their people. Italians, Irish, Poles and Slavs have all elected leaders, from aldermen to members of Congress, who stood up for their rights.

Today's immigrant groups need the same leadership. And not just quiet leadership. they need outspoken public figures and grassroots activism that spread the message of immigrant rights and contributions. Challenging anti-immigrant rhetoric is not very popular these days, but it has never been more important, and more people need to be willing to carry on this struggle.

The Pleasures of Diversity

The Congressional district I represent is a community filled with people who celebrate America while honoring and remembering the nations they have left behind. People express genuine ethnic pride on St. Patrick's Day and Columbus Day, and they receive a day off of work to honor Casimer Polaski, a Polish immigrant who fought heroically during the Revolutionary War.

There are ethnic restaurants, kitchens and grocery stores; you hear Polish and Italian and Korean and of course Spanish being spoken on the streets. And, while the district is full of ethnic diversity and pride, it has its share of ethnic strife, too.

Immigrants made my district, made my city, and their leaders fought for their rights. The Slavs worked to elect Anton Cermak mayor. When an assassin ended his term, their domination gave way to the Irish, who elected decades of mayors from Edmond Kelley to Martin Kennelly to Richard M. Daley—all of whom were careful to slate ethnically backed tickets. Immigrant groups have not always gotten along, but they never lacked leaders who fought for them.

Immigrants to this country are not fighting for a handout; they are fighting to have all the opportunities that they are denied in their native countries.

Now is the time for the leaders of the new immigrants to stand up against division and intolerance, and to explain in clear terms that the new wave of immigrants—from Mexico and Guatemala, from Honduras and Cuba, from El Salvador and Nicaragua—is contributing just as immigrants before them did.

No less a figure than Benjamin Franklin was so frightened of German immigration that he wrote, "all the advantages we have will, in my opinion, be not able to preserve our language, and even our government will become precarious." Franklin was wrong about Germans in 1753. The politicians, talk-radio hosts and editorial writers who fear the new immigrants are just as wrong today.

Leaders must remember our history, learn our facts and find our voices to speak out for fairness and tolerance, and to remind ourselves that new immigrants, like the old ones, are ready to give back as much as and more than they take—economically, culturally, and socially.

Alone Together: Will Being Wired Set Us Free?

Networks will obliterate the industrial model of society. The fear is that they will destroy solitude, and with it human intimacy.

Andrew Kupfer

Imagine, if you can, a small room, hexagonal in shape, like the cell of a bee. An arm-chair is in the centre, by its side a reading-desk—that is all the furniture. And in the arm-chair there sits a swaddled lump of flesh—a woman, about five feet high, with a face as white as a fungus.

An electric bell rang.

"I suppose I must see who it is," she thought. The chair was worked by machinery, and it rolled her to the other side of the room.

"Who is it?" she called. She knew several thousand people; in certain directions human intercourse had advanced enormously . . .

The round plate that she held in her hands began to glow. A faint blue light shot across it, darkening to purple, and presently she could see the image of her son, who lived on the other side of the earth, and he could see her.

—E. M. Forster
"The Machine Stops," 1914

"Come on, honey. Remember those IBM machines. Let's get at it before people go out of style."

—Bobby Darin pickup line
in *State Fair,* 1962

Ever since protohumans with sloping foreheads learned to set things on fire, people have feared and hated technology as much as they have been in its thrall. They have eyed with suspicion the printing press, the automobile, the telephone, and the television as solvents of the glue that binds people together. Each new technology brings a warning: To fall under its spell will be to sacrifice not only simplicity but also community to metamorphose into alienated, isolated, sedentary blobs. In Forster's story, when the machine stops, everybody dies.

This kind of trepidation is sometimes overdrawn—even the advent of the washing machine produced expressions of yearning for simpler times—but it isn't really misplaced. The printing press vanquished the knowledge oligarchy, yet popular culture seems ever more trivial and debased. Modern medicine often prolongs life beyond all reason or desire.

Now information technology is poised to alter the scope of human intercourse, and the familiar combination of promise and dread makes itself felt once again—with an urgency seldom seen in the two centuries since the Industrial Revolution. The new technology holds the potential to change human settlement patterns, change the way people interact with each other, change our ideas of what it means to be human.

Information technology will have the power to reverse what may have been an aberration in human history: the industrial model of society. While people in agrarian societies had for millenniums worked the land around their homes to the rhythm of the sun, industrialization created the time clock and the separate workplace. Wired technology already is assaulting the industrial concept of the workday; as technology brings greater realism to electronic communications, the workplace for many will become untethered from geography, letting people live anywhere. The fear is that in liberating us from geography and the clock, networks will destroy intimacy, both by making solitude impossible and by making physical presence immaterial to communication.

One reason we are wary about information technology is that it is still strange to us, new enough that we notice it all the time. We still marvel at what computers can do, and how we can carry in our laptops enough computing horsepower to have filled an entire laboratory not so many years ago. We view information technology as special, almost magical. Vincent Mosco of the Harvard Center for Information Policy Research, who has written extensively on the history of technology and the way electrification changed population distribution, says people felt the same way about electricity when it was introduced in the 19th century. "Companies used electricity to flash advertisements off the clouds," much in the way that Gothamites summon Batman in times of trouble, says Mosco. "I like that image of people gathering outdoors and watching lights flashing in the sky and seeing that as the spectacle of communications." Today computers, the Internet, and the information superhighway are the magical elements, and even the basic rules of etiquette are unformed, reminiscent of the early days of the telephone. Paul Saffo of the Institute for the Future in

---

REPORTER ASSOCIATE *RAJIV RAO*

Menlo Park, California, says: "Alexander Graham Bell proposed a greeting of 'Hoy! Hoy!'—a variation of 'Ahoy!' It didn't catch on." Instead his great rival Thomas Edison stole a bit of the jam from his crumpet by inventing, as a telephone salutation, the word "hello," a variant of the British exclamation "hallo."

Eventually, though, computer communications—like electricity and telephony—will quite literally fade into the woodwork. When that happens, wired technology will obliterate the significance of two of the great symbols of the Industrial Revolution, the train and the clock, and along with them the idea that society can organize everything to run on set schedules. The temporal shift this technology permits—even demands—is likely to be its most profound and enduring effect.

With an economy that straddles many time zones, the nine-to-five workday will disappear for those whom it hasn't already. People will become accustomed to flitting between their different roles of work, recreation, and repose, constantly prey to interruption, even addicted to it. "The rush and flow of events is like electronic heroin," says Saffo. "And once you get it into your veins it's really hard to stop. You'll figure out a way to interrupt yourself." People may live in bucolic and pastoral settings but not live a pastoral life, competing via cyberspace for work against thousands of others, finishing each job in days or hours, then moving on to the next, like electronic versions of Charlie Chaplin's assembly-line worker in *Modern Times.*

Many assume that people who can leave company headquarters will choose to work in their homes, and wired enthusiasts anticipate a resurgence of familial togetherness. But at least one expert on how the home reflects changes in American society says we may well see less family interaction that we do today. Clifford Clark, an American studies professor at Carleton College in Northfield, Minnesota, predicts: "We will see different family members sitting around different screens in different rooms."

That could touch off domestic turf battles: Our houses aren't suited to these purposes, having evolved over the past century from a large number of little spaces to a small number of big ones. The kitchen was once isolated in the back of the house to keep a continuously fired-up stove from overheating the living quarters, but with the invention of the gas range it moved forward and became a social room as much as a workplace. Today it sometimes flows right into the so-called great room, where families sit in front of the jumbotron to watch surround-sound movies. A shortage of solitary workspace may become just one more source of family disharmony.

Knowledge workers, selling their labor to new species of business that will flourish in the wired economy, may need to be ready to go at a moment's notice. Employers already seek workers via computer networks. But in the future the process will be more pervasive and almost automatic. Professor Thomas Malone of the Center for Coordination Science at MIT says such wired workers will form "overnight armies of intellectual mercenaries."

Imagine a company with a task that needs urgent attention—say, designing a lawnmower or writing a computer program. The company might not maintain a cadre within its ranks to do the job. Instead, it trolls the net for talent, sending out a bulletin that describes the tasks to be done and the skills required of team members. The notice might go directly to qualified applicants, based on résumés filed online. Specialists anywherc in the world instantly submit bids to do a piece of the job, simultaneously triggering a query to their personal references. Winning bidders work together via video hookup, each at his or her home base. The project might last a few weeks or a few days or a few hours. Afterwards the team disbands and the members melt back into the talent pool to bid on new jobs.

Socially, the wired society is likely to bring flip-flops in behavior like the changes wrought by the telephone, which made it acceptable for a man to talk to a strange woman without a formal introduction by a third party. The Internet is making it acceptable for a man to exchange explicit sexual fantasies with a strange woman—or with someone who claims to be a woman but who may really be a trio of male cross-dressers sitting around their screen laughing. At times people breach the bounds of decency and stray into the realm of the allegedly criminal: A college student was recently jailed for distributing via the Internet a depraved story in which he imagined the rape, torture, and murder of a woman he knew, and whose name he disclosed. Another young woman soon replied with an online revenge fantasy of her own.

Many fear that wired communications, by permitting a unique combination of intrusiveness and anonymity, will make people even ruder than they are today. Already people communicating online are rethinking what kind of information they feel comfortable sharing. Mark Weiser, principal scientist at the Computer Science Lab in Xerox's Palo Alto Research Center—and an inventor of the technology that let the Rolling Stones transmit a live video concert over the Internet last November—says that at a business dinner we are likely to talk about our spouses and children but would not usually exchange résumés. Online, though people are guarded about their personal lives since they feel less able to size up, or even identify, their correspondents. Yet they can, in many cases, call up *curricula vitae* that disclose everything their Internet friends have done since high school.

"People are starting to put up different barriers to their interactions," says Weiser, speaking as one who doesn't like barriers very much. He usually has eight video windows open on his computer screen at work, showing his engineering colleagues' offices. Weiser also confesses to being the drummer for a band called Severe Tire Damage that sneaked onto the Internet before the Stones concert as an unscheduled opening act.

In time both the guardedness and the anonymity will evanesce, Weiser says: "As more and more business is conducted online, it will become more of a real place, and real-life expectations will take over. One is that I know who you are. We will stop talking to people we don't know." The wired connection will no longer seem like a strange way of meeting people—which won't be the first time a method that once seemed mad became a part of quotidian routine. And the change in attitude might not take as long as you think. A decade ago, if you telephoned a friend and reached an answering machine, you probably thought, "How rude!" Today you are more likely to be miffed by your thoughtless friends who refuse to buy one.

Despite its potential to free people from geography, the likely effect of wired technology on where people live is murky. While some will be able to leave cities, others won't, and still others won't want to. True, some jobs have already headed for the sticks, particularly back-office operations of financial firms, intensifying a long-term trend that began earlier in the century with improvements in transportation. But many potential movers seem to have sticky feet. Blame this partly on that hobgoblin of managerial minds, force of habit. People might love the idea of sending E-mail to their grandchildren, but as supervisors the same folks don't have the stomach for remote management. People want to see their employees and want to watch them work.

They can't do that via video yet because existing technology is too crude: The picture transmitted by a typical desktop computer videoconference system is a low-resolution, herky-jerky postage stamp. Within the next ten years, though, better devices will be able to send crystalline images with lifelike color and perfectly fluid motion, conveying words, body language, expression. What will it mean when gazing at a face on a video screen is no different than looking at a face through a window? Will the cities empty and the people disperse like leaves in a fall wind?

If history is any guide, wired technology will create forces that pull in the other direction as well. Successive waves of technology, from the telephone to the automobile to rural electrification, have brought predictions of the emptying of cities. Yet the cities endure, and so they will a century from now. The telephone, for example, led to both dispersion and concentration. Not only did it open up remote areas to commerce, but it also helped make possible the most highly concentrated form of living and working space that we know: the skyscraper. Without the telephone to deliver messages, occupants of upper stories would be cut off unless the architect devoted the entire core of the massive structures to elevators and stairways for messengers.

In the information society, expect to see similar pushes and pulls. Most mobile will be the knowledge workers: people whose jobs largely involve talking to others and handling information—in other words, white-collar office workers. For them, electronic links will mostly suffice; they will be able to choose to live by the seashore, say, or near family and friends.

But as if to obey Newtonian laws of motion, information technology will also pull people to the center. By permitting dispersion, information technology promotes the globalization of the economy, guaranteeing a raison d'être for international cities like New York, London, and Tokyo that serve as the nodes for world communications networks—a major reason New York has shown much more resilience that city-bashers predicted. The economic vibrancy of these cities will attract the many people who thirst for amenities like theater, concerts, restaurants, and the continuous paseo of cosmopolitan life.

As they do today, the city dwellers of the information society will depend on a tier of lower-level service workers like barbers and burger flippers, whose work, involving physical contact with other people, cannot be liberated from place by communications technology. (Some higher-level professionals like surgeons will also remain tied to population centers.) Not all the people will be able to follow their bliss to the mountaintops.

Wherever we live, the nature of routine intercourse is likely to be changed by electronic agents—drudges, really, programmed to take over the tedium of interconnectivity. The first commercial prototypes of these agents have recently appeared, including one called Wildfire that acts as an electronic secretary, answering the phone, taking messages, obeying simple verbal commands, and routing phone calls to users wherever they happen to be.

As they become more sophisticated, these software agents will do our shopping, buy our plane tickets, and make our appointments for us, traveling through cyberspace like ghostly echoes of the self. "They won't be intelligent enough to make the clerics nervous," jokes Saffo of the Institute for the Future. "But they will exhibit whimsy and humor, and be interesting enough to convince people to interact with them." Not only will people be talking with these soulless beings, but agents will be interacting with other agents as well. The Hollywood patter of the future may remain "Have your agent call my agent," but people won't be talking about ten-percenters.

Our ghosts may come to haunt us as well. One nightmare scenario not yet on many worry lists is location tracking. With the auctioning off of vast swaths of the radio spectrum for new wireless services and the promise of cheap, lightweight cellular phones, the cellular industry is poised to sweep into the mass market. New low-powered cellular systems will blanket the country with great numbers of closely spaced transmitters. Nearly everyone will be carrying some sort of wireless communications gadget. Whenever they are on—and they are likely to be left on all the time—a signal will travel to the nearest transmitter, letting the network know where to send each user's messages and phone calls.

Cellular companies will be able to use their fine-meshed networks to pinpoint nearly everyone's location and track their movements. This is how the police, with the help of the phone company, tracked down O. J. Simpson as he was driven along the highway in the infamous white Bronco. Anyone with a cellular telephone scanner could also keep tabs on people's locations, even when new digital cellular systems make our conversations secure from eavesdroppers. (Only our words will be encoded; our identification numbers must stay unscrambled so the network can authorize our calls.)

If you don't think anyone really cares where you go from moment to moment, be assured that plenty of companies would pay to find out. Marketers, for example, would love to know who visits which stores, and when, and for how long. They could legally buy this information from the telephone company as easily as they buy mailing lists today. And as with mailing lists, we would have no control over who gets access to this information.

If our ever cosier relationship with wired technology makes us fear for our souls, perhaps that is because the stuff is so seductive. Unlike TV, the new technology requires our participation, drawing us in. As such it is insidious. Management professor Alladi Venkatesh of the University of California at Irvine, an expert on the impact of technology on the household, says: "Television is easy to dismiss. Its limitations are obvious. The danger of the computer is that it gives us the impression that it can do for us what TV has not: make us better people."

It is true that the power to make instant connections anywhere in the world, at any time, can bring inestimable comfort. For the millions who are stuck at home because of age or infirmity or because they are caregivers for young children, for insomniacs who need someone to commune with in the blue hours past midnight, for people who want to find out if their car is a lemon, or how to buy a house, or how to cope with a child's asthma attack, being wired may be the fastest way to connect with others who are willing to share their feelings and knowledge.

But with these gains there is loss. While people may feel just as intensely about friends they make via cyberspace as they do about their face-to-face confreres, the ease with which they form these links means that many are likely to be trivial, short lived, and disposable—junk friends. We may be overwhelmed by a continuous static of information and casual acquaintance, so that finding true soul mates will be even harder than it is today. And the art of quiet repose and contemplation may one day seem as quaint as the 19th-century practice of river gazing—staring at riverscapes to discern their coloristic and picturesque attributes.

MIT's Malone is worried about these risks but tries to remain an optimist. He says he feels closer to some people he has met over the net than he did even to the friends he made growing up in a small town in New Mexico. Those relationships were mere accidents of geography; he and his new friends chose each other through common interest. In an eerie echo of the cautionary tale that E. M. Forster wrote more than 80 years ago, he says, "There must be thousands of people I know personally . . ."

This machine will not stop. In time we will no longer ponder its existence, or be able to imagine a world without its constant hum.

THE PRICE OF FANATICISM

Now that extremists are willing to use weapons of mass destruction, they have crossed a threshold that experts have watched with dread for two decades.

BRUCE W. NELAN

THE IMAGE IS AT ONCE ORDINARY and sinister. Amid the bustle of Kasumigaseki subway station, in downtown Tokyo, three attaché cases stand unattended by the ticket barrier. Suddenly, gas begins hissing ominously out of one of them. When police eventually examine the cases, they discover that each holds containers of clear liquid, a powerful battery-operated vaporizer and a fan to blow the resulting vapor through vents. The cases are rigged to operate as automatic dispensers. But dispensers of what?

They represent the ultimate urban horror. Anonymous, malevolent packages planted by any of the thousands of subway riders and set to kill huge numbers of passersby indiscriminately. The prospective victims are temporarily captives in a subterrancan steel and concrete execution chamber, and they could have died by simply by drawing a breath. The dead would have been selected by sheer chance, depending on petty details like which commuter was on schedule and who had dawdled over breakfast and taken a later train.

Those mysterious attaché cases, perhaps testing devices for a subsequent attack, were found only five days before thousands of riders on the Tokyo subway were felled by nerve gas last week. The liquid inside turned out to be water.

The events in Tokyo were a clear warning to the world. Terrorism has taken a step across a threshold that security experts have been anticipating with dread for decades. It has been known that there are groups out there that are willing to kill at random. There is proof that they are able to use chemical weapons, and possibly biological and radioactive ones as well, that can destroy far more people than conventional bombs and bullets. Now that nerve gas has been used on ordinary citizens, it may possibly happen again: the fact that terrorists are copycats and hungry for publicity makes it a near certainty. With one act, the spectrum of danger has broadened into a threat more terrifying than ever before—and one far more difficult for governments to forestall.

It used to be known who the "terrorists" were: a handful of Middle Eastern or leftist political movements, sponsored and protected by governments, bent on achieving their well-advertised ideological goals through death and intimidation. The next generation of terrorists is more obscure, an assemblage of disparate fanatics pursuing unique or mysterious agendas, with only the capacity for random violence in common. While governments have them under fairly good control and international terrorist incidents are relatively few (321 last year, down from 432 in 1993), it looks to the experts as if the 1990s rise of apocalyptic sects and Islamic extremism has merged with the increasingly easy availability of chemical and biological weapons that can kill thousands in an attack. The potential for random murder and catastrophic governmental disruption lies within reach of small, unsophisticated and irresponsible groups of true believers. "Nightmares are coming true," says Robert Kupperman, a terrorism expert at Washington's Center for Strategic and International Studies. "I think we're in for deep trouble."

Even very sober public officials are deeply concerned. Three weeks ago, Georgia's Senator Sam Nunn sketched a lurid fantasy: how terrorists might wreck the central government of the U.S. On the night of a State of the Union address, when all the top officials are in the Capitol, Nunn said, a handful of fanatics could crash a radio-controlled drone aircraft into the building, "engulfing it with chemical weapons and causing tremendous death and destruction." This scenario, said Nunn, "is not far-fetched," and the technology is all readily available.

Many of the experts say they are surprised that chemical weapons have not been used in a major attack before. The ingredients for making them are available commercially and can be put together by almost any competent chemist. Muslim zealots, for example, are increasingly a younger generation of angry men who have the education and sophistication to construct weapons their fathers and uncles never dreamed of.

Even though radical groups have long had the power to kill more people than they actually did, the fact that they held back somewhat suggests they imposed certain restraints on themselves. Most such groups viewed themselves as political activists rather than wanton killers. They had to appeal to potential supporters of their program and were wary of producing a backlash of revulsion by using the most repellent methods. The cold war and the rules of state-sponsored terrorism curtailed their freedom of action. Governments knew more or less who was sponsoring whom, and the threat of retaliation was always present—as demonstrated when the Reagan Administration sent U.S. bombers to hit Libya in 1986 in retaliation for its support of several terrorist acts. But the end of the cold war and the beginnings of the Middle East peace process have taken Eastern European and some Muslim governments out of the sponsorship business.

At the same time, however, the collapse of the Soviet empire, the creation of new states and the breakup of others have triggered an explosion of ethnic conflicts, with racial and religious hatreds mixed in, giving fresh scope to terrorist free-lancers. Much of the violence committed today in the name of Islam is the work of small, loosely organized cells who emerge for little more than a single act of random vengeance. Sections of Pakistan are ungovernable safe havens for the remnants of 20,000 zealous volunteers from Muslim countries all over the world who went to join the Afghan *mujahedin* in their holy war against the Soviets. An estimated 1,000 fundamentalist fighters still gather in the country's lawless reaches to train and egg each other on. They frequently sally forth aboard international airliners, looking for new places to fight their messianic war.

Some free-lance terrorists have taken up residence in the U.S. They have brought with them a brand of activism previously almost unknown except for occasional episodes of violence among their kind, as when Sikh extremists attacked officials of the Indian government in U.S. cities.

The sense of American immunity was truly swept away in February 1993, when a group of Muslim conspirators detonated a homemade bomb under New York City's World Trade Center. Four months later, nine Islamists were arrested on charges of conspiring to blow up such landmarks as the U.N. and the Lincoln Tunnel. In both cases, the motivation was essentially religious and without any discernible goal: they were simply attacks on the U.S., the Great Satan, in the name of Allah.

The religious ingredient in violence is a dangerous trend the experts have been watching closely. The Irish Republican Army and the Palestine Liberation Organization might have a surface religious orientation, but they and their objectives are political. Some analysts designate even relatively violent Islamic groups "mainstream" terrorists. The Trade Center bombers are in a different category. They do not have clearly visible political motives and seem to have come together rather casually, outside a formal organization, only to inflict punishment on Americans, the infidel enemies of their religion.

In 1968, the first year in which international terrorism seized the headlines, of the eight known groups, all were political, without religious overtones. In 1980, a year after Islamic radicals overthrew the Shah of Iran, overtly religious terrorist groups made their appearance. Of the 48 international groups active in 1992, almost a quarter were religiously motivated. Shi'ite groups, though they commit less than 10% of the attacks worldwide, account for 30% of all the killings.

"Whenever religion is involved, terrorists kill more people," says Bruce Hoffman, director of the Center for the Study of Terrorism and Political Violence at Scotland's University of St. Andrews. Last December a group of Algerian Islamists hijacked an Air France Airbus A300, which they planned to blow up over the center of Paris solely to kill as many people as possible. They would almost certainly have done so if they had not been killed on the ground in Marseilles.

Small, charismatic cults are adopting more violent methods as well. These extremist sects appeal to many people in an antispiritual age because they combine their empowering theology with a warm, supportive environment, at least at first. Those who join become part of a close-knit body of believers who are convinced they understand the meaning of history and what the future holds. That was true of David Koresh's Branch Davidians, and it applies to certain extremist Christian white-supremacist groups bent on "purifying" the U.S.

But once the recruits are in the cult's grip, they encounter a darker side, even if they do not recognize it. Their charismatic leader preaches that they are surrounded by enemies, that nonbelievers are out to crush them and that God commands vengeance. In some sects they are told to commit violent acts, says Hoffman, "because the only way they can hasten redemption or achieve salvation is to eliminate the nonbelievers."

Dehumanization of the enemy is traditional among violent sects. And if the opponents are accepted as children of Satan, killing becomes that much easier. The very basis of their faith makes such killing not only legitimate but also mandatory. In the U.S. there are many shadowy groups lurking—covert militias, survivalists, religious and political cults—with agendas of destruction and a newfound taste for exotic weapons. "You don't hear much about them," says Hugh Stephens of the University of Houston, "but these people are antigovernment and fearful. They are running around with arms and training for the millennium."

Symptoms of the trend have been visible for years. Back in 1972 an American fascist group called the Order of the Rising Sun was grabbed with 80 lbs. of typhoid bacteria cultures that the members planned to dump into the water supplies of Midwestern cities. In 1985 a group of neo-Nazis was arrested with 30 gal. of cyanide they intended to put into the water of New York City and Washington. Now, says Representative Glen Browder, an Alabama Democrat whose district is home to the Army's sole chemical-weapons training base, it appears "the psychological barrier" against the mass use of chemical and biological agents has finally been passed in Tokyo. "It's just a matter of time before it occurs in the U.S.," he says.

Marvin Cetron, president of Forecasting International Ltd., a Virginia-based think tank, last year co-authored an exhaustive study of terrorism for the Pentagon. He thinks a chemical or biological attack on the U.S. is increasingly likely, "perhaps within the next five years." He also predicts that if Sheik Omar Abdel Rahman, the Muslim cleric on trial in New York City, is found guilty on conspiracy charges, there will be "10 or 12 terrorist attacks on U.S. targets" within a few weeks.

If that is so, what is the government doing to prepare? The Pentagon is studying how terrorists might try to spread chemical or biological agents in urban areas and hopes to develop techniques to thwart them. The FBI and CIA are boosting their spending on trying to find and penetrate the groups. and thus catch the plotters before they strike. But to do that, governments must have early and reliable intelligence, which can be almost impossible to obtain about groups that are tiny and disorganized or not yet suspect at all.

Not always, though; there are sometimes early warning signs. Religious cults with apocalyptic ideas frequently publish their violent preachings and often set up their compounds in remote areas. "This filters out the members who are not committed," says Bruce Bueno de Mesquita of the Hoover Institution at Stanford University. Because the groups are isolated, he says, "they don't fit in with the rural community, and they are easy to spot." Law-enforcement agencies also find that while violent cults may start out small and unknown, as they grow and acquire weapons, it becomes harder for them and their potential for violence to remain hidden. Locals become suspicious of them and the purchases they are making and alert the law.

In the world's great cities, the prospect is far more uncertain. In Tokyo the police had a wealth of signals that a major nerve-gas attack might be in the making but were still caught off guard when it came. Some counterterrorism officials are speaking of the Tokyo subway poisoning as a "wakeup call" for governments around the world. But it is also possible that the gas attack in Tokyo was only a preview of what is yet to come.

—Reported by
Barry Hillenbrand/London and J.F.O. McAllister and Mark Thompson/Washington, with other bureaus

The Coming Anarchy

Robert D. Kaplan

Robert D. Kaplan is a contributing editor of The Atlantic Monthly.

How scarcity, crime, overpopulation, tribalism, and disease are rapidly destroying the social fabric of our planet.

The Minister's eyes were like egg yolks, an aftereffect of some of the many illnesses, malaria especially, endemic in his country. There was also an irrefutable sadness in his eyes. He spoke in a slow and creaking voice, the voice of hope about to expire. Flame trees, coconut palms, and a ballpoint-blue Atlantic composed the background. None of it seemed beautiful, though. "In forty-five years I have never seen things so bad. We did not manage ourselves well after the British departed. But what we have now is something worse—the revenge of the poor, of the social failures, of the people least able to bring up children in a modern society." Then he referred to the recent coup in the West African country Sierra Leone. "The boys who took power in Sierra Leone come from houses like this." The Minister jabbed his finger at a corrugated metal shack teeming with children. "In three months these boys confiscated all the official Mercedes, Volvos, BMWs and willfully wrecked them on the road." The Minister mentioned one of the coup's leaders, Solomon Anthony Joseph Musa, who shot the people who had paid for his schooling, "in order to erase the humiliation and mitigate the power his middle-class sponsors held over him."

Tyranny is nothing new in Sierra Leone or in the rest of West Africa. But it is now part and parcel of an increasing lawlessness that is far more significant than any coup, rebel incursion, or episodic experiment in democracy. Crime was what my friend—a top-ranking African official whose life would be threatened were I to identify him more precisely—really wanted to talk about. Crime is what makes West Africa a natural point of departure for my report on what the political character of our planet is likely to be in the twenty-first century.

The cities of West Africa at night are some of the unsafest places in the world. Streets are unlit; the police often lack gasoline for their vehicles; armed burglars, carjackers, and muggers proliferate. . . . Direct flights between the United States and the Murtala Muhammed Airport, in neighboring Nigeria's largest city, Lagos, have been suspended by order of the U.S. Secretary of Transportation because of ineffective security at the terminal and its environs. A State Department report cited the airport for "extortion by law-enforcement and immigration officials." This is one of the few times that the U.S. government has embargoed a foreign airport for reasons that are linked purely to crime. In Abidjan, effectively the capital of the Côte d'Ivoire, or Ivory Coast, restaurants have stick- and gun-wielding guards who walk you the fifteen feet or so between your car and the entrance, giving you an eerie taste of what American cities might be like in the future. An Italian ambassador was killed by gunfire when robbers invaded an Abidjan restaurant. The family of the Nigerian ambassador was tied up and robbed at gunpoint in the ambassador's residence. After university students in the Ivory Coast caught bandits who had been plaguing their dorms, they executed them by hanging tires around their necks and setting the tires on fire. In one instance Ivorian policemen stood by and watched the "necklacings," afraid to intervene. Each time I went to the Abidjan bus terminal, groups of young men with restless, scanning eyes surrounded my taxi, putting their hands all over the windows, demanding "tips" for carrying my luggage even though I had only a rucksack. In cities in six West African countries I saw similar young men everywhere—hordes of them. They were like loose molecules in a very unstable social fluid, a fluid that was clearly on the verge of igniting. . . .

A PREMONITION OF THE FUTURE

West Africa is becoming *the* symbol of worldwide demographic, environmental, and societal stress, in which criminal anarchy emerges as the real "strategic" danger. Disease, overpopulation, unprovoked crime, scarcity of resources, refugee migrations, the increasing erosion of nation-states and international borders, and the empowerment of private armies, security firms, and international drug cartels are now most tellingly demonstrated through a West African prism. West Africa provides an appropriate introduction to the issues, often extremely unpleasant to discuss, that will soon confront our civilization. To remap the political earth the way it will be a few decades hence—as I intend to do in this article—I find I must begin with West Africa.

There is no other place on the planet where political maps are so deceptive—where, in fact, they tell such lies—as in West Africa. Start with Sierra Leone. According to the map, it is a nation-state of defined borders, with a government in control of its territory. In truth the Sierra Leonian government, run by a twenty-seven-year-old army captain, Valentine Strasser, controls Freetown by day and by day also controls part of the rural interior. In the government's territory the national army is an unruly rabble threatening drivers and passengers at most checkpoints. In the other part of the country units of two separate armies from the war in Liberia have taken up residence, as has an army of Sierra Leonian rebels. The government force fighting the rebels is full of renegade commanders who have aligned themselves with disaffected village chiefs. A premodern formlessness governs the battlefield, evoking the wars in medieval Europe prior to the 1648 Peace of Westphalia, which ushered in the era of organized nation-states.

As a consequence, roughly 400,000 Sierra Leonians are internally displaced, 280,000 more have fled to neighboring Guinea, and another 100,000 have fled to Liberia, even as 400,000 Liberians have fled to Sierra Leone. The third largest city in Sierra Leone, Gondama, is a displaced-persons camp. With an additional 600,000 Liberians in Guinea and 250,000 in the Ivory Coast, the borders dividing these four countries have become largely meaningless. Even in quiet zones none of the governments except the Ivory Coast's maintains the schools, bridges, roads, and police forces in a manner necessary for functional sovereignty. . . .

In Sierra Leone, as in Guinea, as in the Ivory Coast, as in Ghana, most of the primary rain forest and the secondary bush is being destroyed at an alarming rate. I saw convoys of trucks bearing majestic hardwood trunks to coastal ports. When Sierra Leone achieved its independence, in 1961, as much as 60 percent of the country was primary rain forest. Now six percent is. In the Ivory Coast the proportion has fallen from 38 percent to eight percent. The deforestation has led to soil erosion, which has led to more flooding and more mosquitos. Virtually everyone in the West African interior has some form of malaria.

Sierra Leone is a microcosm of what is occurring, albeit in a more tempered and gradual manner, throughout West Africa and much of the underdeveloped world: the withering away of central governments, the rise of tribal and regional domains, the unchecked spread of disease, and the growing pervasiveness of war. West Africa is reverting to the Africa of the Victorian atlas. It consists now of a series of coastal trading posts, such as Freetown and Conakry, and an interior that, owing to violence, volatility, and disease, is again becoming, as Graham Greene once observed, "blank" and "unexplored." However, whereas Greene's vision implies a certain romance, as in the somnolent and charmingly seedy Freetown of his celebrated novel *The Heart of the Matter,* it is Thomas Malthus, the philosopher of demographic doomsday, who is now the prophet of West Africa's future. And West Africa's future, eventually, will also be that of most of the rest of the world.

Consider "Chicago." I refer not to Chicago, Illinois, but to a slum district of Abidjan, which the young toughs in the area have named after the American city. ("Washington" is another poor section of Abidjan.) Although Sierra Leone is widely regarded as beyond salvage, the Ivory Coast has been considered an African success story, and Abidjan has been called "the Paris of West Africa." Success, however, was built on two artificial factors: the high price of cocoa, of which the Ivory Coast is the world's leading producer, and the talents of a French expatriate community, whose members have helped run the government and the private sector. The expanding cocoa economy made the Ivory Coast a magnet for migrant workers from all over West Africa: between a third and a half of the country's population is now non-Ivorian, and the figure could be as high as 75 percent in Abidjan. During the 1980s cocoa prices fell and the French began to leave. The skyscrapers of the Paris of West Africa are a façade. Perhaps 15 percent of Abidjan's population of three million people live in shantytowns like Chicago and Washington, and the vast majority live in places that are not much better. Not all of these places appear on any of the readily available maps. This is another indication of how political maps are the products of tired conventional wisdom and, in the Ivory Coast's case, of an elite that will ultimately be forced to relinquish power.

Chicago, like more and more of Abidjan, is a slum in the bush: a checkerwork of corrugated zinc roofs and walls made of cardboard and black plastic wrap. It is located in a gully teeming with coconut palms and oil palms, and is ravaged by flooding. Few residents have easy access to electricity, a sewage system, or a clean water supply. The crumbly red laterite earth crawls with foot-long lizards both inside and outside the shacks. Children defecate in a stream filled with garbage and pigs, droning with malarial mosquitoes. In this stream women do the washing. Young unemployed men spend their time drinking beer, palm wine, and gin while gambling on pinball games constructed out of rotting wood and rusty nails. These are the same youths who rob houses in more prosperous Ivorian neighborhoods at night. . . .

Fifty-five percent of the Ivory Coast's population is urban, and the proportion is expected to reach 62 percent by 2000. The yearly net population growth is 3.6 percent. This means that the Ivory Coast's 13.5 million people will become 39 million by 2025, when much of the population will consist of urbanized peasants like those of Chicago. But don't count on the Ivory Coast's still existing then. Chicago, which is more indicative of Africa's and the Third World's demographic present—and even more of the future—than any idyllic junglescape of women balancing earthen jugs on their heads, illustrates why the Ivory Coast, once a model of Third World success, is becoming a case study in Third World catastrophe.

President Félix Houphouët-Boigny, who died last December at the age of about ninety, left behind a weak cluster of political parties and a leaden bureaucracy that discourages foreign investment. Because the military is small and the non-Ivorian population large, there is neither an obvious force to maintain order nor a sense of nationhood that would lessen the need for such enforcement. The economy has been shrinking since the mid-1980s. Though the French are working assiduously to preserve stability, the Ivory Coast faces a possibility worse than a coup: an anarchic implosion of criminal violence—an urbanized version of what has already happened in Somalia. Or it may become an African Yugoslavia, but one without mini-states to replace the whole.

Because the demographic reality of West Africa is a countryside draining into dense slums by the coast, ultimately the region's rulers will come to reflect the values of these shantytowns. There are signs of this already in Sierra Leone—and in Togo, where the dictator Etienne Eyadema, in power since 1967, was nearly toppled in 1991, not by democrats but by thousands of youths whom the London-based magazine *West Africa* described as "Soweto-like stone-throwing adolescents." Their behavior may herald a regime more brutal than Eyadema's repressive one. . . .

Ali A. Mazrui, the director of the Institute of Global Cultural Studies at the State University of New York at Binghamton, predicts that West Africa—indeed, the whole continent—is on the verge of large-scale border upheaval. Mazrui writes,

> In the 21st century France will be withdrawing from West Africa as she gets increasingly involved in the affairs [of Europe]. France's West African sphere of influence will be filled by Nigeria—a more natural hegemonic power. . . .It will be under those circumstances that Nigeria's own boundaries are likely to expand to incorporate the Republic of Niger (the Hausa link), the Republic of Benin (the Yoruba link) and conceivably Cameroon.

The future could be more tumultuous, and bloodier, than Mazrui dares to say. France *will* withdraw from former colonies like Benin, Togo, Niger, and the Ivory Coast, where it has been propping up local currencies. It will do so not only because its attention will be diverted to new challenges in Europe and Russia but also because younger French officials lack the older generation's emotional ties to the ex-colonies. However, even as Nigeria attempts to expand it, too, is likely to split into several pieces. The State Department's Bureau of Intelligence and Research recently made the following points in an analysis of Nigeria:

> Prospects for a transition to civilian rule and democratization are slim . . . The repressive apparatus of the state security service . . . will be difficult for any future civilian government to control. . . . The country is becoming increasingly ungovernable. . . . Ethnic and regional splits are deepening, a situation made worse by an increase in the number of states from 19 to 30 and a doubling in the number of local governing authorities; religious cleavages are more serious; Muslim fundamentalism and evangelical Christian militancy are on the rise; and northern Muslim anxiety over southern [Christian] control of the economy is intense . . . the will to keep Nigeria together is now very weak.

Given that oil-rich Nigeria is a bellwether for the region—its population of roughly 90 million equals the populations of all the other West African states combined—it is apparent that Africa faces cataclysms that could make the Ethiopian and Somalian famines pale in comparison. This is especially so because Nigeria's population, including that of its largest city, Lagos, whose crime, pollution, and overcrowding make it the cliché par excellence of Third World urban dysfunction, is set to double during the next twenty-five years, while the country continues to deplete its natural resources.

Part of West Africa's quandary is that although its population belts are horizontal [east-west], with habitation densities increasing as one travels south away from the Sahara and toward the tropical abundance of the Atlantic littoral, the borders erected by European colonialists are vertical [north-south], and therefore at cross-purposes with demography and topography. . . . [I]ndeed, the entire stretch of coast from Abidjan eastward to Lagos—is one burgeoning megalopolis that by any rational economic and geographical standard should constitute a single sovereignty, rather than the five (the Ivory Coast, Ghana, Togo, Benin, and Nigeria) into which it is currently divided.

As many internal African borders begin to crumble, a more impenetrable boundary is being erected that threatens to isolate the continent as a whole: the wall of disease. Merely to visit West Africa in some degree of safety, I spent about $500 for a hepatitis B vaccination series and other disease prophylaxis. Africa may today be more dangerous in this regard than it was in 1862, before antibiotics, when the explorer Sir Richard Francis Burton described the health situation on the continent as "deadly." . . . Of the approximately 12 million people worldwide whose blood is HIV-positive, 8 million are in Africa. In the capital of the Ivory Coast, whose modern road system only helps spread the disease, 10 percent of the population is HIV-positive. And war and refugee movements help the virus break through to more-remote areas of Africa. Alan Greenberg, M.D., a representative of the Centers of Disease Control in Abidjan, explains that in Africa the HIV virus and tuberculosis are now "fast-forwarding each other." Of the approximately 4,000 newly diagnosed tuberculosis patients in Abidjan, 45 percent were also found to be HIV-positive. As African birth rates soar and slums proliferate, some experts worry that viral mutations and hybridizations might, just conceivably, result in a form of the AIDS virus that is easier to catch than the present strain.

It is malaria that is most responsible for the disease wall that threatens to separate Africa and other parts of the Third World from more-developed regions of the planet in the twenty-first century. Carried by mosquitoes, malaria, unlike AIDS, is easy to catch. Most people in sub-Saharan Africa have recurring bouts of the disease throughout their entire lives, and it is mutating into increasingly deadly forms. "The great gift of Malaria is utter apathy," wrote Sir Richard Burton, accurately portraying the situation in much of the Third World today. Visitors to malaria-afflicted parts of the planet are protected by a new drug, mefloquine, a side effect of which is vivid, even violent, dreams. But a strain of cerebral malaria resistant to mefloquine is now on the offensive. . . .

And the cities keep growing. I got a general sense of the future while driving from the airport to downtown Conakry, the capital of Guinea. The forty-five-minute journey in heavy traffic was through one never-ending shantytown: a nightmarish Dickensian spectacle to which Dickens himself would never have given credence. The corrugated metal shacks and scabrous walls were coated with black slim. Stores were built out of rusted shipping containers, junked cars, and jumbles of wire mesh. The streets were one long puddle of floating garbage. Mosquitoes and flies were everywhere. Children, many of whom had protruding bellies, seemed as numerous as ants. When the tide went out, dead rats and the skeletons of cars were exposed on the mucky beach. In twenty-eight years Guinea's population will double if growth goes on at current rates. Hardwood logging continues at a madcap speed, and people flee the Guinean countryside for Conakry. It seemed to me that here, as elsewhere in Africa and the Third World, man is challenging nature far beyond its limits, and nature is now beginning to take its revenge.

Africa may be as relevant to the future character of world politics as the Balkans were a hundred years ago, prior to the two Balkan wars and the First World War. Then the threat was the collapse of empires and the birth of nations based solely on tribe. Now the threat is more elemental: *nature unchecked.* Africa's immediate future could be very bad. The coming upheaval, in which foreign embassies are shut down, states collapse, and contact with the outside world takes place through dangerous, disease-ridden coastal trading posts, will loom large in the century we are entering. (Nine of twenty-one U.S. foreign-aid missions to be closed over the next three years are in Africa—a prologue to a consolidation of U.S. embassies themselves.) Precisely because much of Africa is set to go over the edge at a time when the Cold War has ended, when environmental and demographic stress in other parts of the globe is becoming critical, and when the post–First World War system of nation-states—not just in the Balkans but perhaps also in the Middle East—is about to be toppled, Africa suggests what war, borders, and ethnic politics will be like a few decades hence. . . .

Returning from West Africa last fall was an illuminating ordeal. After leaving Abidjan, my Air Afrique flight landed in Dakar, Senegal, where all passengers had to disembark in order to go through another security check, this one demanded by U.S. authorities before they would permit the flight to set out for New York. Once we were in New York, despite the midnight hour, immigration officials at Kennedy Airport held up disembarkation by conducting quick interrogations of the aircraft's passengers—this was in addition to all the normal immigration and customs procedures. It was apparent that drug smuggling, disease, and other factors had contributed to the toughest security procedures I have ever encountered when returning from overseas.

Then, for the first time in over a month, I spotted businesspeople with attaché cases and laptop computers. When I had left New York for Abidjan, all the businesspeople were boarding planes for Seoul and Tokyo, which departed from gates near Air Afrique's. The only non-Africans off to West Africa had been relief workers in T-shirts and khakis. Although the borders within West Africa are increasingly unreal, those separating West Africa from the outside world are in various ways becoming more impenetrable.

But Afrocentrists are right in one respect: we ignore this dying region at our own risk. When the Berlin Wall was falling, in November of 1989, I happened to be in Kosovo, covering a riot between Serbs and Albanians. The future was in Kosovo, I told myself that night, not in Berlin. The same day that Yitzhak Rabin and Yasser Arafat clasped hands on the White House lawn, my Air Afrique plane was approaching Bamako, Mali, revealing corrugated-zinc shacks at the edge of an expanding desert. The real news wasn't at the White House, I realized. It was right below.

Operating in a Period of Punctuated Equilibrium

Lester C. Thurow

THE POLITICAL PROCESS

. . . Governments are in trouble everywhere in the world, since they have no answers to the real problems and worries facing their citizens. The policies being enacted because of the great Republican victory of 1994 don't even address the problems of falling real wages and rising inequality. Eventually those same voters will again become unhappy voters and chase after whatever demagogue happens to be around at the time. The fury that was directed at the Democrats in late 1994 could easily be directed back at the business community in the years ahead if the "right" (wrong?) leaders emerge. A large group of voters with free-floating hostility, not benefiting from the economic system, is not a recipe for economic or political success.

Democracies react well to crises since crises focus everyone's attention on the same issues and demand action. Without a crisis galvanizing public attention, democracies almost never act. To change, democracies need to persuade large numbers of their average citizens (far more than 51 percent) that change is necessary. Majorities are inherently conservative, since change means that the majority must itself abandon old ways. Without a crisis it is difficult to persuade a large majority that something has to change. Without a crisis minorities hurt by change can always block change. Democracies pay much more than proportional attention to one-issue minority groups, since they often swing close elections and can be made into solid supporters quickly—simply support them on their one hot issue.

But current economic events are not a crisis. The real income of nonsupervisory workers is declining less than 1 percent per year. Changes are dramatic over twenty-five-year periods of time, but not in any one year. A dramatic change has occurred in the distribution of income and wealth over the past twenty-five years and absolutely nothing has been done to reverse it. Policies to reverse current trends are not even being debated.

At the same time, these trends are produced by forces so fundamental that it is clear that they are not going to be reversed by marginal reforms in economic policies. Massive structural changes will be required. That is of course what democracies do least well. When democracies are forced to move, instead of making radical changes and moving to the global optimums, democracies tend to move slowly along the line of least resistance to local optimums. With normal evolution, that is the correct strategy. In a period of punctuated equilibrium it is not. Local optimums, the line of least resistance, often leads away from and not toward global optimums.

For a while, using the titles of two recent books, it is possible to have an Age of Contentment for the upper classes while having an Age of Diminishing Expectations for the middle and lower classes.[1] But such a duality is not forever possible. Social systems float on a molten magma of compatible ideologies and technologies. It is not possible to have an ideology of equality (democracy) and an economy that generates ever larger inequalities with absolute income reductions for a majority of the voting population.

American capitalism has strengths and weaknesses when it comes to dealing with these pressures. Its strength is that it has more fundamental political support than capitalism in Europe. The fact that socialist parties have never been a force in American politics says something important about America. Faith in capitalism's ability to deliver rising standards of living will probably die more slowly in the United States than it will in Europe. American capitalism's prime weakness is that it is the major deliverer of what elsewhere would be social welfare benefits (medical care, pensions) to the working middle class. Economically, to lose one's job, and hence one's company-provided fringe benefits, is much less serious in Europe than it is in the United States. There the social welfare state picks up more of what is lost when unemployment strikes.

As middle-class fringe benefits are cut back, the anger of the middle is rising and will rise rapidly. Eventually the middle class will demand that the political process act to stop its benefits and standards of living from being reduced, and it will have less and less interest in politically protecting capitalism. Just such anger led President Clinton and his wife to make health care their primary issue in the first two years of Clinton's administration. The Clintons fumbled the health care reform process, but the middle-class anger at being deprived of health care will return.

Politically, capitalism stands alone in a way that it has not stood alone since the mid-nineteenth century. Then, capitalism survived politically precisely because it coopted groups of workers—middle- and lower-level managers, white-collar workers, skilled blue-collar workers—into thinking of themselves as part of the capitalistic team. With downsizing, capitalism is in effect telling a lot of its past political supporters that they are no longer part of the "team." Having been thrown off capitalism's economic team, it is only a matter of time until those same workers desert capitalism's political team.

In the short run, capitalism can politically afford to be much tougher economically on its workforce than it used to be when socialism or communism threatened it with an internal revolution and an external threat. But at some point, something will arise to challenge capitalism and capitalism will need the political support of more than those small numbers of individuals who are actually owners of substantial amounts of capital. Where is this support to come from?

The facts are clear. Income and wealth inequalities are rising everywhere. Real wages are falling for a large majority. A lumpen proletariat unwanted by the productive economy is growing. The social contract between the middle class and corporate America has been ripped up. The prime remedy for inequalities in the past one hundred years, the social welfare state, is in retreat. Economic plate tectonics is rapidly changing the economic surface of the earth.

The revival of free market survival-of-the-fittest economics is not surprising in that it fills people's need for some kind of social understanding to guide their actions—a return to mythical ancient virtues.[2] It's what people do when they are confused. But something new is going to have to be invented to cope with the current period of punctuated equilibrium and the very different future that will come out of it.

New productive technologies are raising the importance of social investments in infrastructure, education, and research while values are moving toward more individuality with much less social interest in communal investments. A more rugged version of survival-of-the-fittest capitalism is being preached at precisely the moment that the economic system is discovering the productivity gains that can flow from teamwork. The belief that the capitalistic system is perfect and needs no social support has returned just when a new capitalism without ownable capital has to be invented.

A BUILDER'S IDEOLOGY

At this point it is tempting to outline a long list of public policies that would help capitalism get what it needs. What tax and budgetary policies would lead to more long-term investments? What R&D strategies should be employed? Which are the infrastructure projects with long-run spillovers? What are the right reforms in pension and health care programs for the elderly? How does one generate the most skilled labor force in the world? How should formal education be integrated with on-the-job training? Who should pay for what? Some of these questions have been partially answered in the course of this book, but attempting to definitely answer all of these questions now would be a mistake.

Appropriate public policies aren't the current issue. The current issue is persuading ourselves that the world has changed and that we must change with it. Adopting the right policies once the need for change has been recognized intellectually, and more important emotionally, is the easy part of the task. Many public policy possibilities exist for getting from here to there. What we first have to do is figure out where the "there" is and then create a sense of urgency about getting there.

As a boy growing up in Montana before the era of jet aircraft, my flights of fancy still turned to the railroads. At that time one of the fastest trains in the world was the Great Northern's *Empire Builder* as it powered its way across Montana and North Dakota on its way from Minneapolis to Seattle. The name of that train captures the attitudes that must be generated to succeed in the era ahead. As the name implies, there were no empires on the northern Great Plains to be conquered. There was only empty space where James J. Hill envisioned that he and his trains would be the catalyst for building an empire. In the end he was wrong. No empire was ever built. The gold mines that at one time looked so promising petered out. But the name of his train embodies what has to be done.

Today there are no physical empires worth conquering. Holding more physical territory does not make one a better economic competitor. Those who succeed will build the man-made brainpower industries of the future. They will build something where there is today empty intellectual and economic space. Some possibilities that look like economic gold mines will peter out as they did for Mr. Hill, but other technologies that look like wastelands will prove to be economic bonanzas. But to capture those bonanzas one must be there with a commitment to empire building.[3]

No sensible person has ever set the goal of reducing his or her own consumption. Saving is not fun. But participating in the process of using the funds released by saving to build something can be fun. If it is to succeed, the capitalism of the future will have to shift from a consumption ideology to a builder's ideology. Growth is not an automatic process of quietly moving from one equilibrium point to another. The growth path is a noisy process of disequilibrium where a lot of fun is to be had. Technology is not manna from heaven. It is a social process of human creation and innovation. In this context investment must be seen, not as a cost to be avoided, but as a direct generator of utility to be embraced. The individual who invests in what will probably be the most valuable skills that any individual can have, the ability to operate in a global economy, is not being forced to sacrifice consumption but is building a skill set that will bring more enjoyment than an item of consumption.

The savings habits of the self-employed illustrate the attitudes that must be generated. In the United States at every income level the self-employed are much larger savers and investors than those who work for someone else but make the same income. The self-employed directly see what they are building. Building a better business generates more utility for them than having a bigger home or driving a bigger car. As builders their time horizons are much longer than those of either absentee capitalists or consumers.

In big corporations mostly owned by the pension funds and mutual funds, the shareholders are so distant, so diversified, and so amorphous that none of them can get any enjoyment out of creating or building. They only see dividends. If one looks at large corporations still controlled by a dominant family (Mars, Wal-Mart, Miliken, Microsoft), one sees very different behavior patterns and time horizons than those inside equally large businesses owned by institutions. Their personal goals, their family goals, and their business goals are all consistent with a building mentality.

Everyone cannot be self-employed. The economy does not need them in this role and many individuals who should be small savers don't have the necessary personal aptitudes. While everyone cannot participate directly as builders, everyone can participate in the building process in a social sense when government builds the projects that it will need to build. Most Americans did not work on the man-on-the-moon project in the 1960s. Yet all of us took great pride in what was accomplished and I don't remember hearing anyone at the time saying that "too much" money was being spent. In continental Europe the same feeling exists today about high-speed intercity rail travel. Everyone tells you that their trains are the fastest—or soon will be the fastest. Complaints about taxes and government budgets abound, but never once during a year of living in Europe and traveling widely did I hear someone objecting to that part of their government's budgets. . . .

If individuals are to have a builder's mentality, then government must be active visible builders. Some of the building should be physically visible. Deciding to beat the Japanese and Europeans when it comes to having the best intercity high-speed rail network in the world would be a good place to start. But much of the building will be human. The United States should commit to having the most skilled and best-educated labor force in the world. This means being willing to measure objectively where America stands today, finding out who has the best-educated labor force at every level, being willing to chart our progress or lack of progress in catching up with and then passing whoever is best, and committing to doing whatever is necessary to achieve that goal. If something does not work it should be ruthlessly junked and other means adopted—but nothing will be allowed to stop us from reaching the goal.

The real heroes of the future are neither the capitalists of Adam Smith nor the small businessmen politicians love to praise, but those who build new industries.[4] They were willing to live the difficult life outside of the boundaries of routine—to overcome the natural human psychic reluctance to try the new in the face of a social environment that is always attached to the past.[5] They have the ability to dream, the will to conquer, the joy of creating, and the psychic drive to build an economic kingdom.[6]

Joseph Schumpeter thought that capitalism would die out because it would be undercut by the bureaucratization of invention and innovation and by the intellectual scribblers who would point to the nobler goals of other systems such as socialism.[7] He actually predicted the disintegration of the family, since children would cease to be economic assets and parents would refuse to undertake the necessary sacrifices to support them when they became cost centers.[8] Historically, he was wrong about R&D, identified the wrong scribblers, and is looking ever more right about the family.

Research and development, especially fundamental R&D, does have to be bureaucratically funded by big companies or big government. The tinkerer-inventors of nineteenth-century British fame no longer stand at the heart of technological progress. But this still leaves plenty of economic niches for small inventors and innovators to use the basic scientific principles whose discovery was funded by big science to build small firms that eventually become big firms. It also does nothing to reduce the fun of invention for those employed who are funded by the big funders. My institution, MIT, lives or dies with the big funders, but those doing the research have an exciting time and MIT is the country's largest incubator of new firms.

The modern equivalent of Schumpeter's scribbler is the TV set. Officially, it sings hymns of praise for capitalism but unofficially it inculcates a set of antiproduction values. Consumption is the name of the game; no one should postpone immediate gratification. In TV land, creators and builders are conspicuous by their absence. Time horizons become ever shorter based on both the ideologies of the programs and the ways in which that content is presented—moving faster and faster from one scene to another. Put a stopwatch on the evening news programs and measure the longest amount of time they are willing to spend on any one topic no matter how important.

Without an outside threat how does TV man force himself to make the investments and reforms that are essential to the future? Neither his explicit capitalistic ideology nor his implicit TV ideology recognizes sacrificing to build the future. He is the ultimate consumer in the present. Where are the values that support necessary investments in education, R&D, and infrastructure to be generated? If they don't get generated what happens?

The modern scribblers, the TV set, was probably one of the major factors lying behind the fall of the Berlin Wall in 1989. East Germans sat there watching West German television knowing what they were missing. The ideology of socialism could not replace the goods of capitalism. In North Korea, TV sets are built so that they cannot receive the signals of South Korea, and the DMZ still stands. The North Koreans simply don't know they are missing anything that anyone else has.

ADJUSTING TO A NEW GAME

Adjusting to new realities is difficult. Countries fundamentally are what they are and often cannot do what they need to do even though they know what they should do. Everyone in America knows that Americans need to save more, yet the United States can do nothing to reduce its consumption. Europe knows that it cannot forever continue without employment growth, but it cannot give up its fight against the ghosts of inflation and isn't willing to deregulate its labor markets to jump-start its economic engines. Japan knows that its current economy does not work and knows that it uniquely has less residential space per person than any other wealthy society, but it cannot restructure to be a domestically led economy focused on improving the housing stock. Each of the major world players rationally knows some of what it needs to do but cannot act rationally. . . .

Making the necessary changes will be hard for everyone but especially hard for Americans. They are not unique in believing that their social system is the world's best; many citizens of many countries have similar beliefs, but Americans are unique in believing that their social system is perfect—given by founding fathers who are demigods at the very least. The American political system is also now the world's oldest. Both of these factors lead Americans whenever something goes wrong to find fault not with a system that needs institutional repairs but with "bad" individuals—the devil. In American political theology the bad guys never ultimately win. Vietnam was a bigger shock in America than it would have been almost anywhere else precisely because the good guy, we Americans, did not in the end win. In American theology there are no trade-offs between liberty and equality. Americans can have both. The right rules (system) will produce deliverance—and once in place, like Moses' Ten Commandments, those rules are written in stone and never need to be altered. America does not need social planning or elite knowledge. The man on the street knows best. Americans don't have to accept the allocation of losses. Free markets will bring forth not just the best that there is to be had, but perfection at no cost.[9]

Americans will also have to cope with losing their position as the world's dominant economic, political, and military power. Rationality would call for Americans to play an active, but a reduced and different role on the world stage. America has immense powers of persuasion and assimilation; it is the world's only country with global interest and a global reach. But emotionally the loss of leadership is more likely to lead to isolationism.[10] Everyone will deny that they are isolationists ("isolationism" is a bad word, much like "Munich"), but Americans are now saying in the legislation working its way through Congress that they don't want to pay for activities such as the United Nations, that they don't want to pay for regional development banks, and they don't want to send American troops abroad under international auspices—precisely the activities that allow America to be a world military power and exercise leadership. Whatever Americans say, American "isolationism" is in resurgence. . . .

CONCLUSION

The danger is not that capitalism will implode as communism did. Without a viable competitor to which people can rush if they are disappointed with how capitalism is treating them, capitalism cannot self-destruct. Pharaonic, Roman, medieval, and mandarin economies also had no competitors and they simply stagnated for centuries before they finally disappeared.[11] Stagnation, not collapse, is the danger.

Periods of punctuated equilibrium are periods of great optimism and great pessimism. For those very good at playing the old game, the dinosaurs, they are disasters. Millions of years of supremacy disappear in a flash. Evolution along the old lines is impossible. For those who are good at adjusting to new circumstances and can learn to play new games, the mammals, periods of punctuated equilibrium are periods of enormous opportunity. It is precisely the disappearance of the dinosaurs that made it possible for humans to take control of the system. If the dinosaurs had continued to rule, our ancient ancestors probably would have been eaten and we wouldn't be here. But during the transitions in periods of punctuated equilibrium, no one knows who will be a dinosaur and who will be a mammal. That depends upon who is the best at adjusting to a new world—something that can only be known with certainty looking backward.

The intrinsic problems of capitalism visible at its birth (instability, rising inequality, a lumpen proletariat) are still there waiting to be solved, but so are a new set of problems that flow from capitalism's growing dependence upon human capital and man-made brainpower industries. In an era of man-made brainpower industries those who win will learn to play a new game with new rules requiring new strategies. Tomorrow's winners will have very different characteristics than today's winners.

Technology and ideology are shaking the foundations of twenty-first century capitalism. Technology is making skills and knowledge the only sources of sustainable strategic advantage. Abetted by the electronic media, ideology is moving toward a radical form of short-run individual consumption maximization at precisely a time when economic success will depend upon the willingness and ability to make long-run social investments in skills, education, knowledge, and infrastructure. When technology and ideology start moving apart, the only question is when will the "big one" (the earthquake that rocks the system) occur. Paradoxically, at precisely the time when capitalism finds itself with no social competitors—its former competitors, socialism or communism, having died—it will have to undergo a profound metamorphosis.

It is easy to get discouraged and become a pessimist if one looks at what must be done and compares it with the seemingly glacial pace of social change. But to do so would be to make a mistake. Social change occurs in much the same manner as waves hitting rocky cliffs on the Maine coast. On each and every day the rocks win. The waves thunder into them and nothing seemingly happens. But we know with absolute certainty that eventually every one of those rocks will be bits of sand. Every day the waves lose, yet in the long run they win.

Given our new understanding of the tectonic forces altering the economic surface of the earth and the period of punctuated equilibrium that they have created, let's return to the problem of constructing a capitalistic ship that will safely take us into an era. Like Columbus and his men, all of us aboard the good ship "capitalism" are sailing into a new uncertain world. Being smart, Columbus knew that the world was round, but he got his mathematics wrong and thought that the diameter of the world was only three quarters as big as it really is. He also overestimated the eastward land distance to Asia and therefore by subtraction grossly underestimated the westward water distance to Asia. That combination of errors made him think that India (the word for Asia then) was about 3,900 miles from the Canary Islands, more or less where America happened to be. Given the amount of water put on board, without the Americas Columbus and all of his men would have died of thirst and been unknown in our history books.[12]

Columbus goes down in history as the world's greatest explorer, perhaps history's most famous man, because he found the completely unexpected, the Americas, and they happened to be full of gold. One moral of the story is that it is important to be smart, but that it is even more important to be lucky. But ultimately Columbus did not succeed because he was lucky. He succeeded because he made the effort to set sail in a direction never before taken despite a lot of resistance from those around him. Without that enormous effort he could not have been in the position to have a colossal piece of good luck.

With similar persistence and willingness to attempt the unknown, let our journey begin!

Notes

1. Paul Krugman, *The Age of Diminishing Expectations* (Cambridge, Mass.: MIT Press, 1990).
2. Fred Block, *Post-Industrial Possibilities: A Critique of Economic Discourse* (Berkeley: University of California Press, 1990), pp. 2–4.
3. J. L. Baxter, *Social and Psychological Foundations of Economic Analysis* (New York: Harvester Wheatsheaf, 1988).
4. Joseph A. Schumpeter, *The Theory of Economic Development* (New York: Oxford University Press, 1961), p. 92.
5. Ibid. p. 84.
6. Ibid. p. 223.
7. Joseph A. Schumpeter, *Capitalism, Socialism, and Democracy* (New York: Harper Colophon Books, 1975), PP. 132, 139, 143.
8. Ibid. p. 157.
9. Mona Harrington, *The Dream of Deliverance in American Politics* (New York: Alfred A. Knopf, 1986).
10. Fernand Braudel, *A History of Civilization* (New York: Penguin Press, 1994), p. 475.
11. George Brockway, *The End of Economic Man* (New York: W. W. Norton, 1993), p. 253.
12. *Encyclopaedia Britannica,* Vol. 16 (Chicago: 1972), p. 111.

Glossary

Absolute poverty: A condition in which one lacks the essentials of life such as food, clothing or shelter. *See also* Relative poverty.

Achieved position/status: The position of an individual within a system of social stratification based on changeable factors such as occupations, high income, or marriage into higher social strata. *See also* Ascriptive status.

Adaptation: The process by which animal or human species interact with and become fitted to their environments in order to obtain food, shelter, and protection from predation and, ultimately, ensure the biological survival of the species.

Agents of socialization: The people, groups, and organizations who socialize the individual. *See also* Socialization.

Aggression: Some researchers have applied it to any act that inflicts pain or suffering on another individual; others feel that a proper definition must include some notion of intent to do harm.

Alienation: A sense of separation from society. In the context of the bureaucracy, one's feeling of not having control over or responsibility for one's own behavior at work. *See also* Bureaucracy.

Altruism: Behavior motivated by a desire to benefit another individual, or sacrifice by individuals for the benefit of the group as a whole.

Androgyny: A combination of male and female characteristics. The term may be used in a strictly physical sense or it may apply to a wider, social ideal.

Anomie: The loosening of social control over individual behavior that occurs when norms become ineffective.

Anticipatory socialization: The tendency of an individual to adopt the values, attitudes, and behavior that he perceives to be typical of and appropriate for members of a particular group or social category to which he aspires and eventually expects to belong.

Ascriptive status: The position of an individual within a system of social stratification based on factors such as sex, age, race, over which the individual has no control. *See also* Achieved position/status, Social stratification, Status.

Assimilation: The absorption of a subordinate group into the dominant culture.

Authority systems: Systems by which authority is legitimated. According to German sociologist Max Weber, in a traditional system, positions of authority are obtained by heredity. In a charismatic system, leaders are followed because of some extraordinarily appealing personal quality. In a legal-rational system, the office is the source of authority, rather than the officeholder.

Autocratic leader: The type of group leader who is authoritarian and impersonal and who does not participate in group projects. *See also* Democratic leader.

Awareness context: The "total combination of what each interactant in a situation knows about the identity of the other and about his or her own identity in the eyes of the other."

Belief system: Groups of basic assumptions about general concepts such as the existence and nature of God, the meaning of life, or the relationship of the individual and the state held by a culture.

Bias: The theoretical or emotional preconceptions or prejudices of individuals or groups, which may lead to certain subjective interpretations that are radically different from "objective" reality.

Bigotry: The state of mind or behavior of a person of strong prejudice.

Binet-Simon Scale: An intelligence test developed by a French psychologist (Binet) and a psychiatrist (Simon) to assess mental abilities in order to provide special education facilities for the retarded.

Biological determinism: The view of behavior as a product of genetic makeup.

Biological-instinctual theories: Theories of behavior that stress the importance of instinct. *See also* Environmental theories.

Biosocial interaction: The ways in which interrelationships with society influence and are influenced by biological factors. *See also* Biosociology.

Biosocial systems: Systems of social organization such as those among insects, which survive because behavior patterns are biologically controlled.

Biosociology: Biosociology tries to consider the interaction and mutual influences between the social order and the biological makeup of its members.

Birth rate (crude): The number of people born in a single year per 1,000 persons in the population.

Bourgeoisie: The class that owns the means of production. *See also* Proletariat.

Bureaucracy: An authority structure arranged hierarchically for the purpose of efficient operation.

California Psychological Inventory (CPI): One of many questionnaires that has been developed to measure a wide range of personal and social characteristics of an individual.

Case study: A research method that involves intensive examination of a particular social group over time. *See also* Participant observation, Research, Sample survey.

Caste: A rigid form of social stratification, rooted in religious standards in which individuals spend their lives in the stratum into which they were born. *See also* Class, Estate, Social stratification.

Causality: The state in which some condition produces, or always results in, a particular consequence.

Censorship: The act of suppressing or controlling books, plays, films, or other media content or the ideas, values, and beliefs held by certain groups on the grounds that such content is morally, politically, militarily, or otherwise objectionable.

Census: A periodic count and collection of demographic information about an entire population. *See also* Demography.

Central city: The core unit of a metropolitan area. The term is also used to mean "inner city" or "ghetto," with its urban problems of poverty, crime, racial discrimination, poor schools and housing, and so on.

Centralization: Assumption of power and authority by a few within an organized social group.

Central tendency, measure of: A set of statistical measures that are designed to determine a typical case in a particular distribution.

Charisma: Exceptional personal leadership qualities that command authority as contrasted to legal or formal authority. A driving, creative force that attaches both to individuals and to social movements.

Chicago School: The name given to a group of sociologists at the University of Chicago, where the first graduate sociology department in the United States was formed during the 1890s.

Clan: A unilinear descent group whose members trace their common descent to a historically remote, often mystical, ancestor or ancestress. *See also* Kinship.

Class: A category of people who have been grouped together on the basis of one or more common characteristics.

Class conflict: According to Marxist theory, the dynamics for change created by the conflict between ruling classes and subordinate classes in society.

Class consciousness: The awareness, particularly among the working class, or common social, economic, and political conditions. The concept was developed by Karl Marx.

Classless society: According to Marxist theory, the goal of socialism and the state in which all social stratification on the basis of class is eliminated. *See also* Class, Social stratification.

Clinical sociology: The application of sociological skills and methods in a variety of nonacademic settings, including businesses and organizations, social services, and therapeutic practice with individuals and groups.

Closed system: A social stratification system that offers an individual no way to rise to a higher position; based on ascriptive status. *See also* Ascriptive status, Open system.

Coercion: The power to compel people to act against their will, by using force or the threat of force. The constraint of some people by others. *See also* Conflict model, Power.

Cognitive development: A theory of psychology which states that cognitive processes (such as thinking, knowing, perceiving) develop in stages although they function and influence even newborns' behavior.

Cognitive dissonance: A theory developed by American educator and psychologist Leon Festinger concerning the attempt of individuals to achieve consistency between their beliefs and actions, as well as among their beliefs.

Collective behavior: The behavior of a loosely associated group that is responding to the same stimulus. The concept embraces a wide range of group phenomena, including riots, social movements, revolutions, fads, crazes, panics, public opinion, and rumors. All are responses to, as well as causes of, social change.

Communalism: Denotes the degree to which primary relations are confined to one's group (racial, ethnic, religious, and so on). The major emphasis is on the priority of the group over the individual member.

Communism: A political-economic system in which wealth and power are shared harmoniously by the whole community. *See also* Socialism.

Community: The spatial, or territorial, unit in social organization; also the psychological feeling of belonging associated with such units. *See also* Metropolis.

Competitive social system: A social system in which the dominant group views the subordinate group as aggressive and dangerous and thereby in need of suppression. *See also* Paternalistc social system.

Compliance patterns: According to societal theorist Amitai Etzioni, the three ways (coercive, remunerative, and normative) in which formal organizations exercise control over members.

Comte, Auguste (1798–1857): French philosopher who coined the term "sociology" and is considered the founder of the modern discipline.

Concentric-zone theory: Based on the view that cities grew from a central business district outward in a series of concentric circles. Each zone was inhabited by different social classes and different types of homes and businesses. *See also* Multiple-nuclei theory, Sector theory.

Conflict model: The view of society that sees social units as a source of competing values and norms. *See also* Equilibrium model.

Conforming behavior: Behavior that follows the accepted standards of conduct of a group or society. *See also* Deviance.

Conjugal family: A family type in which major emphasis is placed on the husband-wife relationship. *See also* Consanguine family.

Consanguine family: The family type in which the major emphasis is on the blood relationships of parents and children or brothers and sisters.

Consensual validation: A tacit agreement among a group of people as to the meaning of a certain situation or as to how something ought to be done.

Contagion theory: A theory of collective behavior, originated by Gustave LeBon, which states that the rapid spread of common mood among a large number of people is what forms a crowd.

Conventional morality: According to American psychologist Lawrence Kohlberg, the second level of moral development, at which most adults remain. This level involves conformity to cultural or family norms, maintenance of, and loyalty to, the social order. *See also* Postconventional morality, Preconventional morality.

Convergence theory: A theory of collective behavior which states that people with certain tendencies are most likely to come together in a crowd. This theory assumes that crowd behavior is uniform.

Criminal justice system: Authorities and institutions in a society concerned with labeling and punishing criminals according to formal social sanctions.

Criminology: The social science that analyzes crime as a social occurrence; the study of crime, criminality, and the operation of the criminal justice system.

Crowd: A type of social aggregate in which all participants are in the same place at the same time, and they interact in a limited way. *See also* Social aggregate.

Cults: Small groups whose teachings stress ritual, magic, or beliefs widely regarded as false by the dominant culture.

Cultural adaptation: The flexibility of a culture that allows it to change as the environment changes.

Cultural diffusion: The adaptation of a culture as it encounters another and undergoes social change.

Cultural lag: The condition that exists when values or social institutions do not change as rapidly as social practices.

Cultural relativism: The principle of judging a culture on its own terms. *See also* Ethnocentrism.

Culture: The knowledge people need to function as members of the particular groups they belong to; our shared beliefs, customs, values, norms, language, and artifacts.

Death: Termination of life. Though essentially a biological phenomenon, death is also a social process. Most secular societies look upon death as an important event and utilize rituals to prepare the deceased for another life or for the dignity of nonlife.

Death rate (crude): The number of deaths in a single year per 1,000 persons in the population. *See also* Demography.

Democratic leader: A type of group leader who encourages group decision making rather than giving orders. *See also* Autocratic leader.

Democratization: The process of making a party or voluntary organization more responsive to its members.

Demographic transition: The pattern in which death rates fall with industrialization, causing a rise in population and ensuing drop in birth rate which returns the rate of population growth to nearly the same level as before industrialization.

Demography: The study of human population, focusing on birth rate, death rate, and migration patterns.

Dependent variable: The factor that varies with changes in the independent variable. *See also* Independent variable.

Desegregation: Elimination of racial segregation in a society. *See also* Discrimination.

Deterministic: Any theory that sees natural, social, or psychological factors as determined by preceding causes.

Deterrence theory: A theory held by some criminologists that punishment will prevent as well as control crime.

Deviance: The label for all forms of behavior that are considered unacceptable, threatening, harmful, or offensive in terms of the standards or expectations of a particular society or social group. *See also* Conforming behavior, Norm.

Dewey, John (1859–1952): American philosopher and educator, a functionalist, whose ideas about education had a strong effect on schooling. He pressed for a science of education and believed in learning by doing. Individualized instruction and experimental learning can be traced to his theories.

Dialectical materialism: The philosophical method of Karl Marx, who considered knowledge and ideas a reflection of material conditions. Thus the flow of history, for example, can be understood as being moved forward by the conflict of opposing social classes. *See also* Communism.

Discrimination: Unfavorable treatment, based on prejudice, of groups to which one does not belong. *See also* Prejudice.

Disinterestedness: The quality of not allowing personal motives or commitments to distort scientific findings or evaluations of scientific work. *See also* Communalism, Organized skepticism.

Division of labor: The separation of tasks or work into distinct parts that are to be done by particular individuals or groups. Division of labor may be based on many factors, including sex, level of technology, and so on.

Dominance relationships: A system of status within a social organization in which individuals occupy different ranks in respect to one another.

Double standard of sexual behavior: A moral judgment by which sexual activity of men is considered appropriate or excused while that of women is considered immoral. *See also* Sex role.

Dramaturgical perspective: The point of view, favored by American sociologist Erving Goffman, that social interaction can be compared to a dramatic presentation.

Durkheim, Emile (1858–1917): French sociologist and one of the founders of modern sociology. Deeply influenced by the positivism of French philosopher Auguste Comte, Durkheim's major concern was with social order, which he believed to be the product of a cohesion stemming from a common system of values and norms.

Ecological determinism: The point of view stressing how environment affects behavior.

Economic determinism: The doctrine, supported by Karl Mark, that economic factors are the only bases for social patterns.

Economic modernization: Shift from an agricultural-based economy to an industrial one.

Egalitarianism: Emphasis within a society on the concept of equality among members of social systems.

Egocentricity: The characteristic quality of very young children, their awareness of only their own point of view.

Electra complex: A love-hate relationship of parents that was postulated by Sigmund Freud as occurring during a girl's childhood.

Elite groups: Members of the top ranks of society in terms of power, prestige, and economic or intellectual resources.

Empiricism: A philosophical school of thought that holds that all knowledge is grounded in sense experience. This view denies the existence on innate principles or ideas.

Encounter groups: Groups of individuals who meet to change their personal lives by confronting each other, discussing personal problems, and talking more honestly and openly than in everyday life. *See also* Group therapy.

Endogamy: Marriage within one's social group. *See also* Exogamy.

Environmental theories: Theories of behavior that stress the influence of learning and environment. *See also* Biological-instinctual theories.

Equilibrium model: A view of society as a system of interdependent parts that function together to maintain the equilibrium of the whole system. *See also* Conflict model, Functionalism.

Erikson, Erik (1902–1994): Danish-born psychoanalytic theorist who theorized that individuals move through a series of psychosocial stages throughout life, with the integrity of the personality depending largely on the individual's success in making appropriate adaptations at previous stages.

Estate: A form of social stratification based on laws, usually about one's relationship to land. *See also* Social stratification.

Ethnic group: A social group distinguished by various traits, including language, national or geographic origin, customs, religion, and race.

Ethnicity: The act or process of becoming or being a religious, racial, national, cultural, or subcultural ethnic group. *See also* Ethnic group.

Ethnocentrism: The tendency to judge other groups by the standards of one's own culture and to believe that one's own group values and norms are better than others'. *See also* Cultural relativism.

Ethology: The comparative study of animal behavior patterns as they occur in nature.

Etzioni, Amitai W. (1920–   ): Contemporary societal theorist, especially active in the study of political sociology, complex organizations, and social change.

Eugenics: The science of controlling heredity.

Evolution: A process of change by which living organisms develop, and each succeeding generation is connected with its preceding generation.

Evolutionary change: A gradual process of social change. *See also* Revolutionary change.

Exchange theory: The viewpoint that stresses that individuals judge the worth of particular interactions on the basis of costs and profits to themselves.

Exogamy: Marriage outside one's social group. *See also* Endogamy.

Experiment: A research method in which only one factor is varied at a time and efforts are made to keep other variables constant in order to isolate the causal or independent variable. *See also* Independent variable, Research.

Extended family: A family type consisting of two or more nuclear families. Also characterized as three or more generations, who usually live together.

Facilitating conditions: In a model of suburban growth, those factors that make movement from city to suburb possible. Such factors include commuter transportation systems and communications technology. *See also* Motivating conditions.

Family: A set of people related to each other by blood, marriage, or adoption. Family membership is determined by a combination of biological and cultural factors that vary among societies.

Family life cycle: The process of characteristic changes that a family's task (such a childrearing) undergo over time.

Family planning: The theory of population control that assumes that parents should be able to determine and produce the number of children they want, spaced at the intervals they think best. *See also* Population control.

Folk Taxonomy: Classification system used by a culture to organize its cognitive categories.

Formal and complex organization: Large social units purposely set up to meet specific, impersonal goals. *See also* Informal organization.

Freud, Sigmund (1856–1939): Viennese founder of modern psychology and originator of psychoanalysis. Basic to Freud's theories are the beliefs that much of human behavior is unconsciously motivated and that neuroses often have their origins in early childhood wishes or memories that have been repressed. He developed an account of psychosexual development in which he said that sexuality was present even in infants, although the nature of this sexuality changed as the individual progressed through a sequence of stages to mature adult sexuality. Freud also proposed a division of the self into the *id* (instinctual desires), the *ego* (the conscious self), and the *superego* (conscience arising from socialization). The ego mediates between the pressures of the other two parts in an effort to adapt the individual to the demands of society, and personality formation is largely the result of this process. *See also* Psychoanalytic theory.

Functionalism: A dominant school in modern sociology which assumes that each part of the social structure functions to maintain the society and which views social change according to the equilibrium model; also called structural-functionalism. *See also* Equilibrium model.

Game theory: The study of situations in which the outcome of interaction depends on the joint action of the partners.

Gemeinschaft/Gesellschaft: Simple, close-knit communal form of social organization/impersonal bureaucratic form. Typology of social organization devised by German sociologist Ferdinand Töennies and used to understand variety and changes in societies' social structure.

Gender identity: A child's awareness of being either male or female. *See also* Sex role.

Gene pool: The total of genes present in the population.

Generalized other: According to American social psychologist and philospher George Herbert Mead, the developmental stage in which children adopt the viewpoint of many other people or, in short, of society in general. *See also* Significant other.

Genetic engineering: Altering the reproductive process in order to alter the genetic structure of the organism.

Generic load: The presence of genes in a population that are capable of reducing fitness.

Genocide: Deliberate destruction of a racial or ethic group.

Genotype/phenotype: Genotype is the entire structure of genes that are inherited by an organism from its parents. Phenotype is the observable result of interaction between the genotype and the environment.

Gerontology: The study of the problems of aging and old age.

Group marriage: Marriage among two or more women and two or more men at the same time.

Group process: The dynamics of group functioning and decision making and of the interactions of group members.

Group space: A concept of sociologist Robert Bales, from his research on social groups. Bales correlated many factors and then constructed dimensions, such as dominance, likeability, task orientation, along which group members could be placed. When these dimensions are combined in three dimensions, they form the group space.

Group therapy: A form of psychotherapy in which interaction among group members is the main therapeutic mode.

Hierarchy: The relative positions of individuals or groups within a body or society and their relationship to power and control. *See also* Social sciences.

Hobbes, Thomas (1588–1679): British philosopher and writer who theorized about social order and social conflict. He was the first social conflict theorist.

Hobbesian question: The term referring to seventeenth-century philosopher Thomas Hobbes' question of how society could establish and maintain social order. Today, sociologists apply this question to the problem of conformity within the social order.

Humanistic sociology: The branch of sociology that is people-centered, focuses upon human problems, and tries to serve broad human interests rather than merely those of some elite.

Ideal type: A conceptual model or tool used to help analyze social occurrences. It is an abstraction based on reality, although it seldom, if ever, occurs in exactly that form.

Identity: According to American psychoanalyst Erik Erikson, a person's sense of who and what he or she is.

Imperialism: The expansionist and empire-building policies of nation-states, involving a subordinate-superordinate relationship in which one political community achieves sovereignty over another.

Independent variable: The causal variable, or factor that changes. *See also* Dependent variable.

Individuation: The development and recognition of the individual as a distinct being in the group.

Industrialization: The systematic organization of production through the use of machinery and a specialized labor force.

Infant mortality rate: The number of children per thousand live births who die during their first year of life.

Influence: A subtle form of power involving the ability to sway people to do what they might not otherwise do. *See also* Power.

Informal norms: The rules governing behavior generally set by an informal group instead of the formal requirements of an organization. *See also* Informal organization.

Informal organization: In contrast to and within a formal organization, those groups of people or roles they play that cut across the official bureaucratic pattern. *See also* Formal and complex organization.

Instinct: An unlearned fixed action pattern that occurs in response to specific stimuli as a result of complex hormonal and neurological processes.

Institution: Complex and well-accepted way of behaving aimed at meeting broad social goals. The major social institutions are government, family, religion, education, and the economy. *See also* Organization.

Intelligence: A capacity for knowledge. There is no agreement on a precise definition, although intelligence has come to refer to higher-level abstract processes.

Intelligence quotient (IQ): The ration of a child's mental age to his chronological age multiplied by a constant of 100. Adult IQ indicates the intelligence of a particular individual in relation to the statistically average person, who is arbitrarily given an IQ of 100.

Interest groups: Political factions made up of citizens who associate voluntarily and who aim to influence communal action. *See also* Pluralism.

Intergenerational learning: Learning by one generation from another. It is found generally among nonhuman primates as well as among humans.

Internalization: Involves the incorporation of the role-behavior of others as it is perceived in social interactions. The internalization of norms is responsible for the development of the superego and constitutes the central core of personality.

Kin selection: A process in which individuals cooperate, sacrifice themselves, or do not reproduce so that their kin can survive and reproduce.

Kinship: A system of organizing and naming relationships that arise through marriage (affinal kinship) and through birth (consanguine kinship).

Labeling theory: The school of thought that sees deviance or criminality as a status imposed by societal reaction. *See also* Opportunity theory.

Learning theory: An attempt to account for the manner in which the response of an organism is modified as a result of experience. Learning is to be distinguished from remembering, which is not necessary for learning and involves only the recall of previous experience.

Leisure, sociology of: A subdiscipline concerned with the analysis of patterns of nonwork behavior and their relationship to other social and economic variables.

Level of interaction: The way in which people relate to one another. Interactions may be subtle and nearly undetectable, or they may be clear and obvious. People may relate on a number of different levels with each statement or gesture. *See also* Group process.

Linguistic relativity: The concept that different languages analyze and portray the universe in different ways.

Locke, John (1632–1704): British philosopher and political theorist who put forward a social contract theory of government, which saw people as rational and dignified and entitled to overthrow any government that grew tyrannical. *See also* Social contract.

Macrosociology: The sociological study of relations between groups. Some sociologists consider it the study of the entire society or social system. *See also* Microsociology.

Malthus, Thomas (1766–1834): British economic and demographic theorist who predicted that population increases would outrun increases in food production, with starvation as a result.

Marriage: The social institution that sanctions, or gives approval to, the union of husband and wife and assumes some permanence and conformity to social custom. Marriage patterns differ among societies.

Marx, Karl (1818–1883): The German-born economic, political, and social thinker whose ideas provided the inspiration for modern communism.

Marxism: The revolutionary social-science tradition founded by Karl Marx and Friedrich Engels. Its basic concepts include materialism, alienation, the labor theory of value, class struggle, the dictatorship of the proletariat, and the revolutionary role of the vanguard party and of national liberation movements.

Mass communications: The organized transmission of a message to an audience by means of a technological medium. The term is usually applied to the activities of broadcasting, the press, and publicly screened films.

Mass culture: The way of life produced in advanced industrial countries (also called Pop or Popular culture). Mass culture involves standardized material goods, art, lifestyles, ideas, tastes, fashions, and values. It is the homogenized product of the mass media.

Mass society: A term used to characterize complex industrial societies. It draws attention to the uniformity of material goods, ideas, and lifestyles. Mass societies are often contrasted with traditional, case, or elitist class societies.

Materialism: Essentially economic determinism, this outlook upon human behavior suggests that the evolution of human societies is caused by the condition of material agencies—that is, by technology and the products of technology, which together comprise the economic institution.

Mead, George Herbert (1863–1931): American social psychologist and philosopher whose theories of mind, self, and society had a major influence on sociological approaches such as role theory and symbolic interactionism.

Mead, Margaret (1901–1978): An American anthropologist, Mead conducted field research in Samoa, the Manus, the Admiralty Islands, with the Omaha Indian tribe, and the Arapesh, Mundugumor, and Tchambuli in New Guinea. She also studied the American character and food habits and helped to develop the national-character approach to the study of complex societies.

Measures of central tendency: Descriptive statistical techniques used to measure the central tendency of distribution of group scores or results.

Mechanical solidarity: A concept developed by French sociologist Emile Durkheim to describe the form of social cohesion that exists in small-scale societies that have minimal division of labor.

Median age: The age that divides the population in half. Half of the population is older and half younger than the median age.

Megalopolis: Urban areas made up of more than one metropolis, "supercities." (For example, the area between New Hampshire and northern Virginia is one megalopolis.) *See also* Metropolis.

Mental age: The average age of an individual who obtains a certain score on an intelligence test.

Methodology: The logic of applying the scientific perspective and the set of rules for conducting research. *See also* Scientific method.

Metropolis: Urban area made up of separate cities, towns, and unincorporated areas that are interrelated.

Microsociology: The sociological study of interaction between individuals. *See also* Macrosociology.

Migration: The movement of people, a variable affecting the size and composition of population. Migration may be internal, within a country, or international, between countries. *See also* Demography.

Mills, C. Wright (1916–1962): The leader of mid-twentieth-century American sociological thought, who attempted to develop a radical sociological critique of capitalist society. His social-interactionist position, derived from Max Weber and Herbert Spencer, also influenced his thinking.

Miscegenation: Marriage or sexual relations between two persons of different races.

Modeling: A term used mainly by psychologists to describe the process of imitation, also known as learning by observation. Modeling is considered a very important aspect of socialization.

Modernization: The process of gradual change in a society from traditional social, economic, and political institutions to those characteristic of modern urban, industrial societies. *See also* Industrialization, Social modernization.

Monasticism: An organized system of withdrawal from everyday life and devotion to religious principles.

Moral absolutism: The idea that one's own moral values are the only true ones and that they are the proper basis for judging all others. *See also* Cultural relativism.

Moral development: The growth of a child into an adult who is willing to make the sacrifices necessary for social living. Study of moral development has focused on how people come to adopt their culture's standards of right and wrong and how they resist the temptation to defy the rules of acceptable conduct.

Mores: Folkways or customs to which group members attach social importance or necessity; standards of behavior that carry the force of right and wrong. *See also* Socialization.

Motivating conditions: In a model of suburban growth, factors that stimulate the shift of population from city to suburb. Such factors include deteriorating conditions in the cities and rising economic productivity. *See also* Facilitating conditions.

Multiple-nuclei theory: Theory of urban development stating that a city grows from a number of centers rather than from a single point. *See also* Concentric-zone theory, Sector theory.

Natural increase: Births minus deaths per 1,000 population.

Natural selection: The evolutionary process by which those individuals of a species with the best-adapted genetic endowment tend to survive to become parents of the next generation. *See also* Evolution.

Negative rites: According to French sociologist Emile Durkheim, rites that maintain taboos or prohibitions. *See also* Piacular rites.

Nonparticipant observations: A research method used in case studies by social scientists who come into contact with others but do not interact and behave primarily as a trained observer. *See also* Participant observation.

Norm: A shared standard for judging the behavior of an individual. Norms are elements of culture.

Normative organization: According to societal theorist Amitai Etzioni, a formal organization to which people belong because of personal interest or commitment to the organization's goals.

Nuclear family: The smallest family type, consisting of parents and their children. In Western society, custom has broadened the basic definition to include childless couples and single parents.

Oedipus complex: Refers to the idea, which originated with Sigmund Freud, that at a particular moment in the development of the child, his or her identifications with both mother and father undergo a radical transformation. Until the Oedipal moment, children, regardless of sex, have primary identification with the mother. In the Oedipal moment the father interposes himself between the child and the mother.

Open system: A social stratification system that allows an individual to rise to a higher position based on achieved status. *See also* Achieved position/status, Closed system.

Opportunity theory: The school of criminology that sees criminality as conduct. It is based on the writings of American sociological theorist Robert Merton, who reasoned that deviance results from pressures within the social structure. *See also* Labeling theory.

Organization: A deliberately formed group of people who achieve the aims of a social institution. *See also* Institution.

Organizational development: A field of endeavor that seeks to help organizations adapt to a difficult and changing environment by techniques such as sensitivity training, and which aims to humanize and democratize bureaucracies. *See also* Formal and complex organization.

Organized skepticism: The suspension of judgment until all relevant facts are at hand and the analysis of all such facts according to established scientific standards. *See also* Communalism, Disinterestedness.

Paradigm: A collection of the major assumptions, concepts, and propositions in a substantive area. Paradigms serve to orient research and theorizing in the area, and in this respect they resemble models.

Parsons, Talcott (1902–1979): American sociologist, his career passed through a number of phases, ranging from a substantive approach to social data involving a moderate level of abstraction to an analytic approach of almost metaphysical abstraction. *See also* Functionalism.

Participant observation: A research method used in case studies by social scientists who interact with other people and record relatively informal observations. *See also* Case study, Nonparticipant observations.

Paternalistic social system: A social system in which people or groups are treated in the manner in which a father controls his children. *See also* Competitive social system.

Pathological behavior: Conduct that results from some form of physical or mental illness or psychological problem. *See also* Deviance.

Pecking order: A hierarchical relationship of dominance and submission within a flock, herd, or community.

Peer: An associate (playmate, classmate, friend, etc.). Although socialization begins in the family, as the child matures peers or groups influence the child's behavior.

Personality: The individual's pattern of thoughts, motives, and self-concepts.

Phenomenology: A scientific method that attempts to study an individual's awareness of experience without making assumptions, theories, or value judgment that would prejudice the study. *See also* Relativism.

Piacular rites: According to French sociologist Emile Durkeim, religious rites that comfort or console individuals, help the community in times of disaster, and ensure the piety of the individual. *See also* Negative rites.

Piaget, Jean (1896–1980): Swiss biologist and psychologist who demonstrated the developmental nature of children's reasoning processes. He believes that humans pass through a universal, invariant development sequence of cognitive stages.

Pluralism: A state of society in which a variety of groups and institutions retain political power and distinctive cultural characteristics.

Pluralistic society: A society in which power is distributed among a number of interest groups that are presumed to counter balance each other.

Political modernization: The shift in loyalty or administrative structure from traditional authorities, such a tribal and religious leaders, to large-scale government organizations or from regional to national government. *See also* Social modernization.

Political sociology: The sociological study of politics, which, in turn, involves the regulation and control of the citizenry. A close relative of political sociology is political science.

Population control: Lowering the rate of natural increase of population. *See also* Natural increase.

Population explosion: A sudden, dramatic growth in the rate of natural increase of population. *See also* Natural increase.

Positivism: A philosophy that rejects abstract ideas in favor of a factual, scientific orientation to reality.

Postconventional morality: According to American psychologist Lawrence Kohlberg, the final level of moral development, which few people ever attain. This level is concerned with the moral values and individual rights apart from the group or society. *See also* Conventional morality, Preconventional morality.

Poverty culture: The belief that poverty-stricken people anywhere in the world share certain values, beliefs, and attitudes toward the world.

Power: The ability of people to realize their will even against others' opposition. *See also* Elite groups.

Pragmatism: A philosophical view generally credited to American psychologist William James and American philosopher and educator John Dewey, which stresses that concepts should be analyzed and evaluated in terms of practical consequences.

Preconventional morality: According to American psychologist Lawrence Kohlberg, the first level of moral development. At this level, children know cultural labels of good and bad, although they judge behavior only in terms of consequences. *See also* Conventional morality, Postconventional morality.

Prejudice: A biased prejudgement; an attitude in which one holds a negative belief about members of a group to which one does not belong. Prejudice is often directed at minority ethnic or racial groups. *See also* Stereotype.

Primary groups: Groups such as the family, work group, gang, or neighborhood, which are characterized by face-to-face contact of members and which are thought to significantly affect members' personality development. *See also* Secondary group.

Products of culture: Religion, art, law, architecture, and all the many material objects used and produced by a given cultural group.

Projection: According to Sigmund Freud, the tendency for people to attribute to others beliefs or motives that they have but cannot bring themselves to recognize or admit consciously. *See also* Prejudice.

Proletariat: According to Karl Marx, the industrial working class. *See also* Bourgeoisie.

Protestant ethic: According to Max Weber, the belief that hard work and frugal living would ensure future salvation.

Psychoanalytic theory: A theory of personality development, based on the work of Sigmund Freud. which maintains that the personality develops through a series of psychosexual stages as it experiences tension between demands of society and individual instincts for self-indulgence and independence. *See also* Personality.

Race: Biologically, the classification of people by observed physical characteristics; cultures, the meaning we give to physical characteristics and behavior traits when identifying in- and out-groups.

Race and ethnic relations: Social interactions among members of different groups that are based on, or affected by, an awareness of real or imagined racial or ethnic differences. *See also* Race.

Racism: A belief in racial superiority that leads to discrimination and prejudice toward those races considered inferior. *See also* Discrimination, Prejudice, Race.

Randomization: A technique for controlling all other variables that might confound the relationship between the independent variable(s) and the dependent variable in an experimental situation. Randomization is accomplished by assigning the objects (including people) to be observed at random to the experimental and control groups. This procedure guarantees that the characteristics associated with the objects will be uniformly distributed in all of the experimental and control conditions.

Rationalization: According to German sociologist Max Weber, the systematic application of impersonal and specific rules and procedures to obtain efficient coordination within modern organizations. *See also* Formal and complex organization.

Recidivism: The return to criminal behavior after punishment has been administered. *See also* Deterrence theory.

Relative poverty: Poverty of the lower strata of society as compared to the abundance enjoyed by members of higher strata. *See also* Absolute poverty.

Relativism: The idea that different people will have different experiences and interpretations of the same event. *See also* Phenomenology.

Reliability: A criterion for evaluating research results that refers to how well the study was done. A reliable study can be duplicated and its results found by other researchers. *See also* Validity.

Religion: A communally held system of beliefs and practices that are associated with some transcendent supernatural reality. *See also* Sect.

Replacement level: The rate of population increase at which individuals merely replace themselves. *See also* Zero population growth.

Research: In the application of scientific method, the process by which an investigator seeks information to verify a theory. *See also* Scientific method, Theory.

Resocialization: Major changes of attitudes or behavior, enforced by agents of socialization, that are likely to occur in institutions in which people are cut off from the outside world, spend all day with the same people, shed all possessions and identity, break with the past, and lose their freedom of action. *See also* Socialization.

Restricted code: According to British sociologist Basil Bernstein, the kind of ungrammatical, colloquial speech available to both middle-class and working-class people.

Revolutionary change: Violent social changes, most likely to occur when the gap between rising expectations and actual attainments becomes too frustrating for people to bear. *See also* Evolutionary change, Rising expectations.

Rising expectations: The tendency of people to expect and demand improved social, economic, and political conditions as social change progresses within a society.

Rite of passage: A ceremony that dramatizes a change in an individual's status. Weddings and funerals are examples.

Role: The behavior of an individual in relation with others. Also, the behavior considered acceptable for an individual in a particular situation or in the performance of a necessary social function. *See also* Role label, Role performance.

Role convergence: A growing similarity in roles that were formerly segregated and distinct. As men and women come to share domestic tasks, for example, their roles converge. *See also* Sex role.

Role label: The names assigned to an individual who acts in a particular way. Role labels may be broad ("laborer") or specific ("people who get colds easily").

Role performance: The actual behavior in individuals in a particular role.

Role portrayal: The adapting of roles to fit one's style of interaction.

Rumor: Unconfirmed stories and interpretations. They are the major form of communication during the milling process in collective behavior. *See also* Collective behavior.

Rural areas: Settlements of fewer that 2,500 residents or areas of low population density, such as farmlands. *See also* Urban areas.

Salience: The degree of importance of a group to its members; its impact on members. Generally, the smaller the group, the more salient it can become. *See also* Small group.

Sample survey: A research method in which a representative group of people is chosen from a particular population. Sample surveys may be conducted by interview or questionnaire. *See also* Case study, Experiment.

Scapegoat: A person or community that is made the undeserving object of aggression by others. The aggression derives from the need to allocate blame for any misfortune experienced by the aggressors. *See also* Prejudice.

Scientific method: The process used by scientists to analyze phenomena in a systematic and complete way. It is based on an agreement that criteria must be established for each set of observations referred to as fact and involves theory, research, and application. *See also* Research, Theory.

Secondary group: A social group characterized by limited face-to-face interaction, relatively impersonal relationships, goal-oriented or task-oriented behavior, and possibly formal organization. *See also* Primary groups.

Sect: A relatively small religious movement that has broken away from a larger church. A sect generally is in opposition to the larger society's values and norms.

Sectarianism: Having characteristics of sects, such as opposition to and withdrawal from, the larger society. *See also* Sect.

Sector theory: Theory of urban development which states that urban growth tends to occur along major transportation routes and that new residential areas are created at the edges of older areas of the same class. These developments produce more or less homogeneous pie-shaped sectors. *See also* Concentric-zone theory, Multiple-nuclei theory.

Secularization: The displacement of religious beliefs and influences by worldly beliefs and influences.

Segmental roles: Specialized duties by people in a bureaucratic society and over which they have little control. *See also* Role, Specialization.

Segregation: Involuntary separation of groups, on the basis of race, religion, sex, age, class, nationality, or culture.

Sex role: The culturally determined set of behavior and attitudes considered appropriate for males and females. *See also* Gender identity.

Shaman: The individual in a tribal or nonliterate society who is priest, sorcerer, and healer all in one. The shaman treats diseases, exorcizes evil spirits, and is considered to have supernatural powers.

Significant other: Those other people who are most important for an individual in determining his or her behavior. *See also* Generalized others.

Simmel, Georg (1858–1918): German sociologist and conflict theorist who proposed that a small number of stable forms of interaction underlie in the superficial diversity of manifest social occurrences. *See also* Conflict model.

Skinner, B. F. (1904–1990): American psychologist most noted for his rigorous adherence to the principles of behaviorism. One of Skinner's most important contributions is the distinction between classical *Pavlovian conditioning,* in which behavior is elicited (for example, reflex), and *operant conditioning,* in which behavior is emitted.

Small group: An interaction system in which members have face-to-face contact and which tends to have important effects on members' behavior. *See also* Primary groups.

Social aggregate: A relatively large number of people who do not know one another or who interact impersonally. Aggregates have loose structures and brief lives. *See also* Collective behavior, Crowd.

Social bonding: The quality of forming relatively permanent associations, found in both human and some animal and insect societies.

Social change: An alteration of stable patterns or social organization and interaction, preceded or followed by changes in related values and norms.

Social conflict: Disagreement over social values and competing interests. *See also* Conflict model.

Social constraints: Factors that produce conformity to the behavioral expectations of society, such as ridicule, expulsion from a group or punishments. Knowledge of social constraints is taught during socialization. *See also* Socialization.

Social contract: An agreement binding all parties that sets up rights, responsibilities, powers, and privileges and forms a basis for government.

Social control: Techniques and strategies for regulating human behavior.

Social Darwinism: The view which sees society as an organism that grows more perfect through the natural selection of favored individuals. In this view, the wealthier and better-educated classes are more "fit" because they have competed their way to success. Social Darwinism applies Darwin's theory of biological evolution to social groups. *See also* Evolution, Natural selection.

Social disorganization: The breakdown of institutions and communities, which results in dislocation and breakdown of ordinary social controls over behavior.

Social distance: The relative positions of members or groups in a stratified social system; the degree of social acceptance that exists between certain social groups in a society.

Social dynamics: All the forces and processes involved in social change.

Social engineering: Systematic planning to solve social problems.

Social epidemiology: The study of illness rates in a population within a specific geographic area. *See also* Sociology of medicine.

Social group: A collection of interrelating human beings. A group may consist of two or more people. The interaction may involve performing a complex task—a surgical team—or simple proximity—all the drivers on a road during rush hour. Groups may be classified as primary or secondary. *See also* Primary groups, Secondary group, Small group.

Social interaction: The effect that two or more people have on each other's behavior, thoughts, and emotions through symbolic and nonsymbolic modes of expression.

Social isolates: Children who have had minimal human contact because of abandonment or parental neglect. Also refers to people cut off from social contact voluntarily or involuntarily.

Social mobility: The movement of people up or down a social hierarchy based on wealth, power, and education.

Social modernization: A process of change in social institutions, usually viewed as a movement from traditional or less-developed institutions to those characteristic of developed societies. *See also* Economic modernization.

Social movement: A long-term collective effort to resist or to promote social change. *See also* Collective behavior.

Social organization: A general term used in different contexts, but usually referring to organizational aspects of societies, communities, institutions, and groups.

Social relations perspectives: A view that emphasizes factors other than intelligence, such as family, in determining an individual's economic position. *See also* Technocratic perspective.

Social sciences: Branches of learning concerned with the institutions of human societies and with human behavior and interrelationships. Social sciences draw their subject matter from the natural sciences.

Social stratification: A system of social inequality in which groups are ranked according to their attainment of socially valued rewards.

Social system: The arrangement or pattern of organization of any social group. A system is a whole make up of interacting parts.

Socialism: An economic system in which means of production (land, equipment, materials) are collectively owned and controlled by the state rather than by private individuals. *See also* Communism.

Socialization: The complex process by which individuals learn and adopt the behavior patterns and norms that enable them to function appropriately in their social environments. *See also* Agents of socialization, Personality.

Society: A social group that is relatively large, self-sufficient, and continues from generation to generation. Its members are generally recruited through the process of socialization. *See also* Conflict model, Functionalism, Socialization, Sociology.

Sociobiology: A relatively new field which is a branch of behavioral biology that studies the biological bases of the social behavior and social organization of all animal species. *See also* Biosociology.

Sociocultural: Social organization in which patterns of behavior are largely governed by a network of learned values, norms, and beliefs. *See also* Culture, Norm.

Sociogram: A diagram showing the interaction among group members. A sociogram of a group might show, for example, who is most liked and who is least liked. *See also* Group process.

Sociological perspective: The point of view of the sociologist. It aims at precision and objectivity through the scientific method. *See also* Scientific method.

Sociology: The social science concerned with the systematic study of human society and human social interaction. *See also* Society.

Sociology of death: The inquiry into the impact of dying on a patient's relationship to self, to others, and to the social structure as a whole. *See also* Thanatology.

Sociology of education: The scientific analysis of both formal and informal learning in a society.

Sociology of medicine: The study of the definition, causes, and cure of disease in different societies and social groups. The sociology of medicine also studies the social organization of modern medical care and the social roles of staff and patients at various medical facilities.

Sociology of work: A study of the relations of production and consumption and the influence of work on social organization and social change. *See also* Social change.

Specialization: A concentration of work in a specific area. According to German sociologist Max Weber, specialization is a characteristic of an ideal type of bureaucratic organization. *See also* Bureaucracy, Ideal type.

Spencer, Herbert (1820–1913): British philosopher whose descriptive sociology was very influential and formed the basis for Social Darwinism. *See also* Social Darwinism.

Standard Metropolitan Statistical Area (SMSA): A Census Bureau concept for counting population in core cities, their suburbs, satellite communities, and other closely related areas. SMSAs ignore usual political divisions, such as state boundaries. *See also* Metropolis.

State: The political-legal system that represents a whole country, its territory, and people. A state is a more formal legal and technical entity than the broader concept, "society." *See also* Society.

Statistics: A method for analyzing data gathered from samples of observations in order to: describe the amount of variation in each of the variables; describe hypothetical relationships among variables; to make inferences from the results to the larger population from which the sample was drawn.

Status: The position of the individual (actor) in a system of social relationships. *See also* Achieved position/status, Ascriptive status.

Status group: According to German sociologist Max Weber, people with similar lifestyles and social standing.

Stereotype: An exaggerated belief associated with some particular category, particularly of a national, ethnic, or racial group. *See also* Sex role.

Stigmatization: The labeling of individuals in such a way that they are disqualified from full social acceptance and participation. Criminalization is a part of this process. *See also* Deviance.

Stratification: A system of social inequality based on hierarchy orderings of groups according to their members' share in socially valued rewards. The nature of these rewards varies from society to society but usually consists of wealth, power, and status.

Structural differentiation: The specialization of institutions, social roles, and functions that accompanies social change.

Structural-functionalism: *See* Functionalism.

Structuralism: An intellectual approach that emphasizes studying the underlying structures of human behavior rather than obvious surface events.

Subcultures: Various groups within the society who share some elements of the basic culture but who also possess some distinctive folkways and mores. *See also* Culture.

Surrogate religion: A belief system that substitutes for a traditional religion. Communism is an example.

Symbol: Anything that stands for something else. For example, words may be symbols of objects, ideas, or emotions.

Symbolic interaction: The process of interaction between human beings conducted at the symbolic level (for example, through language). The school of social psychology known as *symbolic interactionism* has stressed the implications of the process for socialization.

Target population: In a model of suburban growth, a group of people who are affected both by facilitating and motivating conditions. This population consisted of young to middle-age white married couples. *See also* Facilitating conditions and Motivating conditions.

Task segregation: A division of labor based on a feature such as the sex or age of the participants. Task segregation is common in most societies. *See also* Division of labor.

Taxonomy: A classification system of cognitive categories. *See also* Folk taxonomy.

Technocracy: The domination of an industrial society by a technical elite. *See also* Elite groups, Technocratic perspective.

Technocratic perspective: The view that sees the hierarchical division of labor as a result of the need to motivate the ablest individuals to undertake the most extensive training, which will allow them to perform the most difficult and important occupations in a society. *See also* Technocracy.

Thanatology: The study of theories, causes, and conditions of death.

Theory: A set of generalized, often related, statements about some phenomenon. A theory is useful in generating hypotheses. Middle-range theories interrelate two or more empirical generalizations. Grand theory organized all concepts, generalizations, and middle-range theories into an overall explanation. *See also* Research.

Third World: Those non-Western peoples who have experienced modern capitalism in the form of imperialism. Originally referring specifically to the colonized societies of Asia, Africa, and Latin America, the term has come to be applied as well as to national minorities within the United States—Chicanos, blacks, Puerto Ricans, Asian Americans, and Indians.

Töennies, Ferdinand (1855–1936): Classical German sociologist who was the first to recognize the impact of the organic point of view on positivism. He identified the social organization concepts of *Gemeinschaft* and *Gesellschaft. See also* Gemeinschaft/Gesellschaft.

Totemism: Religious belief in which a totem—a representation of some natural object in the environment—figures prominently. Totems serve as symbols of clans and sacred representations. *See also* Clan.

Traditional society: Rural, agricultural, homogeneous societies characterized by relatively simple means of production.

Tylor, Edward Burnett (1832–1917): British pioneer anthropologist upon whose central ideas about culture all modern definitions are based.

Typology: A classification scheme containing two or more categories (types) based on characteristics of the things being classified that are considered by the classifier to be of importance.

Universalism: A rule for scientific innovation, according to American sociological theorist Robert Merton. It refers to an objectivity that does not allow factors such as race, religion, or national origin to interfere with scientific inquiry. *See also* Communalism, Disinterestedness, Organized skepticism.

Urban areas: According to Census Bureau definitions, settlements of 2,500 or more persons. *See also* Rural areas.

Urban society: A form of Social organization in which: (1) economic exchange and markets are very important; (2) social roles are highly specialized; (3) centralized administrative and legal agencies provide political direction; and (4) interaction tends to be impersonal and functional. *See also* Urbanization.

Urbanism: The ways in which the city affects how people feel, think, and interact.

Urbanization: The movement of people from country to city as well as the spread of urban influence and cultural patterns to rural areas. Also refers to the greater proportion of the population in urban areas than in rural areas. *See also* Urban society.

Utilitarian organization: According to societal theorist Amitai Etzioni, a formal organization that people join for practical reasons, mainly jobs and salaries.

Utilitarianism: A political philosophy based on the principle of "the greatest good of the greatest number." The concept was developed by the British philosopher Jeremy Bentham, who based it on the principle that pleasure was preferable to pain. Public policy, then, ought to be aimed at what he termed *maximum utility*—that is, the maximization of pleasure for the maximum number of citizens.

Validity: A criterion for evaluating research results that refers to how well the data actually reflect the real world. *See also* Reliability, Research.

Value-added theory: Neil Smelser's theory that postulates five stages in the development of collective behavior. *Social conduciveness* describes situations that permit collective behaviors to occur. *Structural strain* refers to problems in the social environment. The growth of a *generalized belief* involves the interpretation of structural conduciveness and strain in a way that favors collective behavior. *Precipitating factors* are events that trigger collective behavior. *Mobilization for action* is the "organizational" component and usually involves explicit instruction and/or suggestions. *See also* Collective bchavior.

Values: Individual or collective conceptions of what is desirable. This conception usually has both emotional and symbolic components. *See also* Norm.

Variables: Factors that can change. Researchers must state the specific variables they intend to measure, An independent variable is causal. A dependent variable changes according to the independent variable's behavior. *See also* Research, Scientific method.

Verstehen: Subjective understanding that, according to German sociologist Max Weber, must be employed in sociological investigation. *See also* Positivism.

Weber, Max (1864–1920): German sociologist whose work profoundly influences Western sociological thought and method. The key to Weber's analysis of the modern world is his concept or *rationalization*—the substitution of explicit formal rules and procedures for earlier spontaneous, rule-of-thumb methods and attitudes. The result of this process was a profound "disenchantment of the world," which has been carried to its ultimate form in capitalist society, where older values were being subordinated to technical methods. The prime example of the rationalized institution was bureaucracy. *See also* Rationalization, Verstehen.

Woman suffrage: The right of women to vote. *See also* Women's movement.

Women's movement: A social movement by women to gain equal social, economic, and legal status with men. *See also* Social movement.

Zero population growth: A point at which population stops increasing. *See also* Population control, Replacement level.

SOURCES

The Study of Society, Second Edition. © 1977 by Dushkin Publishing Group/Brown & Benchmark Publishers, Guilford, CT 06437.

The Encyclopedic Dictionary of Sociology, Fourth Edition. © 1991 by Dushkin Publishing Group/Brown & Benchmark Publishers, Guilford, CT 06437.

(1995–1996)

Index

Credits/Acknowledgments

Cover design by Charles Vitelli

1. Culture
Facing overview—Photo by Louis P. Raucci. 12-14, 16—Photos by Colin M. Turnbull.

2. Socialization and Social Control
Facing overview—United Nations photo by John Robaton.

3. Groups and Roles in Transition
Facing overview—EPA-Documerica Color Slide.

4. Stratification and Social Inequalities
Facing overview—United Nations photo by W. A. Graham.

5. Social Institutions: Issues, Crises, and Change
Facing overview—Photo by Pamela Carley.

6. Social Change and the Future
Facing overview—Digital Stock photo.

*PHOTOCOPY THIS PAGE!!!**

ANNUAL EDITIONS ARTICLE REVIEW FORM

■ NAME: _______________ DATE: _______________

■ TITLE AND NUMBER OF ARTICLE: _______________

■ BRIEFLY STATE THE MAIN IDEA OF THIS ARTICLE: _______________

■ LIST THREE IMPORTANT FACTS THAT THE AUTHOR USES TO SUPPORT THE MAIN IDEA:

■ WHAT INFORMATION OR IDEAS DISCUSSED IN THIS ARTICLE ARE ALSO DISCUSSED IN YOUR TEXTBOOK OR OTHER READINGS THAT YOU HAVE DONE? LIST THE TEXTBOOK CHAPTERS AND PAGE NUMBERS:

■ LIST ANY EXAMPLES OF BIAS OR FAULTY REASONING THAT YOU FOUND IN THE ARTICLE:

■ LIST ANY NEW TERMS/CONCEPTS THAT WERE DISCUSSED IN THE ARTICLE, AND WRITE A SHORT DEFINITION:

*Your instructor may require you to use this ANNUAL EDITIONS Article Review Form in any number of ways: for articles that are assigned, for extra credit, as a tool to assist in developing assigned papers, or simply for your own reference. Even if it is not required, we encourage you to photocopy and use this page; you will find that reflecting on the articles will greatly enhance the information from your text.

We Want Your Advice

ANNUAL EDITIONS revisions depend on two major opinion sources: one is our Advisory Board, listed in the front of this volume, which works with us in scanning the thousands of articles published in the public press each year; the other is you—the person actually using the book. Please help us and the users of the next edition by completing the prepaid article rating form on this page and returning it to us. Thank you for your help!

ANNUAL EDITIONS: SOCIOLOGY 97/98

Article Rating Form

Here is an opportunity for you to have direct input into the next revision of this volume. We would like you to rate each of the 44 articles listed below, using the following scale:

1. **Excellent: should definitely be retained**
2. **Above average: should probably be retained**
3. **Below average: should probably be deleted**
4. **Poor: should definitely be deleted**

Your ratings will play a vital part in the next revision. So please mail this prepaid form to us just as soon as you complete it.
Thanks for your help!

Rating	Article
____	1. Tribal Wisdom
____	2. The Mountain People
____	3. Overworked Americans or Overwhelmed Americans?
____	4. A De-Moralized Society: The British/American Experience
____	5. The Decline of Bourgeois America
____	6. Guns and Dolls
____	7. Children of the Universe
____	8. Crime in America: Violent and Irrational—and That's Just the Policy
____	9. Moral Credibility and Crime
____	10. When Violence Hits Home
____	11. Legalization Madness
____	12. The Way We Weren't: The Myth and Reality of the "Traditional" Family
____	13. Where's Papa?
____	14. Modernizing Marriage
____	15. Now for the Truth about Americans and Sex
____	16. Ending the Battle between the Sexes
____	17. Men: Tomorrow's Second Sex
____	18. Crisis of Community: Make America Work for Americans
____	19. The Strange Disappearance of Civic America
____	20. Are Today's Suburbs Really Family-Friendly?
____	21. Winner Take All
____	22. Working Harder, Getting Nowhere

Rating	Article
____	23. Poverty's Children: Growing Up in the South Bronx
____	24. Upside-Down Welfare
____	25. Dismantling the Welfare State: Is It the Answer to America's Social Problems?
____	26. Whites' Myths about Blacks
____	27. Affirmative Action: It Benefits Everyone and Let's Get Rid of It
____	28. The Longest Climb
____	29. Violence against Women
____	30. Money Changes Everything
____	31. Hyper Democracy
____	32. The Death of Common Sense
____	33. Reinventing the Corporation
____	34. The Revolution in the Workplace: What's Happening to Our Jobs?
____	35. Seeking Abortion's Middle Ground
____	36. Health Unlimited
____	37. A New Vision for City Schools
____	38. Can Churches Save America?
____	39. Carrying Capacity: Earth's Bottom Line
____	40. The New Assault on Immigrants
____	41. Alone Together: Will Being Wired Set Us Free?
____	42. The Price of Fanaticism
____	43. The Coming Anarchy
____	44. Operating in a Period of Punctuated Equilibrium

(Continued on next page)

ABOUT YOU

Name ______________________________ Date ______________

Are you a teacher? ❑ Or a student? ❑

Your school name ______________________________

Department ______________________________

Address ______________________________

City ______________________ State __________ Zip __________

School telephone # ______________________________

YOUR COMMENTS ARE IMPORTANT TO US !

Please fill in the following information:

For which course did you use this book? ______________________________

Did you use a text with this *ANNUAL EDITION*? ❑ yes ❑ no

What was the title of the text? ______________________________

What are your general reactions to the *Annual Editions* concept?

Have you read any particular articles recently that you think should be included in the next edition?

Are there any articles you feel should be replaced in the next edition? Why?

Are there any World Wide Web sites you feel should be included in the next edition? Please annotate.

May we contact you for editorial input?

May we quote your comments?

No Postage Necessary if Mailed in the United States

ANNUAL EDITIONS: SOCIOLOGY 97/98

BUSINESS REPLY MAIL

First Class Permit No. 84 Guilford, CT

Postage will be paid by addressee

Dushkin/McGraw·Hill
Sluice Dock
Guilford, Connecticut 06437